Design of Experiments: A Modern Approach

First Edition

Bradley Jones
JMP Division of SAS, Inc

and

Douglas C. Montgomery
Arizona State University

WILEY

VP AND EDITORIAL DIRECTOR	Laurie Rosatone
SENIOR DIRECTOR	Don Fowley
EDITOR	Jennifer Brady
EDITORIAL MANAGER	Judy Howarth
CONTENT MANAGEMENT DIRECTOR	Lisa Wojcik
CONTENT MANAGER	Nichole Urban
SENIOR CONTENT SPECIALIST	Nicole Repasky
PRODUCTION EDITOR	Indirakumari S
COVER PHOTO CREDIT	Courtesy of Bradley Jones, © elen_studio / Shutterstock

This book was set in 10/12 pt ITC New Baskerville Std by SPi Global and printed and bound by Quad Graphics.

Founded in 1807, John Wiley & Sons, Inc. has been a valued source of knowledge and understanding for more than 200 years, helping people around the world meet their needs and fulfill their aspirations. Our company is built on a foundation of principles that include responsibility to the communities we serve and where we live and work. In 2008, we launched a Corporate Citizenship Initiative, a global effort to address the environmental, social, economic, and ethical challenges we face in our business. Among the issues we are addressing are carbon impact, paper specifications and procurement, ethical conduct within our business and among our vendors, and community and charitable support. For more information, please visit our website: www.wiley.com/go/citizenship.

ISBN: 978-1-119-74601-0 (PBK)
ISBN: 978-1-119-61132-5 (EVALC)

Library of Congress Cataloging-in-Publication Data

LCCN: 2019025994

The inside back cover will contain printing identification and country of origin if omitted from this page. In addition, if the ISBN on the back cover differs from the ISBN on this page, the one on the back cover is correct.

V10019004_061620

PREFACE

There are several textbooks covering material in design of experiments (DOE). It is a fair question, then, to ask, "Why write another DOE textbook?"

One answer is based on the observation that in 2018 over a quarter of the DOE courses taught at the university level rely on course notes rather than a text. We view this as evidence of pent-up demand for a different kind of textbook than is currently available.

A characteristic of many DOE textbooks is that they focus as much or more on analysis than on design. A student might get the impression that there is only one appropriate design for any scenario and this design should be orthogonal. Orthogonal designs have the desirable feature that the analysis of the data generated after running the experiment is less demanding than the analysis of observational data.

We hasten to say that orthogonal designs are good designs, in general. In fact, many of the example designs in this book are the familiar orthogonal designs that appear in other textbooks. However, we use design optimization as our design construction approach, which distinguishes this textbook from others. Many orthogonal designs appear here not because they are orthogonal but because they are optimal, sometimes by many criteria.

The problem with focusing primarily on orthogonal design is that orthogonal designs do not always exist for a given scenario and set of resource constraints. By contrast, it is possible to generate an optimal or highly efficient design in many situations where an orthogonal design does not exist. This makes the optimal design construction approach more practically applicable due to its flexibility in accommodating the unique features of a design problem.

Two-level screening designs appear in many textbooks as well as this one. We have strong anecdotal evidence that engineers are uncomfortable with these designs due to their strong belief that the relationship of most factors with a response is not linear. For this reason, we introduce definitive screening designs as a three-level alternative to standard screening designs.

Many traditional DOE textbooks devote considerable space to the development and illustration of the manual computing techniques for the analysis of variance and other related methods. The speed, capabilities, and widespread availability of computer software has increased exponentially in the last few decades. There is no longer a need for practitioners of DOE to be able to analyze experiments using hand calculation. This is virtually never done in practice. Therefore, except for development of the basic analysis of variance computations for the simplest cases, we have relied on software for analysis. This allows us to streamline the presentation of the important design material. We have also chosen not to include an appendix of statistical tables. Percentage points of the standard reference distributions (normal, t, chi-square, and F) can be found directly from both statistics and spreadsheet software. Statistics software routinely computes P-values as well, so tables of these distributions are not very useful.

We intend this book to be used as a text for a first course in design of experiments, probably at the upper division or first year graduate level. The prerequisites are modest; anyone who has completed a first course in statistics and has familiarity with matrix algebra will be comfortable reading the book. The organization of the book follows a traditional outline for a design of experiments course. Chapter 1 is an introduction to the topic and contains some historical notes, a recommended "process" for designing and conducting experiments and concludes with a review of some basic statistics topics. Chapter 2 treats simple comparative experiments, that is, single-factor experiments where the factor has only two levels. The two-sample t-test is introduced as the analysis technique. Chapter 3 introduces the analysis of variance in the context of a single categorical factor experiment with multiple levels. Chapter 4 continues the treatment of single-factor experiments, but now the factor is continuous and linear regression is the analysis technique. Chapter 5 introduces two-factor factorial experiments and extends the analysis of variance to that situation. Chapter 6 deals with blocking, including the familiar randomized complete block design. We also show how many blocking problems can be conveniently solved using optimal design methodology. Chapter 7 treats 2^k factorial designs, including blocked versions of these designs. We show that these are optimal designs. Chapter 8 discusses experiments for factor screening, including the familiar 2^{k-p} fractional factorial designs, some nonregular fractional factorial designs, and definitive screening designs. Chapter 9 treats the case of blocked designs where the blocks are random. The important case of split-plot experiments is presented in Chapter 10. Chapter 11 discusses response surface methodology, including response surface optimization techniques, the classical response surface designs, and the use of optimal designs in this framework. Finally, Chapter 12 introduces designs for nonlinear models, including the important case of designs for binomial response data and other members of the family of generalized linear models. Throughout the book, the use of design optimality is stressed as the technique for design construction and selection.

We assume students and other users of the book will have access to software for both design and analysis of experiments. There are many such software applications. To our knowledge, however, only JMP is capable of generating optimal designs and embedding a proper analysis for all the scenarios introduced in this textbook. The interested student can find a substantial amount of information about JMP at the URL, www.jmp.com.

About the Authors

Dr. Bradley Jones is the Distinguished Research Fellow in the JMP division of SAS where he is responsible for the development of new methods in design of experiments (DOE). He built the JMP Custom Designer, a general and powerful tool for generating optimal experimental designs. He holds a patent on the use of DOE for minimizing registration errors in the manufacture of laminated circuit boards and is the inventor of the prediction profile plot for interactive exploration of multiple input and output response surfaces.

He is the co-discoverer with Christopher Nachtsheim of Definitive Screening Designs. With Peter Goos, he wrote *Optimal Design of Experiments: A Case Study Approach*, which won the 2012 Ziegel Prize.

He received the Brumbaugh Award for the paper making the largest contribution to industrial quality control in 2009 and again in 2011. He is the recipient of the Lloyd S. Nelson Award for the article having the greatest immediate impact to practitioners in 2010 and again in 2012. In 2012, he received the Statistics in Chemistry Award from the American Statistical Association for outstanding collaborative work with a chemist. He also won the Youden Prize for the best expository paper in *Technometrics* in 2012. He is a former Editor-in-Chief of the *Journal of Quality Technology* and a Fellow of the American Statistical Association.

Dr. Douglas C. Montgomery is Regent's Professor of Industrial Engineering and ASU Foundation Professor of Engineering at Arizona State University. Previously he was the John M. Fluke Distinguished Professor of Engineering, Director of Industrial Engineering and Professor of Mechanical Engineering at the University of Washington in Seattle and Professor of Industrial and Systems Engineering at the Georgia Institute of Technology. He holds BSIE, MS, and Ph.D. degrees from Virginia Tech. Dr. Montgomery's industrial experience includes engineering assignments with Union Carbide Corporation and Eli Lilly and Company. He also has extensive consulting experience with over 100 organizations, mostly in the areas of industrial statistics, quality improvement, and product/process improvement/design.

Dr. Montgomery's professional interests focus on industrial statistics, including design of experiments, quality and reliability engineering, applications of linear models, and time-series analysis and forecasting. The Office of Naval Research, the National Science Foundation, the United States Army, the Department of Defense, and private industry have sponsored his research. He has supervised 68 doctoral dissertations and over 40 MS theses and MS Statistics Projects.

Dr. Montgomery is an author of 12 books and over 270 archival journal papers. His textbooks on design of experiments and statistical quality control are the most widely used texts on these subjects worldwide. Google Scholar indicates that he has over 100,000 citations to his published work and his h-index is 80. He is currently one of the Chief Editors of *Quality and Reliability Engineering International* and is a former Editor of the *Journal of Quality Technology*.

He is a Fellow of the American Statistical Association, an Honorary Member of the American Society for Quality, a Fellow of the Royal Statistical Society, a Fellow of the Institute of Industrial Engineers, an Elected Member of the International Statistical Institute and an Academician of the International Academy for Quality. His recognition awards include the Shewhart Medal, the Distinguished Service Medal, the William G. Hunter Award, the Brumbaugh Award, the Lloyd S. Nelson Award, and the Shewell Award (twice) from ASQ, the Deming Lecture Award from ASA, the George Box Medal from ENBIS (European Network for Business and Industrial Statistics), the Greenfield Medal from the Royal Statistical Society, and the Ellis R. Ott Award. He has also received several outstanding teaching awards, including the Arizona State University Engineering College Graduate Teaching Excellence Award in 1994. He was named an ASU Outstanding Doctoral Mentor in 2004.

Contents

SS Student solution available in interactive e-text

Experimental Design: Principles and Practices and Statistics Review

1

1.1 The Strategy of Experimentation

Observing a system or process while it is in operation is an important part of the learning process and is an integral part of understanding how systems and processes work. The great New York Yankees catcher Yogi Berra said that "... you can observe a lot just by watching." However, to understand what happens to a process when you change certain input factors, you have to do more than just watch—you actually have to change the factors. This means that to really understand cause-and-effect relationships in a system, you must deliberately change the input variables to the system and observe the changes in the system output that these changes to the inputs produce. In other words, you need to conduct **experiments** on the system. Observations on a system or process can lead to theories or hypotheses about what makes the system work, but experiments of the type described above are required to demonstrate that these theories are correct.

Experiments are conducted in virtually all fields of science, engineering, industry, and business, usually to discover something about a particular process, product, or system. Each experimental **run** is really just a **test**. Obviously, testing plays a vital role in how products are designed and improved, how processes are developed and optimized, and how systems perform. We define an **experiment** as a series of tests in which deliberate changes are made to the controllable input variables of a process or system with the objective of identifying the reasons for the observed changes in the output or response variables of the system. Specifically, we usually want to determine which input variables are responsible for the observed changes in response to develop a model relating the response to the important input variables and to use this model for making decisions about the system. The person conducting the experiment is called the **experimenter**.

This book is about planning and conducting experiments as well as analyzing the resulting data so that valid and objective conclusions are obtained. Our focus is on experiments in engineering, science, and business. Experimentation plays an important role in **technology commercialization** and **product realization** activities, which consist of new product design and formulation, manufacturing process development, and process improvement. The objective in many cases may be to develop a **robust** process, that is, a process affected minimally by external sources of variability. There are also many applications of designed experiments in a nonmanufacturing or non-product-development setting, such as marketing, service operations, e-commerce, and general business operations.

As an example of an experiment, suppose that a biomedical engineer is designing a new pump for the intravenous delivery of a drug. The pump should deliver a constant quantity or dose of the drug over a specified period of time. The primary design parameter of the pump that the engineer is considering at this stage is the diameter of the pump cylinder. Two diameters are feasible: 3/16th in. and 1/4th in. Here, the objective of the experimenter is to determine which diameter produces the appropriate flow rate for the particular type of pump. In this type of experiment, a common approach would be to build several prototypes and subject each prototype to a standard test to determine the effect of the cylinder diameter on the drug delivery rate. The average delivery rate that is observed for each cylinder diameter would be used to determine the best pump design to use in the product.

This is a relatively simple experiment. It only has one **factor**, the cylinder diameter, and the factor only has two **levels**. However, with a little thought, a number of important questions come to mind:

1. Are these two diameters the only ones that should be considered?
2. Are there any other factors that might potentially affect the drug delivery rate? Should those be investigated in this experiment (such as the length of the cylinder, or the length of the plunger, or the pressure at which the pump operates)?
3. How many prototypes of each cylinder should be tested?
4. How should the experiment actually be conducted; that is, in what order should the prototypes be tested?
5. What method of data analysis should be used?
6. What difference in average drug delivery rate between the two cylinder diameters will be considered important?

These are all important questions. All of them, and perhaps many others, need to be answered satisfactorily before conducting the experiment.

As a second example, consider the design of the home page for an e-commerce business. A great deal of business activity is generated through this website, so its design is a very important aspect of business operations. Suppose that the website has the following components: (1) a photo flash image, (2) a main headline, (3) a sub headline, (4) a main text copy, and (5) a main image on the right. The response variable is the click-through rate, that is, the number of visitors who click through into the site divided by the total number of visitors to the site. The click-through rate can be tracked automatically for a particular website. Suppose that there are five choices for the photo flash image, four choices for the main headline, three choices for the sub headline, two choices for the main text copy, and two choices for the main image.

This is a more complex experiment involving five factors. The most conservative classical approach to a multifactor experiment is to test all possible combinations of factor levels. This is running a **factorial design**. If we use a factorial design for this experiment, Web pages for all possible combinations of these factor levels must be constructed and tested (240 Web pages). Even if a website design can be tested in a few hours once it is in place, the logistical aspects of designing 240 Web pages are formidable. Fortunately, in multifactor experiments, we often do not need to run a complete factorial; a **fractional factorial** experiment that uses a small number of the possible Web page designs would likely be successful. This experiment would require a fractional factorial where the factors have different numbers of levels.

Finally, consider designing an aircraft engine. The engine is a commercial turbofan, intended to operate in the cruise configuration at 40,000 ft and 0.8 Mach. The design parameters that need to be studied include inlet flow, fan pressure ratio, overall pressure, stator outlet temperature, and many other factors. The output response variables in this system are specific fuel consumption and engine thrust. In designing this system, it would be prohibitively expensive to build prototypes or actual test articles early in the design process, so the engineers use a **computer model** of the system that allows them to focus on the key design parameters of the engine and to vary them in an effort to optimize the performance of the engine. Designed experiments can be employed with the computer model of the engine to determine the most important design parameters and their optimal settings. Engineers frequently employ various types of computer models (finite element models, computational fluid dynamic models, electrical circuit simulators, traffic flow models, and so on) for design activities, and experimental design techniques are an integral part of how to employ those computer models to make the engineering design and development activities more effective and efficient.

Experimentation is a vital part of the **scientific** (or **engineering**) **method**. Of course, there are situations where the scientific phenomena are so well understood that useful results including mathematical models can be developed directly by applying these well-understood principles. The models of such phenomena that follow directly from the physical mechanism are usually called **mechanistic models**. A simple example is the familiar equation for current flow in an electrical circuit, Ohm's law, $E = IR$. However, most problems in science, engineering, and business require **observation** of the system at work and **experimentation** to elucidate information about why and how it works. Well-designed experiments can often lead to a model of system performance; such experimentally determined models are called **empirical models**. The focus of this book is on designing an experiment so that the experimenter can obtain a valid and useful empirical model of the system under study. These empirical models can be manipulated and analyzed just as a mechanistic model can.

A well-designed experiment is important because the results and conclusions that can be drawn from the experiment depend, to a large extent, on the method of data collection. To illustrate this point, suppose that the bioengineer in the drug delivery pump experiment used all of the smaller diameter cylinders in morning tests and all of the larger diameter cylinders in afternoon tests. She plans to compare the sample average drug delivery rates from the two sets of tests to draw conclusions. It is possible that there is a temperature difference between the morning and afternoon test period and that temperature could have some impact on the performance of the pump. Now, when the average drug delivery rates are compared, she is unable to say how much of the observed difference is the result of the two different time periods and how much is the result of inherent differences between the cylinder diameters. Thus, the method of data collection has directly impacted the conclusions that can be drawn from the experiment.

In general, experiments are used to study the performance of processes and systems.

The process or system can be represented by the model shown in Figure 1.1. We can usually visualize the process as a combination of resources (which could include operations, machines, methods, people, computer code and software, but are not limited to those explicit items) that transforms some set of input variables or design factors (which could be component parts, raw materials, personnel or operator actions, or information, for example) into an output that has one or more observable response variables. Some of the design factors $x_1, x_2, \ldots, x_p$ are controllable, whereas other variables $z_1, z_2, \ldots, z_q$ are uncontrollable (although they may be controllable for purposes of a test). The objectives of the experiment may include the following:

1. Determining which factors are most influential on the response y
2. Determining where to set the influential xs so that the response y is almost always at or near its most desired value
3. Determining where to set the influential xs so that variability in y is small
4. Determining where to set the influential xs so that the effects of the uncontrollable variables $z_1, z_2, \ldots z_q$ are minimized.

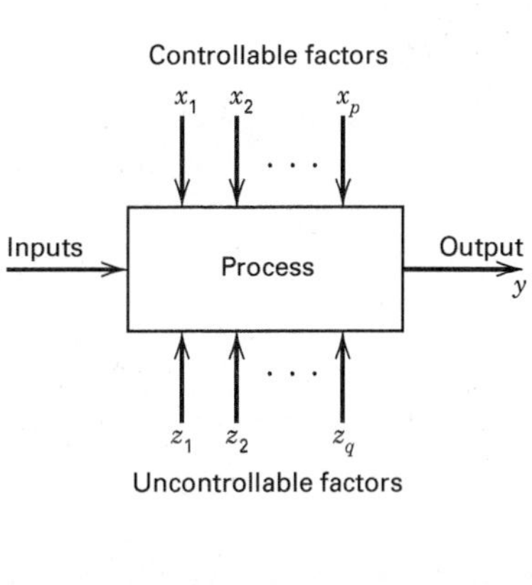

FIGURE 1.1 General Model of a System or Process

As you can see from the examples that we presented, experiments often involve several factors. The general approach to planning and conducting the experiment is called the **strategy of experimentation**. We will illustrate three strategies with an extension of the drug delivery pump experiment described earlier. The performance of this pump depends on more than the cylinder diameter. Suppose that the other factors include the length of the cylinder, the fit or clearance between the cylinder wall and the plunger, the plunger length, and the diameter of the tubing that connects the pump to the patient's blood vessel. This is a total of five design factors. There are other factors that could be considered, such as the choice of material for the tubing and the diameter of the needle that is inserted into the patient's vein, but the bioengineer decides that these factors can be ignored; that is, these factors are not important because their effects are likely to be so small that they have no practical value. Engineers, scientists, and business analysts often make these types of decisions about some of the factors they are considering.

Now let us consider how the five design factors that we have identified as potentially important could be experimentally tested to determine their effect on the drug delivery rate. Each trial of the experiment requires assembling and testing a prototype with a specific configuration of the five factors. Suppose that a maximum of eight prototypes can be tested over the course of the experiment. One approach would be to select an essentially arbitrary combination of these factors, test them, and see the result. Then, depending on the observed results, another combination of factors is selected, a second prototype assembled, and another test performed. This approach could be continued, switching the values of one or more design factors from test to test until all eight prototypes have been built and tested. This strategy of experimentation, called the **best-guess approach**, is frequently used in practice. It can work reasonably well, too, because the experimenters often have a great deal of technical (theoretical) knowledge or insight about the system they are studying, as well as considerable practical experience.

The best-guess approach has at least two disadvantages. First, suppose the initial "best guess" does not produce the desired results. Now the experimenter has to take another guess at the correct combination of factor levels. This could continue for a long time, without any guarantee of success. It is certainly possible that the entire budget for the experiment (eight prototypes) could be exhausted without finding a satisfactory solution. In fact, the optimal solution could be some set of factor values that lie in between the specific values of the factors that are being tested. Second, suppose the initial best guess (or one of the early ones) produces an acceptable result. Now the experimenter is tempted to stop testing, although there is no guarantee that the *best* solution has been found. A competitor that uses a better experimental approach can possibly develop a product that will outperform yours.

Another strategy of experimentation that is used extensively in practice is the **one-factor-at-a-time (OFAT)** approach. This method consists of selecting a starting point, or **baseline** set of levels, for each factor and then successively varying each factor over its range with the other factors held constant at the baseline level. After all tests are performed, a series of graphs are usually constructed showing how the response variable is affected by varying each factor with all other factors held constant.

To illustrate the OFAT approach, suppose that each one of the five factors in the drug delivery pump has two values (or levels) that we want to test. We will call these levels "low" and "high" and denote that by -1 and $+1$ respectively. Table 1.1 shows an OFAT experiment for this problem. The first thing we notice about the OFAT is that it only requires six runs. This is an apparent efficiency, as we were prepared to perform eight tests. But further examination of the experiment shows that the six runs are used very poorly. For example, only run 1 and run 2 tell you anything about changing the cylinder diameter. Furthermore, only run 1 and run 3 tell you anything about changing the cylinder length. In fact, the effect of changing any one variable is determined by using the results from only two of the six runs. This is very inefficient use of the data—every conclusion is based on only two tests even though six tests were performed. Furthermore, if there is some variability in the drug delivery rate observations (as there always will be), then this variability could have an effect on the conclusions.

Table 1.1 A One-factor-at-a-time Experiment for the Drug Delivery Pump

	Factor				
Run	**Cylinder diameter**	**Cylinder length**	**Fit**	**Plunger length**	**Tubing diameter**
1	−1	−1	−1	−1	−1
2	+1	−1	−1	−1	−1
3	−1	+1	−1	−1	−1
4	−1	−1	+1	−1	−1
5	−1	−1	−1	+1	−1
6	−1	−1	−1	−1	+1

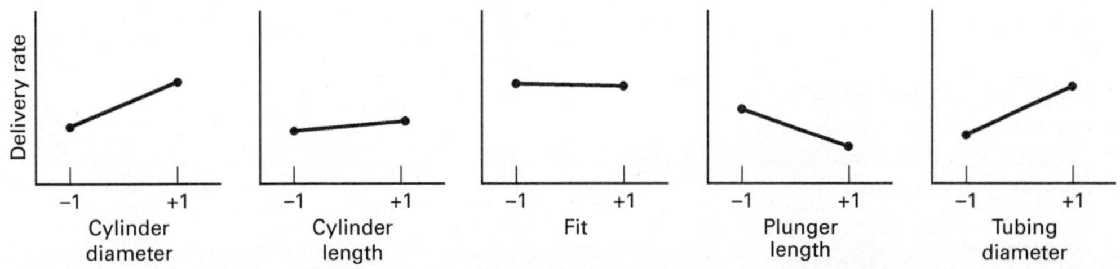

FIGURE 1.2 Results from the OFAT in Table 1.1

Figure 1.2 shows a set of hypothetical graphs that represent the results from the OFAT in Table 1.1. A straight line connects the observed response at the high and low levels of each factor. The slope of the line determines the effect of the factor; a positive slope means that increasing the level of the factor from low to high increases the drug delivery rate, while a slope that is zero (or nearly so) indicates the factor has no effect. Each one of the data values that determine the slope in each graph is the result of a single run.

Another major disadvantage of the OFAT strategy is that it fails to consider any possible **interaction** effect of the factors (an interaction is the failure of one factor to produce the same effect on the response at different levels of another factor). Figure 1.3 shows an interaction between the factors cylinder diameter and cylinder length for the drug delivery pump experiment. Notice that if the engineer uses the smaller diameter cylinder, the length of the cylinder has virtually no effect on the drug delivery rate, but if she uses the larger diameter cylinder, much different results are obtained by changing the cylinder length. Interactions between factors are very common, and if they occur, the OFAT strategy will usually produce poor results. Many engineers do not recognize this, and, consequently, OFAT experiments are run frequently in practice. (Some individuals actually think that this strategy is related to the scientific method or that it is a "sound" engineering principle.) OFAT experiments are always less efficient than other methods based on a statistical approach to design.

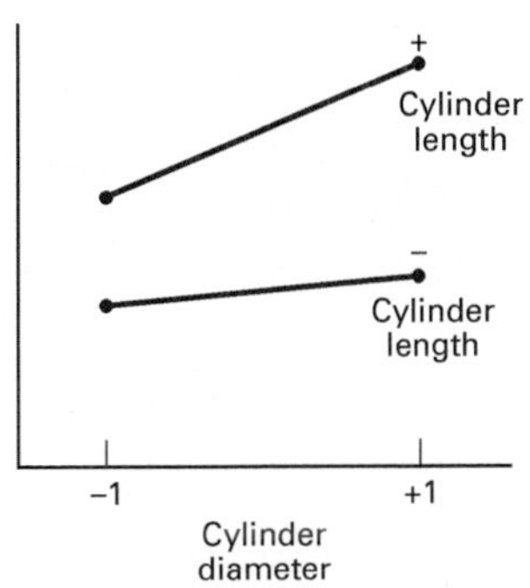

FIGURE 1.3 An Interaction Between Cylinder Diameter and Cylinder Length

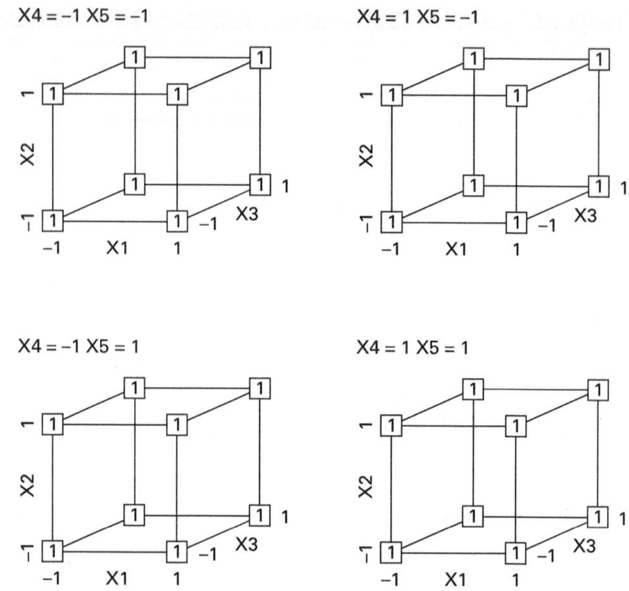

FIGURE 1.4 The 2^5 Full-factorial Design (the Symbol 1 at Each Cube Corner Indicates That a Run Should Be Made at That Location)

The correct approach to dealing with several factors is to conduct a **factorial** experiment. This is an experimental strategy in which factors are varied together, instead of one at a time. The factorial experimental design concept is extremely important, and we will illustrate how it can be applied to the drug delivery pump experiment. We will continue to assume that each factor has two levels, low and high, represented symbolically by -1 and $+1$, respectively. A complete or full-factorial experiment would require that we run every possible combination of the five factors across two levels each. This design has $2^5 = 32$ runs (see Figure 1.4). Because the experimenter only has a budget that will allow building eight prototypes, the full-factorial strategy will not work here. Fortunately, if there are four to five or more factors, a full factorial is usually not necessary and we can use a smaller fraction of the original runs. Figure 1.5 shows a one-half fraction of

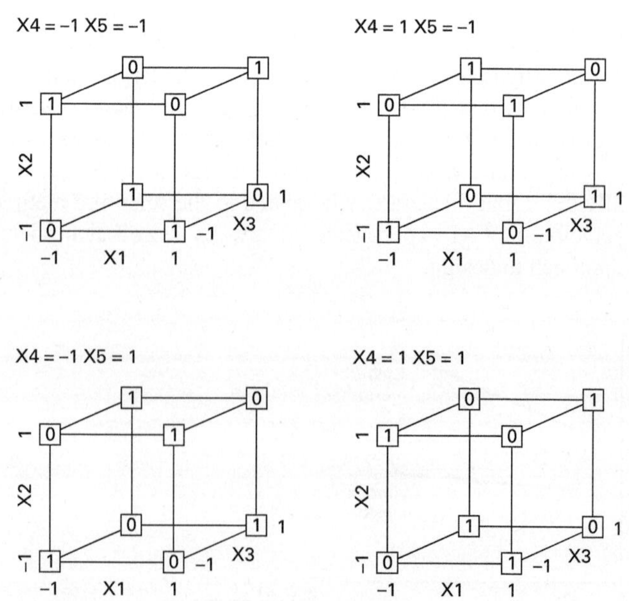

FIGURE 1.5 The One-half Fraction of the 2^5 Design (the Cells with 1 Indicate Where Runs Should Be Made)

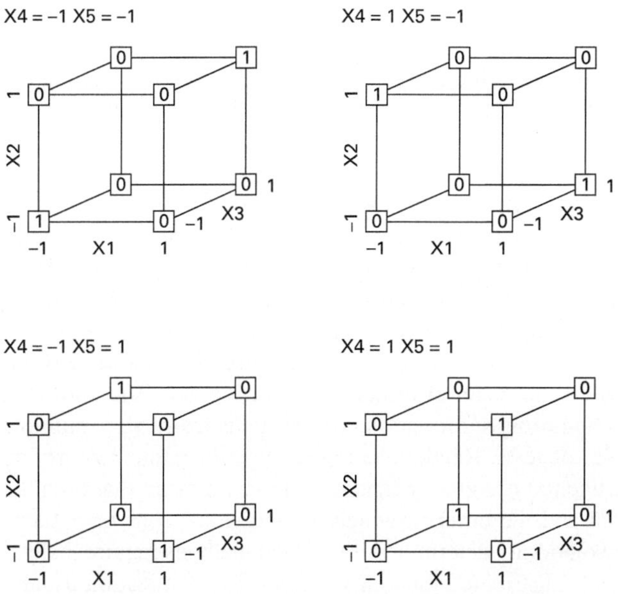

FIGURE 1.6 The One-quarter Fraction of the 2^5 Design (the Cells with 1 Indicate Where Runs Should Be Made)

the 2^5. This design has 16 runs, so it still is not feasible. This is an excellent design, however, because it allows the experimenter to estimate the effects of all five factors and all ten of the two-factor interactions involving these five factors. There is also an eight-run fraction of the 2^5, called a one-quarter fraction. This design is shown in Figure 1.6 and Table 1.2.

One very important feature of the factorial experiment is evident from this simple example; namely, factorials and fractional factorials make the most efficient use of the experimental data. Notice from Table 1.2 that this experiment has eight observations, and for every factor there are four observations on the drug delivery rate with the factor at the high level (+1) and four observations with the factor at the low level (−1). So if we take the difference in the averages of these groups of four observations as a measure of the effect of a factor on the drug delivery rate, every factor effect is estimated using all eight runs. No other strategy of experimentation makes such efficient use of the data. This is an important and useful feature of factorials.

Table 1.2 A One-quarter Fraction of the 2^5 Design for the Drug Delivery Pump Experiment

	Factor				
Run	**Cylinder diameter**	**Cylinder length**	**Fit**	**Plunger length**	**Tubing diameter**
1	−1	−1	−1	+1	+1
2	+1	−1	−1	−1	−1
3	−1	+1	−1	−1	+1
4	+1	+1	−1	+1	−1
5	−1	−1	+1	+1	−1
6	+1	−1	+1	−1	+1
7	−1	+1	+1	−1	−1
8	+1	+1	+1	+1	+1

1.2 Basic Principles

If an experiment is going to be performed most efficiently, a scientific approach to planning the experiment must be employed. **Statistical design of experiments** refers to the process of planning the experiment so that appropriate data will be collected and analyzed by statistical methods, resulting in valid and objective conclusions. A statistical approach to experimental design is necessary if we wish to draw meaningful conclusions from the data. When the problem involves data that are subject to experimental errors, statistical methods are the only **objective** approach to analysis. Thus, there are two aspects to any experimental problem: the design of the experiment and the statistical analysis of the data. These two aspects are closely related because the possibilities for analysis depend directly on the design employed, both of which are addressed in this book.

The four basic principles of experimental design are **the factorial principle**, **randomization**, **replication**, and **blocking**. We discussed the factorial principle in Section 1.1. The core idea here is that the most efficient way of acquiring information is to systematically vary multiple factors in a carefully constructed series of tests. Randomization is the cornerstone underlying the use of statistical methods in experimental design. By randomization, we mean that both the allocation of the experimental material and the order in which the individual runs or trials of the experiment are to be performed are randomly determined. Statistical methods require that the observations (or errors) be independently distributed random variables. Randomization usually makes this assumption valid. By properly randomizing the experiment, we also assist in "averaging out" the effects of extraneous factors that may be present. For example, suppose that the pressure in the drug delivery pump experiment increases throughout the day and the drug delivery rate may be affected by pump pressure. If all the smaller cylinder diameter prototypes are tested in the morning and all of the larger diameter cylinder prototypes are tested in the afternoon, we may be introducing systematic bias into the experimental results. This bias handicaps one of the cylinder diameters and consequently invalidates our results. Testing the prototypes in random order throughout the day is the solution to this problem.

Computer software programs are widely used to assist experimenters in selecting and constructing experimental designs. These programs often present the runs in the experimental design in random order. This random order is created by using a random number generator. Even with such a computer program, it is still often necessary to assign units of experimental material, operators, gauges or measurement devices, and so forth for use in the experiment. Sometimes, experimenters encounter situations where randomization of some aspect of the experiment is difficult. For example, in a chemical process, temperature may be a very hard-to-change variable as we may want to change it less often than we change the levels of other factors. In an experiment of this type, **complete randomization** would be difficult because it would add time and cost. There are statistical design methods for dealing with **restricted randomization**. Some of these approaches are discussed in subsequent chapters.

By **replication** we mean an independent repeat of each factor combination in the experiment. In the drug delivery pump experiment discussed in Section 1.1, replication would consist of testing a prototype with the smaller cylinder diameter and a prototype with the larger cylinder diameter. Thus, if four prototypes of each type are tested, we say the experiment has been replicated four times, or that four **replicates** have been obtained. Each of the eight total observations should be run in random order. Replication has two important properties. First, it allows the experimenter to obtain an estimate of the experimental error. This estimate of error becomes a basic unit of measurement for determining whether observed differences in the data are really **statistically significantly** different. Second, if the sample mean ($\bar{y}$) is used to estimate the true mean response for one of the factor levels in the experiment, replication permits the experimenter to obtain a more precise estimate of this parameter. For example, if σ^2 is the variance of an individual observation and there are n replicates, the variance of the sample mean is

$$\sigma_{\bar{y}}^2 = \frac{\sigma^2}{n}$$

The practical implication of this is that replication improves the precision of estimation of the true mean of the response variable. For example, if we had only $n = 1$ replicate and observed $y_1 = 10.1$ ml/h for the smaller diameter cylinder and $y_2 = 10.3$ ml/h for the larger diameter cylinder, we would be unable to make a statistically sound decision about the effect of the cylinder diameter on the drug delivery rate—that is, the observed difference from this experiment could be the result of experimental error. The point is that without replication we have no way of knowing why the two observations are different. On the other hand, if n was reasonably large and the experimental error was sufficiently small and if we observed $\bar{y}_1 < \bar{y}_2$, we would be reasonably safe in concluding that the larger cylinder diameter results in a higher drug delivery rate than the smaller cylinder diameter.

Often when the runs in an experiment are randomized, two (or more) consecutive runs will have exactly the same levels for some of the factors. For example, suppose we have three factors in an experiment: pressure, temperature, and time. When the experimental runs are randomized, we find the following:

Run number	**Pressure (psi)**	**Temperature (°C)**	**Time (min)**
i	30	100	30
$i+1$	30	125	45
$i+2$	40	125	45

Notice that between runs i and $i + 1$, the levels of pressure are identical and between runs $i + 1$ and $i + 2$, the levels of both temperature and time are identical. To obtain a true replicate, the experimenter needs to "twist the pressure knob" to an intermediate setting between runs i and $i + 1$, and reset pressure to 30 psi for run $i + 1$. Similarly, temperature and time should be reset to intermediate levels between runs $i + 1$ and $i + 2$ before being set to their design levels for run $i + 2$. Part of the experimental error is the variability associated with hitting and holding factor levels.

There is an important distinction between **replication** and **repeated measurements**. For example, suppose that a silicon wafer is etched in a single-wafer plasma etching process, and a critical dimension on this wafer is measured three times. These measurements are not replicates; they are a form of repeated measurements, and in this case, the observed variability in the three repeated measurements is a direct reflection of the inherent variability in the measurement system or gauge. As another illustration, suppose that as part of an experiment in semiconductor manufacturing four wafers are processed simultaneously in an oxidation furnace at a particular gas flow rate and time and then a measurement is taken on the oxide thickness of each wafer. Once again, the measurements on the four wafers are not replicates but repeated measurements. In this case, they reflect differences among the wafers and other sources of variability within that particular furnace run. Replication reflects sources of variability both **between** runs and (potentially) **within** runs.

Blocking is a design technique used to improve the precision with which comparisons among the factors of interest are made. Often blocking is used to reduce or eliminate the variability transmitted from **nuisance factors**—that is, factors that may influence the experimental response but in which we are not directly interested. For example, an experiment in a pharmaceutical manufacturing may require two batches of the active drug to make all the required runs. However, there could be differences between the batches due to variability in the process that produces the active drug, and if we are not specifically interested in this effect, we would think of the batches of active drug as a nuisance factor. Generally, a block is a set of relatively homogeneous experimental conditions. In the pharmaceutical manufacturing example, each batch of the active drug would form a block, because the variability within a batch would be expected to be smaller than the variability between batches. Typically, as in this example, each level of the nuisance factor becomes a block. Then the experimenter divides the observations from the statistical design into groups that are run in each block. We study blocking in detail in several places in the text.

The four basic principles of experimental design, the factorial principle, randomization, replication, and blocking, deserve consideration in every experiment. We illustrate and emphasize them repeatedly throughout this book.

1.3 Practical Guidelines for Designing an Experiment

To use the statistical approach in designing and analyzing an experiment, it is necessary for everyone involved in the experiment to have a clear idea in advance of exactly what is to be studied, how the data are to be collected, and at least a qualitative understanding of how these data are to be analyzed. An outline of the recommended procedure is given in Table 1.3. We now give a brief discussion of this outline and elaborate on some of the key points. For more information about practical aspects of design, see Coleman and Montgomery (1993) and the references therein.

1.3.1 Recognition of and Statement of the Problem

This may seem to be a rather obvious point, but in practice it is often neither simple to realize that a problem requiring experimentation exists nor is it easy to develop a clear and generally accepted statement of this problem. It is necessary to develop all ideas about the objectives of the experiment. Usually, it is important to solicit input from all concerned parties: engineering, quality assurance, manufacturing, marketing, management, operating personnel (who usually have much insight and who are too often ignored), and when possible the customer. For this reason, a **team approach** to designing experiments is recommended.

It is usually helpful to prepare a list of specific problems or questions that need to be addressed by the experiment. A clear statement of the problem often contributes substantially to better understanding of the phenomenon being studied and the final solution of the problem. It is also important to keep the overall objective in mind; for example, is this a new process or system—in which case the initial objective is likely to be **characterization** or **factor screening**—or is it a mature or reasonably well-understood system that has been previously characterized—in which case the objective might be **optimization**? There are many possible objectives of an experiment, including **confirmation** (is the system performing the same way now that it did in the past?), **discovery** (what happens if we explore new materials, variables, operating conditions, etc.?), and **stability** or **robustness** (under what conditions do the response variables of interest seriously

Table 1.3 Guidelines for Designing an Experiment

Step	
1. Recognition of and statement of the problem	Preexperimental planning
2. Selection of the response variable[a]	Preexperimental planning
3. Choice of factors, levels, and ranges[a]	Preexperimental planning
4. Choice of experimental design	
5. Performing the experiment	
6. Statistical analysis of the data	
7. Conclusions and recommendations	

[a] In practice, steps are often done simultaneously or in reverse order.

degrade? or, how can we reduce the variability in the response variable that arises from sources that we cannot directly control?). Obviously, the specific questions to be addressed in the experiment relate directly to the overall objectives. An important aspect of problem formulation is the recognition that one large comprehensive experiment may not answer the key questions satisfactorily. A single comprehensive experiment requires the experimenters know the answers to a lot of questions, and if they are wrong, the results may be disappointing. This leads to wasting time, materials, and other resources and may result in never answering the original research questions satisfactorily. A **sequential** approach employing a series of smaller experiments, each with a specific objective, such as factor screening, is a better strategy.

1.3.2 Selection of the Response Variable

In selecting the response variable, the experimenter should be certain that this variable really provides useful information about the process under study. Most often, the average or standard deviation (or both) of the measured characteristic will be the response variable. Multiple responses are not unusual. Gauge capability (or measurement error) is also an important factor. If gauge capability is inadequate, only relatively large factor effects will be detected by the experiment or perhaps additional replication will be required. In some situations where gauge capability is poor, the experimenter may decide to measure each experimental unit several times and use the average of the repeated measurements as the observed response. It is usually critically important to identify issues related to defining the responses of interest and how they are to be measured *before* conducting the experiment. Sometimes, designed experiments are employed to study and improve the performance of measurement systems. (As noted in Table 1.3, steps 2 and 3 are often done simultaneously or in the reverse order.)

1.3.3 Choice of Factors, Levels, and Ranges

When considering the factors that may influence the performance of a process or system, the experimenter usually discovers that these factors can be classified as either **potential design factors** or nuisance factors. The potential design factors are those factors that the experimenter may wish to vary in the experiment. Often we find that there are a lot of potential design factors, and some further classification of them is helpful. Some useful classifications are **design factors**, **held-constant factors**, and **allowed-to-vary** factors. The design factors are the factors actually selected for study in the experiment. Held-constant factors are variables that may exert some effect on the response, but for purposes of the present experiment these factors are not of interest, so they will be held at a specific level. For example, in an etching experiment in the semiconductor industry, there may be an effect that is unique to the specific plasma etch tool used in the experiment. However, this factor would be very difficult to vary in an experiment, so the experimenter may decide to perform all of the experimental runs on one particular (ideally "typical") etcher. Thus, this factor has been held constant. As an example of allowed-to-vary factors, the experimental units or the "materials" to which the design factors are applied are usually nonhomogeneous, yet we often ignore this unit-to-unit variability and rely on randomization to balance out any material or experimental unit effect. We often assume that the effects of held-constant factors and allowed-to-vary factors are relatively small.

Nuisance factors, on the other hand, may have large effects that must be accounted for, yet we may not be interested in them in the context of the present experiment. Nuisance factors are often classified as **controllable**, **uncontrollable**, or **noise factors**. A controllable nuisance factor is one whose levels may be set by the experimenter. For example, the experimenter can select different batches of raw material or different days of the week when conducting the experiment. The blocking principle, discussed in the previous section, is often useful in dealing with controllable nuisance factors. If a nuisance factor is uncontrollable in the experiment, but it can be measured, an analysis procedure called the **analysis of covariance** can often be used to compensate for its effect. For example, the relative humidity in the process environment may affect process performance, and if the humidity cannot be controlled, it can be measured and treated as a covariate. When a factor that varies naturally and uncontrollably in the process can be controlled for purposes of an experiment, we often call it a noise factor. In these kinds of experimental situations, our objective is usually to find the settings of the controllable design factors that minimize the variability transmitted from the noise factors. This is sometimes called a process robustness study or a robust design problem. Blocking, analysis of covariance, and process robustness studies are discussed later in the text.

Once the experimenter has selected the design factors, he or she must choose the ranges over which these factors will be varied and the specific levels at which runs will be made. Thought must also be given to how these factors are to be controlled at the desired values and how they are to be measured. For instance, in the drug delivery pump experiment, the engineer has defined five factors that may affect the drug delivery rate. The experimenter will also have to decide on a region of interest for each factor (that is, the range over which each factor will be varied) and on how many levels of each variable to use. **Process knowledge** is required to do this. This process knowledge is usually a combination of practical experience and theoretical understanding. It is important to investigate all factors that may be of importance and to be not overly influenced by past experience, particularly when we are in the early stages of experimentation or when the process is not very mature.

When the objective of the experiment is **factor screening** or **process characterization**, it is usually best to keep the number of factor levels low. Often, two levels are adequate in factor screening studies. Choosing the region of interest is also important. In factor screening, the region of interest should be relatively large—that is, the range over which the factors are varied should be broad. As we learn more about which variables are important and which levels produce the best results, the region of interest will usually become narrower.

The **cause-and-effect diagram** can be a useful technique for organizing some of the information generated in preexperimental planning. Figure 1.7 is the cause-and-effect diagram constructed while planning an experiment to resolve problems with wafer charging (charge accumulation on the wafers) encountered in an etching tool used in semiconductor manufacturing. The cause-and-effect diagram is also known as a **fishbone diagram** because the "effect" of interest or the response variable is drawn along the spine of the diagram and the potential causes or design factors organized in a series of ribs. The cause-and-effect diagram uses the traditional

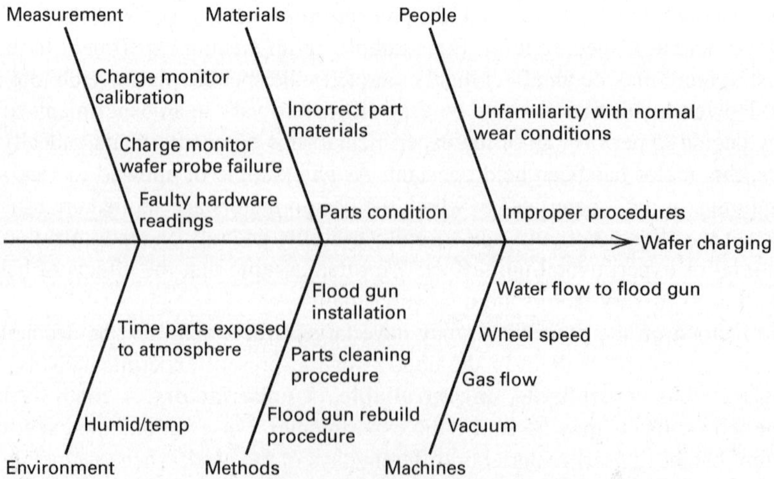

FIGURE 1.7 A Cause-and-effect Diagram for the Wafer Charging Experiment

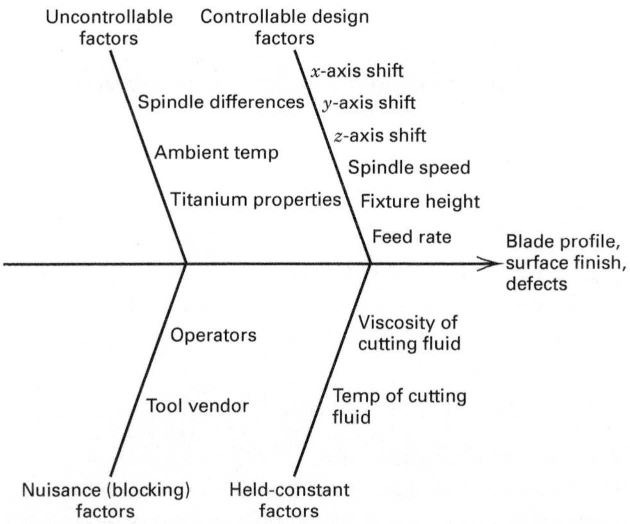

FIGURE 1.8 A Cause-and-effect Diagram for the CNC Experiment

causes of measurement, materials, people, environment, methods, and machines to organize the information and potential design factors. Notice that some of the individual causes will probably lead directly to a design factor that will be included in the experiment (such as wheel speed, gas flow, and vacuum), while others represent potential areas that will need further study to turn them into design factors (such as operators following improper procedures), and still others may lead to either factors that will be held constant during the experiment or blocked (such as temperature and relative humidity). Figure 1.8 is a cause-and-effect diagram for an experiment to study the effect of several factors on the turbine blades produced on a computer-numerical-controlled (CNC) machine. This experiment has three response variables: blade profile, blade surface finish, and surface finish defects in the finished blade. The causes are organized into groups of controllable factors from which the design factors for the experiment may be selected, uncontrollable factors whose effects will probably be balanced out by randomization, nuisance factors that may be blocked, and factors that may be held constant when the experiment is conducted. It is not unusual for experimenters to construct several different cause-and-effect diagrams to assist and guide them during preexperimental planning.

We reiterate how crucial it is to bring out all points of view and process information in steps 1 through 3. We refer to this as **preexperimental planning**. Coleman and Montgomery (1993) provided worksheets that can be useful in preexperimental planning. It is unlikely that one person has all the knowledge required to do this adequately in many situations. Therefore, we advocate for a team effort in planning the experiment. Most of your success will hinge on how well the preexperimental planning is done.

1.3.4 Experimental Design Generation

If the above preexperimental planning activities are done correctly, this step is relatively easy. Generation of the design considers several decision variables, including the sample size (number of replicates), selection of a suitable run order for the experimental trials, and determination of whether or not blocking or other randomization restrictions are involved. It also considers selecting a tentative **empirical model** to describe the results. The model is a quantitative relationship (equation) between the response and the important design factors. In many cases, a low-order polynomial model will be adequate. A **first-order** model in two variables is

$$y = \beta_0 + \beta_1 x_1 + \beta_2 x_2 + \varepsilon$$

where y is the response, the xs are the design factors, the βs are unknown parameters that will be estimated from the data in the experiment, and ε is a random error term that accounts for the experimental error in the system that is being studied. First-order models are used extensively in screening or characterization experiments. A common extension of the first-order model is to add an **interaction** term, say

$$y = \beta_0 + \beta_1 x_1 + \beta_2 x_2 + \beta_{12} x_1 x_2 + \varepsilon$$

where the cross-product term $x_1 x_2$ represents the two-factor interaction involving the design factors. Because such interactions are relatively common, the first-order model with interaction is widely used. Higher-order interactions can also be included in experiments with more than two factors if necessary. Another widely used model is the **second-order** model:

$$y = \beta_0 + \beta_1 x_1 + \beta_2 x_2 + \beta_{12} x_1 x_2 + \beta_{11} x_1^2 + \beta_{22} x_2^2 + \varepsilon$$

Second-order models are often used in optimization experiments.

There are several interactive statistical software packages that support this phase of experimental design. The experimenter can enter information about the number of factors, levels, and ranges, and the model that is anticipated to be appropriate, and these programs will then allow the experimenter to generate a design that is, in some sense, the "best" design for the problem. The software will also provide diagnostic information about how the design will perform for the anticipated model. This allows the experimenter to generate more than one design for his or her problem and compare the design choices available. These programs will usually also provide a worksheet (with the order of the runs randomized) for use in conducting the experiment.

In selecting the design, it is important to keep the experimental objectives in mind. In many engineering experiments, we already know at the outset that some of the factor levels will result in different values for the response. Consequently, we are interested in identifying *which* factors cause this difference and in estimating the **magnitude** of the response change. In other situations, we may be more interested in verifying **uniformity**. For example, two production conditions A and B may be compared, A being the standard and B being a more cost-effective alternative. The experimenter will then be interested in demonstrating that using B does not lower the yield relative to the use of A.

1.3.5 Performing the Experiment

When running the experiment, it is vital to monitor the process carefully to ensure that everything is being done according to planning. Errors in experimental procedure at this stage will usually destroy experimental validity. Up-front planning is crucial to success. It is easy to underestimate the logistical and planning aspects of running a designed experiment in a complex manufacturing or research and development environment. Coleman and Montgomery (1993) suggested that prior to conducting the experiment a few trial runs or pilot runs are often helpful. These runs provide information about the consistency of experimental material, a check on the measurement system, a rough idea of experimental error, and a chance to practice the overall experimental technique. This also provides an opportunity to revisit the decisions made in steps 1–4, if necessary.

1.3.6 Statistical Analysis of the Data

Statistical methods should be used to analyze the data so that results and conclusions are **objective** rather than judgmental in nature. If the experiment has been designed correctly and performed according to the design, the statistical methods required are not elaborate. Software packages can assist in data analysis, and most of the programs used in step 4 to select the design provide a seamless, direct interface to the statistical analysis. Often we find that simple **graphical methods** play an important role in data analysis and interpretation, and computer software plays an important role in visualizing the results of the experiment. Because many of the questions that the experimenter wants to answer can be cast into a hypothesis-testing framework, statistical hypothesis testing and confidence interval estimation procedures are very useful in analyzing the data from a designed experiment. Fitting the **empirical model** for the experiment is also important. Residual analysis and model adequacy checking are also well-advised analysis techniques. We discuss these issues in detail later. Remember that statistical methods cannot prove that a factor (or factors) has a particular effect. They only provide guidelines as to the reliability and validity of results. When properly applied, statistical methods do not allow anything to be proved experimentally, but they do allow us to measure the likely error or risk in a conclusion. The primary advantage of statistical methods is that they add **objectivity** to the decision-making process. Statistical techniques coupled with good engineering or process knowledge and common sense generally lead to sound conclusions.

1.3.7 Conclusions and Recommendations

Once the data have been analyzed, the experimenter must draw *practical* conclusions about the results and recommend a course of action. Graphical methods are often useful in this stage, particularly in presenting the results to others. **Follow-up runs** and **confirmation testing** should also be performed to validate the conclusions from the experiment. Throughout this entire process, it is important to keep in mind that experimentation is an important part of the learning process, where we tentatively formulate hypotheses about a system, perform experiments to investigate these hypotheses, and on the basis of the results formulate new hypotheses, and so on. This suggests that experimentation is **iterative**. It is often a major mistake to design a single, large, comprehensive experiment at the start of a study. A successful experiment requires knowledge of the important factors, the ranges over which these factors should be varied, the appropriate number of levels to use, the proper units of measurement for these variables, and the correct form of the model for the results.

Generally, we do not perfectly know the answers to these questions, but we learn about them as we go along. As an experimental program progresses, we often drop some input variables, add others, change the region of exploration for some factors, or add new response variables. Consequently, it is desirable to experiment **sequentially**. This will ensure that sufficient resources are available to perform confirmation runs and ultimately accomplish the final objective of the experiment.

1.4 A Brief History of Designed Experiments

There have been four eras in the modern development of statistical experimental design. The agricultural era was led by the pioneering work of Sir Ronald A. Fisher in the 1920s and early 1930s. During that time, Fisher was responsible for statistics and data analysis at the Rothamsted Agricultural Experimental Station near London, England. Fisher recognized that a poor method for generating data often hampered the data analysis. By interacting with scientists and researchers in many fields, he developed insights that led to the four basic principles of experimental design that we discussed in Section 1.2: the factorial principle, randomization, replication, and blocking. Fisher systematically introduced statistical thinking and principles into designing experimental investigations, including the likelihood concept and the analysis of variance. His two books (the most recent editions are Fisher (1958, 1966)) had profound influence on the use of statistics, particularly in agricultural and related life sciences. It is not an overstatement to say that Fisher's work transformed agricultural science. For an excellent biography of Fisher, see Box (1978).

Although applications of statistical design in industrial settings certainly began in the 1930s, the second, or industrial, era was catalyzed by the development of response surface methodology (RSM) by Box and Wilson (1951). They recognized and exploited the fact that many industrial experiments are fundamentally different from their agricultural counterparts in two ways: (1) the response variable can usually be observed (nearly) immediately and (2) the experimenter can quickly learn crucial information from a small group of runs that can be used to plan the next experiment. Box (1999) calls these two features of industrial experiments **immediacy** and **sequentiality**. Over the next 30 years, RSM and other design techniques spread throughout the chemical and the process industries, mostly in research and development work. George Box was the intellectual leader of this movement. However, the application of statistical design at the plant or manufacturing process level was still not extremely widespread. Some of the reasons for this include an inadequate training in basic statistical concepts and methods for engineers, scientists, and business personnel, the lack of computing resources and user-friendly software to support its implementation.

It was during the second industrial era that work on **optimal** design of experiments began. Kiefer (1959, 1961) and Kiefer and Wolfowitz (1959) proposed a formal approach to selecting a design based on specific objective optimality criteria. Their initial approach was to select a design that would result in the model parameters being estimated with the best possible precision. This approach did not find much application because of the lack of computer tools for its implementation. However, there have been great advances in both algorithms for generating optimal designs and computing capability over the last 20 years.

The increasing interest of Western industry in quality improvement that began in the late 1970s ushered in the third era of statistical design. The work of Genichi Taguchi (Taguchi and Wu (1980), Kackar (1985), and Taguchi (1987, 1991)) had a significant impact on expanding the interest in and the use of designed experiments. Taguchi advocated using designed experiments for what he termed **robust parameter design**, or

1. Making processes insensitive to environmental factors or other factors that are difficult to control
2. Making products insensitive to variation transmitted from components
3. Finding levels of the process variables that force the mean to a desired value while reducing variability around this value.

Taguchi suggested highly fractionated factorial designs and other orthogonal arrays along with some novel statistical methods to solve these problems. The resulting methodology generated much discussion and controversy. Part of the controversy arose because Taguchi's methodology was advocated in the West initially (and primarily) by entrepreneurs, and the underlying statistical science had not been adequately peer reviewed. By the late 1980s, the results of peer review indicated that although Taguchi's engineering concepts and objectives were well founded, there were substantial problems with his experimental strategy and methods of data analysis. For specific details of these issues, see Box (1988), Box, Bisgaard, and Fung (1988), Hunter (1985, 1989), Pignatiello and Ramberg (1992), and Myers, Montgomery, and Anderson-Cook (2016). Many of these concerns are also summarized in the extensive panel discussion in the May 1992 issue of *Technometrics* (see Nair et al. (1992)).

There were several positive outcomes of the Taguchi controversy. First, designed experiments became more widely used in the discrete parts of industries, including automotive and aerospace manufacturing, electronics and semiconductors, and many other industries that had previously made little use of the technique. It was also one of the drivers of the fourth era of statistical design, because not only did applications of designed experiments increase but also these applications spawned new research problems and a lot of highly useful new methodology appeared. An important aspect of this new methodology was renewed interest in optimal design. Because computers were more powerful and widely available, algorithms to construct optimal designs were developed and began to be implemented in commercial software.

This fourth era of statistical design has included an expanded general interest in optimal design by researchers and software developers, as well as practitioners. Today the optimal design approach is a viable and attractive way to approach experimental design problems in the industrial and business world. We use the optimal design approach extensively throughout this book. This era has also produced other new methodology, including alternatives to Taguchi's technical methods that allow his engineering concepts to be carried into practice efficiently and effectively. Formal education in statistical experimental design is becoming part of many engineering programs in universities, at both undergraduate and graduate levels. More students in business are being exposed to the concepts as well. The successful integration of good experimental design practice into engineering, science, and business is a key factor in future competitiveness. Applications of designed experiments have grown far beyond the agricultural origins. There is not a single area of science, engineering, or business that has not successfully employed statistically designed experiments. There has also been considerable utilization of designed experiments in many other areas, including the service sector of business, financial services, government operations, e-commerce, and many nonprofit business sectors. An article appeared in *Forbes* magazine on March 11, 1996, entitled The New Mantra: MVT," where MVT stands for "multivariable testing," a term that the authors use to describe factorial designs. The article describes the many successes that a diverse group of companies have had through their use of statistically designed experiments.

1.5 A Review: Using Statistical Techniques in Experimentation

Much of the research in engineering, science, business, and industry is empirical and makes extensive use of experimentation. Statistical methods can greatly increase the efficiency of these experiments and often strengthen the conclusions so obtained. The proper use of statistical techniques in experimentation requires that the experimenter keep the following points in mind:

1. **Use your nonstatistical knowledge of the problem.** Experimenters are usually highly knowledgeable in their fields. For example, a civil engineer working on a problem in hydrology typically has considerable practical experience and formal academic training in this area, and a marketing research analyst has a lot of knowledge about the specific business problem under investigation. In some fields, there is a large body of physical theory from which relationships between factors and responses are drawn. This type of nonstatistical knowledge is invaluable in choosing factors, determining factor levels, deciding how many replicates to run, interpreting the results of the analysis, and so forth. Using a designed experiment is no substitute for thinking about the problem.

2. **Keep the design and analysis as simple as possible.** Don't be overzealous in the use of complex, sophisticated statistical techniques. Relatively simple design and analysis methods are adequate and easier to explain. This is a good place to reemphasize steps 1–3 of the procedure recommended in Section 1.3. If you do the preexperimental planning carefully and select a reasonable design, the analysis will almost always be relatively straightforward. In fact, a well-designed experiment will sometimes almost analyze itself! However, if you botch the preexperimental planning and execute the experimental design badly, it is unlikely that even the most complex and elegant statistics can save the day.

3. **Recognize the difference between practical and statistical significance.** Just because two experimental conditions produce mean responses that are statistically significantly different, there is no assurance that this difference is large enough to have any practical value.

For example, an engineer may determine that a modification to an automobile fuel injection system may produce a true mean improvement in gasoline mileage of 0.1 mi/gal and be able to determine that this is a statistically significant result. However, if the cost of the modification is $1000, the 0.1 mi/gal difference will be too small to be of any practical value.

4. **Experiments are usually iterative.** Remember that in most situations it is unwise to design too comprehensive an experiment at the start of a study. Successful design requires the knowledge of important factors, the ranges over which these factors are varied, the appropriate number of levels for each factor, and the proper methods and units of measurement for each factor and response. Generally, we are not well equipped to answer these questions at the beginning of the experiment, but we learn the answers as we go along. This argues in favor of the **iterative**, or **sequential**, approach discussed previously. Of course, there are situations where comprehensive experiments are entirely appropriate, but as a general rule most experiments should be iterative. Often these first efforts are primarily learning experiences, and some resources must be available to accomplish the final objectives of the experiment.

1.6 Review of Some Basic Statistical Concepts and Methods

1.6.1 Data Description

Statistics is the **science of data**. Collecting, describing, displaying, analyzing, and drawing conclusions from data are some of the many aspects of statistics. As an example, the United States Golf Association tests golf balls to ensure that they conform to the rules of golf. Balls are tested for weight, diameter, roundness, and conformance to an overall distance standard. The overall distance test is conducted by hitting balls with a driver swung by a mechanical device nicknamed Iron Byron, after the legendary great player Byron Nelson, whose swing the machine is said to emulate. The distances in yards achieved for 10 balls of a particular brand are as follows: 292, 270, 271, 278, 279, 268, 284, 271, 262, and 283. These distances are the total of the carry in the air and the distance the ball bounces or rolls after contact with the ground. We think of these observations as a **sample** of the possible distances that golf balls of this particular type could possibly travel. Much of statistics is involved with drawing conclusions about a population from sample data that is collected from that population.

Notice that the individual observations in the sample differ, so there is fluctuation, or **noise**, in the observed distances. This noise is usually called **experimental error** or simply **error**. It is a statistical error, meaning that it arises from variation that is uncontrolled and generally unavoidable. In this case, the error arises from ball-to-ball differences, differences in Iron Byron's swing from shot to shot, the differences in performance of the golf club from shot to shot (such as the flex in the shaft or the point of contact of the ball on the club face), the point at which the ball contacts the ground, which impacts the amount of roll and other factors. The presence of error or noise implies that the response variable, overall distance, is a **random variable**. A random variable may be either discrete or continuous. If the set of all possible values of the random variable is either finite or countably infinite, then the random variable is discrete, whereas if the set of all possible values of the random variable is an interval, then the random variable is continuous.

1.6.1.1 Graphical and Numerical Description of Variability

We often use simple graphical methods to assist in analyzing the data from an experiment. The dot diagram, illustrated in Figure 1.9 for the golf ball distance data, is a very useful device for displaying a small body of data (say up to about 20 observations). The dot diagram enables the

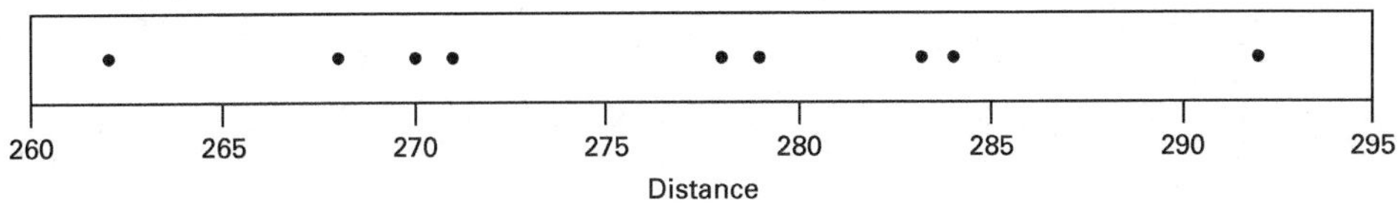

FIGURE 1.9
A Dot Diagram of the Golf Ball Distance Data

experimenter to see quickly the general location or central tendency of the observations and their spread. For example, in the golf ball overall distance data, the dot diagram reveals that the center of the data is about 275 yards, but that there is a lot of variability, with some balls traveling as little as 262 yards and some balls traveling as far as 292 yards, a range of 30 yards.

We can also describe data numerically. The **sample average** or **sample mean** is a very good way to measure the central tendency in sample data. The sample mean is defined as

$$\bar{y} = \frac{\sum_{i=1}^{n} y_i}{n} \tag{1.1}$$

where y_i are the observations in the sample and n is the sample size. The sample average for the golf ball distance data is

$$\bar{y} = \frac{\sum_{i=1}^{n} y_i}{n} = \frac{292 + 270 + 271 + \cdots + 283}{10} = \frac{2758}{10} = 275.8 \text{ yards}$$

The sample average is a measure of the middle of the data in the sense that if you view the dot diagram as a system of unit weights the diagram balances exactly at the sample average. It is analogous to the first moment in mechanics.

Variability is usually measured by the **sample variance** or **sample standard deviation**. The sample variance is

$$S^2 = \frac{\sum_{i=1}^{n} (y_i - \bar{y})^2}{n-1} = \frac{\sum_{i=1}^{n} y_i^2 - \frac{\left(\sum_{i=1}^{n} y_i\right)^2}{n}}{n-1} \tag{1.2}$$

A large sample variance implies large variability in the sample data. The sample standard deviation S is the positive square root of the sample variance. Many individuals prefer to work with the standard deviation because it is expressed in the same units as the observations y (yards in the golf ball distance data), while the sample variance is in the square of the original units. The sample variance and standard deviation for the golf ball distance data are

$$S^2 = \frac{\sum_{i=1}^{n} y_i^2 - \frac{\left(\sum_{i=1}^{n} y_i\right)^2}{n}}{n-1} = \frac{761{,}384.0 - \frac{(2758)^2}{10}}{9} = 80.84$$

$$S = 8.99 \text{ yards}$$

If the data are fairly numerous, the dots in a dot diagram become difficult to distinguish and a histogram may be preferable. Figure 1.10 presents a histogram for 200 observations on the metal recovery, or yield, from a smelting process. The histogram shows the central tendency, spread, and

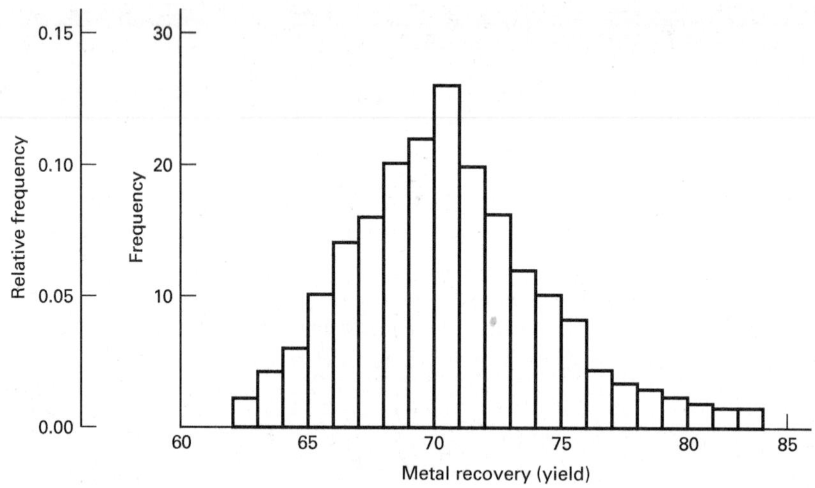

FIGURE 1.10 Histogram of 200 Observations on Metal Recovery (Yield) from a Smelting Process

general shape of the distribution of the data. Recall that a histogram is constructed by dividing the horizontal axis into bins (usually of equal length) and drawing a rectangle over the jth bin with the area of the rectangle proportional to n_j, the number of observations that fall in that bin. If the bins are of equal length, then the height of the rectangle will equal the number of sample observations in that bin. Histograms are large-sample tools and should generally not be used unless there are at least 30 or more observations.

The **box plot** (or **box-and-whisker plot**) is a useful way to display data. A box plot displays the minimum, the maximum, the lower and upper quartiles (the 25th percentile and the 75th percentile, respectively), and the median (the 50th percentile) on a rectangular box aligned either horizontally or vertically. The box extends from the lower quartile to the upper quartile, and a line is drawn through the box at the median. Lines (or whiskers) extend from the ends of the box to (typically) the minimum and maximum values. (There are several variations of box plots that have different rules for denoting the extreme sample points; see Montgomery and Runger (2018) for more details.) Figure 1.11 presents the box plot for the golf ball distance data. The median distance is 274.5 yards, which is very similar to the sample average (275.8 yards), and

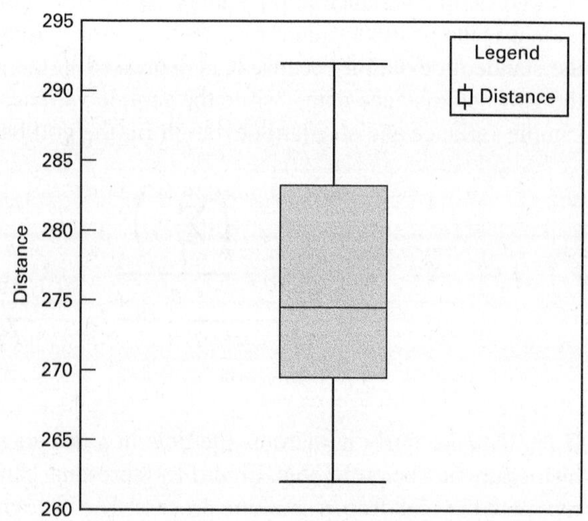

FIGURE 1.11 Box Plot of the Golf Ball Distance Data

both whiskers are similar in length. These are indications that the sample of data we have collected has a reasonably symmetric shape.

Dot diagrams, histograms, and box plots are useful graphical tools for summarizing the information in a sample of data, and the sample mean and sample variance or standard deviation are widely used numerical descriptive summaries. To describe the observations that might occur in a sample more completely, we can also use the concept of a probability distribution.

1.6.1.2 Probability Distributions

A random variable, say y, is characterized by its probability distribution. If y is discrete, we often call the probability distribution of y, say $p(y)$, the probability mass function of y. If y is continuous, the probability distribution of y, say $f(y)$, is often called the probability density function for y.

Figure 1.12 illustrates hypothetical discrete and continuous probability distributions.

Notice that in the discrete probability distribution (Figure 1.12a), it is the height of the function $p(y_j)$ that represents probability, whereas in the continuous case (Figure 1.12b), it is the area under the curve $f(y)$ associated with a given interval that represents probability. The properties of probability distributions may be summarized quantitatively as follows:

$$y \text{ discrete}: \quad 0 \le p(y_j) \le 1 \quad \text{all values of } y_j$$

$$P(y = y_j) = p(y_j) \quad \text{all values of } y_j$$

$$\sum_{\substack{\text{all values}\\ \text{of } y_j}} p(y_j) = 1$$

$$y \text{ continuous}: \quad 0 \le f(y)$$

$$P(a \le y \le b) = \int_a^b f(y)\, dy$$

$$\int_{-\infty}^{\infty} f(y)\, dy = 1$$

1.6.1.3 Mean, Variance, and Expected Values

The **mean**, μ, of a probability distribution is a measure of its central tendency or location. Mathematically, we define the mean as

$$\mu = \begin{cases} \int_{-\infty}^{\infty} y f(y)\, dy & y \text{ continuous} \\ \sum_{\text{all } y} y p(y_j) & y \text{ discrete} \end{cases} \tag{1.3}$$

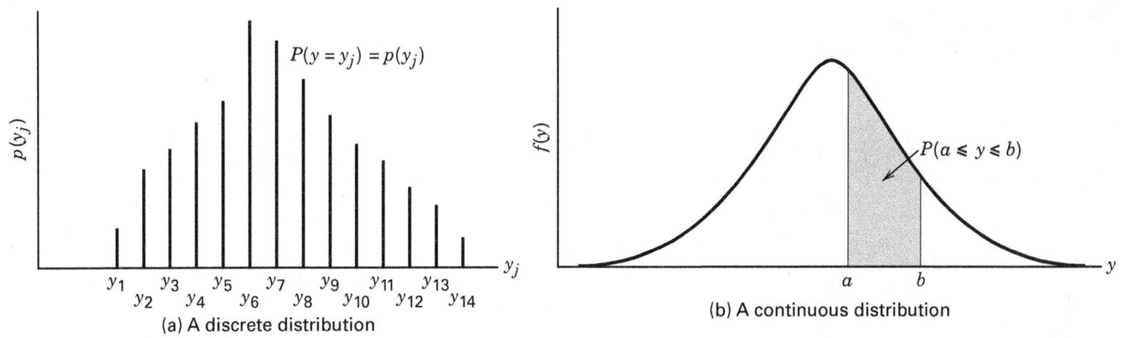

FIGURE 1.12 (a) Discrete and (b) Continuous Probability Distributions

We may also express the mean in terms of the **expected value** or the long-run average value of the random variable y as

$$\mu = E(y) = \begin{cases} \int_{-\infty}^{\infty} yf(y)\,dy & y \text{ continuous} \\ \sum_{\text{all } y} yp(y_j) & y \text{ discrete} \end{cases} \tag{1.4}$$

where E denotes the **expected value operator**.

The variability or dispersion of a probability distribution can be measured by the **variance**, defined as

$$\sigma^2 = \begin{cases} \int_{-\infty}^{\infty} (y-\mu)^2 f(y)\,dy & y \text{ continuous} \\ \sum_{\text{all } y} (y-\mu)^2 p(y_j) & y \text{ discrete} \end{cases} \tag{1.5}$$

Note that the variance can be expressed entirely in terms of expectation because

$$\sigma^2 = E[(y-\mu)^2] \tag{1.6}$$

Finally, the variance is used so extensively that it is convenient to define a **variance operator** V such that

$$V(y) = E[(y-\mu)^2] = \sigma^2 \tag{1.7}$$

The concepts of expected value and variance are used extensively throughout this book, and it may be helpful to review several elementary results concerning these operators. If y is a random variable with mean μ and variance σ^2 and c is a constant, then

1. $E(c) = c$
2. $E(y) = \mu$
3. $E(cy) = cE(y) = c\mu$
4. $V(c) = 0$
5. $V(y) = \sigma^2$
6. $V(cy) = c^2V(y) = c^2\sigma^2$

If there are two random variables, say, y_1 with $E(y_1) = \mu_1$ and $V(y_1) = \sigma_1^2$ and y_2 with $E(y_2) = \mu_2$ and $V(y_2) = \sigma_2^2$, we have

7. $E(y_1 + y_2) = E(y_1) + E(y_2) = \mu_1 + \mu_2$

It is possible to show that

8. $V(y_1 + y_2) = V(y_1) + V(y_2) + 2\text{Cov}(y_1, y_2)$

where

$$\text{Cov}(y_1, y_2) = E[(y_1 - \mu_1)(y_2 - \mu_2)] \tag{1.8}$$

is the **covariance** of the random variables y_1 and y_2. The covariance is a measure of the linear association between y_1 and y_2. More specifically, we may show that if y_1 and y_2 are independent, then $\text{Cov}(y_1, y_2) = 0$. We may also show that

9. $V(y_1 - y_2) = V(y_1) + V(y_2) - 2\text{Cov}(y_1, y_2)$

If y_1 and y_2 are **independent**, we have

10. $V(y_1 \pm y_2) = V(y_1) + V(y_2) = \sigma_1^2 + \sigma_2^2$

and

11. $E(y_1 \cdot y_2) = E(y_1) \cdot E(y_2) = \mu_1 \cdot \mu_2$

However, note that, in general

12. $E\left(\frac{y_1}{y_2}\right) \neq \frac{E(y_1)}{E(y_2)}$

regardless of whether or not y_1 and y_2 are independent.

1.6.2 Random Samples, Statistics and Sampling Distributions

The objective of statistical inference is to draw conclusions about a population using a sample from that population. Most of the methods that we will study assume that **random samples** are used. That is, if the population contains N elements and a sample of n of them is to be selected, and if each of the $N!/[(N-n)!n!]$ possible samples has an equal probability of being chosen, then the procedure employed is called **random sampling**. A key feature of a random sample is that the observations in the random sample are independent random variables. In practice, it is sometimes difficult to obtain random samples, and random numbers generated by a computer program may be helpful.

Statistical inference makes considerable use of quantities computed from the observations in the sample. We define a **statistic** as any function of the observations in a sample that does not contain unknown parameters. For example, the sample mean $\overline{y}$ and the sample variance S^2 are both statistics. The sample mean is a **point estimator** of the population mean μ, and the sample variance S^2 is a **point estimator** of the population variance σ^2. In general, an estimator of an unknown parameter is a statistic that corresponds to that parameter. Note that a point estimator is a random variable. A particular numerical value of an estimator, computed from sample data, is called an **estimate**. For example, suppose we wish to estimate the mean and variance of the breaking strength of a particular type of textile fiber. A random sample of $n = 25$ fiber specimens is tested, and the breaking strength is recorded for each. The sample mean and variance are computed according to Eqs. (1.1) and (1.2), respectively, and are $\overline{y} = 21.7$ and $S^2 = 1.20$. Therefore, the estimate of μ is 21.7, and the estimate of σ^2 is 1.20.

We know that the variance of the sample mean is $\sigma_{\overline{y}}^2 = \sigma^2/n$. The square root of this quantity, $\sigma_{\overline{y}} = \sigma/\sqrt{n}$, is called the **standard error** of the mean. A small value of the standard error indicates that the mean has been estimated with high precision. When the population variance is unknown, we can replace σ by the sample standard deviation S, and $\widehat{\sigma}_{\overline{y}} = S/\sqrt{n}$ is called the **estimated standard error**.

Several properties are required of good point estimators. Two of the most important are the following:

1. An **unbiased** point estimator is often desirable. That is, the long-run average or expected value of the point estimator should be equal to the parameter that is being estimated.

   Although unbiasedness is desirable, this property alone does not always make an estimator a good one.

2. An unbiased estimator should have minimum variance. This property states that the minimum variance point estimator has a variance that is smaller than the variance of any other estimator of that parameter.

We may easily show that $\bar{y}$ and S^2 are unbiased estimators of μ and σ^2, respectively. First consider $\bar{y}$. Using the properties of expectation, we have

$$\begin{aligned}
E(\bar{y}) &= E\left(\frac{\sum_{i=1}^{n} y_i}{n}\right) \\
&= \frac{1}{n}\sum_{i=1}^{n} E(y_i) \\
&= \frac{1}{n}\sum_{i=1}^{n} \mu \\
&= \mu
\end{aligned}$$

because the expected value of each observation y_i is μ. Thus, $\bar{y}$ is an unbiased estimator of μ.

Now consider the sample variance S^2. We have

$$\begin{aligned}
E(S^2) &= E\left[\frac{\sum_{i=1}^{n} (y_i - \bar{y})^2}{n-1}\right] \\
&= \frac{1}{n-1} E\left[\sum_{i=1}^{n} (y_i - \bar{y})^2\right] \\
&= \frac{1}{n-1} E(SS)
\end{aligned}$$

where $SS = \sum_{i=1}^{n} (y_i - \bar{y})^2$ is the **corrected sum of squares** of the observations y_i. Now

$$\begin{aligned}
E(SS) &= E\left[\sum_{i=1}^{n} (y_i - \bar{y})^2\right] \\
&= E\left[\sum_{i=1}^{n} y_i^2 - n\bar{y}^2\right] \qquad (1.9)
\end{aligned}$$

$$\begin{aligned}
&= \sum_{i=1}^{n} (\mu^2 + \sigma^2) - n(\mu^2 + \sigma^2/n) \\
&= (n-1)\sigma^2 \qquad (1.10)
\end{aligned}$$

Therefore,

$$E(S^2) = \frac{1}{n-1} E(SS) = \sigma^2$$

Therefore, S^2 is an unbiased estimator of σ^2.

1.6.2.1 Degrees of Freedom

The quantity $n - 1$ in Eq. (1.10) is called the number of degrees of freedom of the sum of squares SS. This is a very general result; that is, if y is a random variable with variance σ^2 and $SS = \sum_{i=1}^{n} (y_i - \bar{y})^2$ has v degrees of freedom, then

$$E\left(\frac{SS}{v}\right) = \sigma^2 \tag{1.11}$$

The number of degrees of freedom of a sum of squares is equal to the number of independent elements in that sum of squares. For example, SS in Eq. (1.9) consists of the sum of squares of the n elements $y_1, y_2, \ldots, y_n$. These elements are not all independent because $\sum_{i=1}^{n}(y_i - \bar{y}) = 0$; in fact, only $n - 1$ of them are independent, implying that SS has $n - 1$ degrees of freedom.

1.6.2.2 The Normal and Other Sampling Distributions

Often we are able to determine the probability distribution of a particular statistic if we know the probability distribution of the population from which the sample was drawn. The probability distribution of a statistic is called a sampling distribution. We now briefly discuss several useful sampling distributions.

One of the most commonly used sampling distributions is the normal distribution. If y is a normal random variable, the probability distribution of y is

$$f(y) = \frac{1}{\sigma\sqrt{2\pi}} e^{-(1/2)[(y-\mu)/\sigma]^2} \quad -\infty < y < \infty \tag{1.12}$$

where $-\infty < \mu < \infty$ is the mean of the distribution and $\sigma^2 > 0$ is the variance. The normal distribution is shown in Figure 1.13.

Because sample runs that differ as a result of experimental error often are well described by the normal distribution, it plays a central role in the analysis of data from designed experiments. Many important sampling distributions may also be defined in terms of normal random variables. We often use the notation $y \sim N(\mu, \sigma^2)$ to denote that y is distributed normally with mean μ and variance σ^2.

An important special case of the normal distribution is the **standard normal distribution**; that is, $\mu = 0$ and $\sigma^2 = 1$. We see that if $y \sim N(\mu, \sigma^2)$, the random variable

$$z = \frac{y - \mu}{\sigma} \tag{1.13}$$

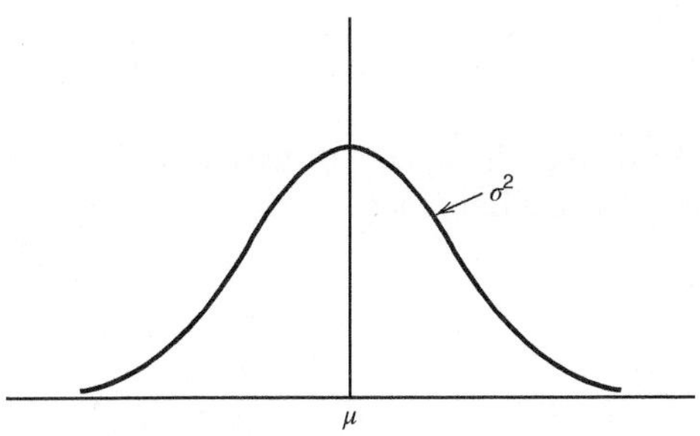

FIGURE 1.13 The Normal Distribution

follows the standard normal distribution, denoted $z \sim N(0, 1)$. The operation demonstrated in Eq. (1.13) is often called **standardizing** the normal random variable. It is often useful to be able to calculate probabilities associated with the standard normal random variable.

Many statistical techniques assume that the random variable is normally distributed. The central limit theorem is often a justification of approximate normality.

The Central Limit Theorem

If $y_1, y_2, \ldots, y_n$ is a sequence of n independent and identically distributed random variables with $E(y_i) = \mu$ and $V(y_i) = \sigma^2$ (both finite) and $x = y_1 + y_2 + \cdots + y_n$, then the limiting form of the distribution of

$$z_n = \frac{x - n\mu}{\sqrt{n\sigma^2}}$$

as $n \to \infty$, is the standard normal distribution.

This result states essentially that the sum of n independent and identically distributed random variables is approximately normally distributed. In many cases, this approximation is good for very small n, say $n < 10$, whereas in other cases large n is required, say $n > 100$. Frequently, we think of the error in an experiment as arising in an additive manner from several independent sources; consequently, the normal distribution becomes a plausible model for the combined experimental error.

An important sampling distribution that can be defined in terms of normal random variables is the **chi-square** or **χ^2 distribution**. If $z_1, z_2, \ldots, z_k$ are normally and independently distributed random variables with mean 0 and variance 1, abbreviated NID(0, 1), then the random variable

$$x = z_1^2 + z_2^2 + \cdots + z_k^2$$

follows the **chi-square distribution with k degrees of freedom**. The density function of chi-square is

$$f(x) = \frac{1}{2^{k/2}\Gamma\left(\frac{k}{2}\right)} x^{(k/2)-1} e^{-x/2} \quad x > 0 \tag{1.14}$$

Several chi-square distributions are shown in Figure 1.14. The distribution is asymmetric, or **skewed**, with the mean and variance

$$\mu = k$$
$$\sigma^2 = 2k$$

respectively. The appendix provides JMP scripting language (JSL) commands for calculating the percentage points of the chi-square distribution.

As an example of a random variable that follows the chi-square distribution, suppose that $y_1, y_2, \ldots, y_n$ is a random sample from an $N(\mu, \sigma^2)$ distribution. Then

$$\frac{SS}{\sigma^2} = \frac{\sum_{i=1}^{n} (y_i - \bar{y})^2}{\sigma^2} \sim \chi^2_{n-1} \tag{1.15}$$

That is, SS/σ^2 is distributed as chi-square with $n - 1$ degrees of freedom.

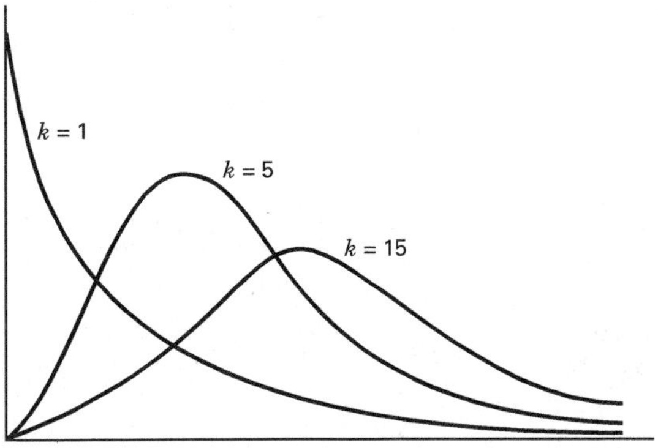

FIGURE 1.14 The Chi-square Distribution for Several Degrees of Freedom

Many of the techniques used in this book involve the computation and manipulation of sums of squares. The result given in Eq. (1.15) is extremely important and occurs repeatedly; a sum of squares in normal random variables when divided by σ^2 follows the chi-square distribution.

The sample variance can be written as

$$S^2 = \frac{SS}{n-1} \tag{1.16}$$

If the observations in the sample are NID(μ, σ^2), then the distribution of S^2 is $[\sigma^2/(n-1)]\chi^2_{n-1}$. Thus, the sampling distribution of the sample variance is a constant times the chi-square distribution if the population is normally distributed.

If z and χ^2_k are independent standard normal and chi-square random variables, respectively, the random variable

$$t_k = \frac{z}{\sqrt{\chi^2_k/k}} \tag{1.17}$$

follows the t-**distribution with k degrees of freedom**, denoted t_k. The density function of t is

$$f(t) = \frac{\Gamma[(k+1)/2]}{\sqrt{k\pi}\Gamma(k/2)} \frac{1}{[(t^2/k)+1]^{(k+1)/2}} \qquad -\infty < t < \infty \tag{1.18}$$

and the mean and variance of t are $\mu = 0$ and $\sigma^2 = k/(k-2)$ for $k > 2$, respectively. Several t-distributions are shown in Figure 1.15. Note that if $k = \infty$, the t-distribution becomes the standard normal distribution. The appendix provides JSL commands for calculating the percentage points of the t-distribution.

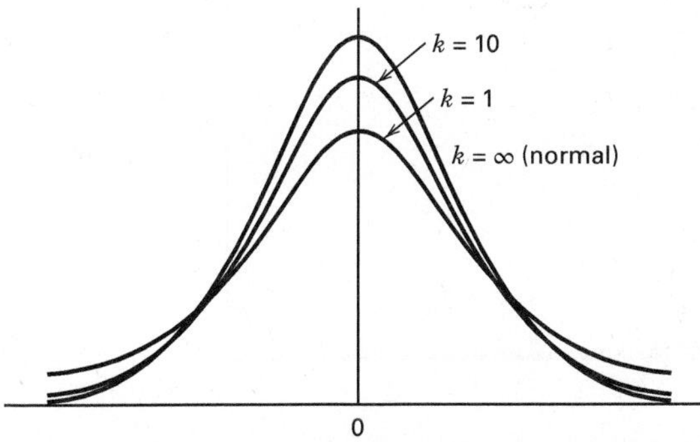

FIGURE 1.15 The t-distribution for Several Different Degrees of Freedom

If $y_1, y_2, \ldots, y_n$ is a random sample from the $N(\mu, \sigma^2)$ distribution, then the quantity

$$t = \frac{\overline{y} - \mu}{S/\sqrt{n}} \tag{1.19}$$

is distributed as t with $n - 1$ degrees of freedom.

The final sampling distribution that we will consider is the ***F*-distribution**. If χ_u^2 and χ_v^2 are two independent chi-square random variables with u and v degrees of freedom, respectively, then the ratio

$$F_{u,v} = \frac{\chi_u^2/u}{\chi_v^2/v} \tag{1.20}$$

follows the ***F*-distribution with *u numerator* degrees of freedom** and ***v denominator* degrees of freedom**. If x is an F random variable with u numerator and v denominator degrees of freedom, then the probability distribution of x is

$$h(x) = \frac{\Gamma\left(\frac{u+v}{2}\right)\left(\frac{u}{v}\right)^{u/2} x^{(u/2)-1}}{\Gamma\left(\frac{u}{x}\right)\Gamma\left(\frac{v}{2}\right)\left[\left(\frac{u}{v}\right)x + 1\right]^{(u+v)/2}} \qquad 0 < x < \infty \tag{1.21}$$

Several F-distributions are shown in Figure 1.16. This distribution is very important in the statistical analysis of designed experiments. The appendix provides JSL commands for calculating the percentage points of the F-distribution.

As an example of a statistic that is distributed as F, suppose we have two independent normal populations with common variance σ^2. If $y_{11}, y_{12}, \ldots, y_{1n_1}$ is a random sample of n_1 observations from the first population, and if $y_{21}, y_{22}, \ldots, y_{2n_2}$ is a random sample of n_2 observations from the second, then

$$\frac{S_1^2}{S_2^2} \sim F_{n_1-1, n_2-1} \tag{1.22}$$

where S_1^2 and S_2^2 are the two sample variances. This result follows directly from Eqs. (1.15) and (1.20).

1.6.3 Statistical Intervals and Tests of Hypotheses

The field of statistical inference consists of those methods used to make decisions or to draw conclusions about a population. These methods use the information contained in a random sample

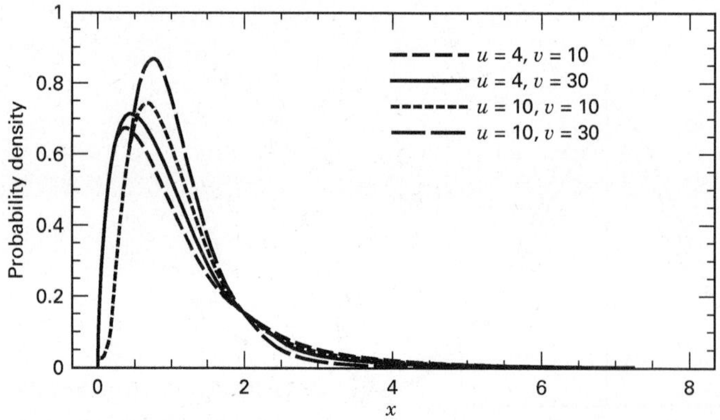

FIGURE 1.16 The F-distribution for Several Different Choices of the Two Parameters

from the population in drawing conclusions. Statistical inference may be divided into two major areas: parameter estimation and hypothesis testing. As an example of a parameter estimation problem, suppose that an engineer is analyzing the tensile strength of a component used in an automobile chassis. Because variability in tensile strength is naturally present between the individual components because of differences in raw material batches, manufacturing processes, and measurement procedures (for example), the engineer is interested in estimating the mean tensile strength of the components. Knowledge of the statistical sampling properties of the estimator used would enable the engineer to establish the precision of the estimate.

Now consider a different situation involving the strength of the automobile chassis components. The engineer conjectures that the mean strength of these components is greater than 1000 psi. Statistical hypothesis testing is a framework for solving problems of this type. In this case, the hypothesis would be that the mean strength exceeds 1000 psi. There is no emphasis on estimating strength; instead, the focus is on drawing conclusions about the stated hypothesis.

1.6.3.1 Comparing a Single Mean to a Specified Value, Variance Known

Some experiments involve comparing only one population mean μ to a specified value, say μ_0. The hypotheses are

$$\begin{aligned} H_0 &: \mu = \mu_0 \\ H_1 &: \mu \neq \mu_0 \end{aligned} \tag{1.23}$$

The statement H_0 is called the null hypothesis and H_1 is called the alternative hypothesis. The alternative hypothesis here is two-sided (or two-tailed) because we are interested in rejecting the null hypothesis for any value of the mean μ that differs from the value μ_0.

If the population is normal with known variance, or if the population is nonnormal but the sample size is large enough so that the central limit theorem applies, then the hypothesis may be tested using a direct application of the normal distribution. The **one-sample Z-test** statistic is

$$Z_0 = \frac{\bar{y} - \mu_0}{\sigma/\sqrt{n}} \tag{1.24}$$

If the null hypothesis is true, the distribution of the test statistic is standard normal, that is, $N(0,1)$. Notice that the test statistic is just the ratio of the difference between the sample mean and the hypothesized value of the population mean divided by the standard error. It is a dimensionless quantity that shows how far $\bar{y}$ is from μ_0 in standard deviation units.

The standard normal distribution is called the **reference distribution** for the test. This implies that when we take a random sample from the population of interest and compute the test statistic we expect Z_0 to be close to zero. That is, a small positive or small negative value of the test statistic is evidence in support of the null hypothesis. However, if Z_0 is not very close to zero, this is evidence against H_0. Suppose that we are willing to incorrectly reject the null hypothesis 5% of the time; that is, reject H_0 and conclude that the mean μ is not equal to μ_0 when it is actually equal to μ_0. Let $\Phi(Z)$ be the cumulative standard normal probability associated with the value Z; that is, $\Phi(Z)$ is the area (probability) to the left of Z on the standard normal distribution. Suppose that we set this probability equal to 0.975. That is, $\Phi(1.96) = 0.975$. We refer to this value of Z as the **upper 2½ percent point of the standard normal distribution** and we denote it by $Z_{0.025}$. Now, the standard normal distribution is symmetric around zero, so $-Z_{0.025} = -1.96$ would be the **lower 2½ percent point of the standard normal distribution** because the probability below this value is 0.025. These two Z-values, $Z_{0.025} = 1.96$ and $-Z_{0.025} = -1.96$, are the **critical values** of the statistical test and values of Z above 1.96 and below -1.96 are called the **critical region** for the test statistic. If the value of the test statistic Z_0 in Eq. (1.24) is greater than 1.96 or less than -1.96, we would reject the null hypothesis. This choice of critical values ensures that if the null hypothesis is true we will only wrongly reject H_0 5% of the time. Rejecting the null hypothesis when it is true is called a **type I error**, and in our example the probability of a type I error was selected to be 0.05. In general, we use the symbol α to represent the type I error probability. The probability of making a type I error is also called the **significance level** of the statistical test.

There is nothing "magic" about the choice of 0.05 as the probability of a type I error. The experimenter can select any value that he or she thinks is appropriate. However, values in the range 0.01–0.10 are widely used in practice, with 0.05 being a "default" choice in much engineering and scientific research. You should consider the consequences (economic or otherwise) when choosing the type I error rate.

To generalize, suppose that the type I error probability is α. Let $Z_{\alpha/2}$ be the upper $\alpha/2$ percentage point of the standard normal distribution. Then to test the hypotheses in Eq. (1.23), we would take a random sample of size n from the population of interest and compute the value of the test statistic Z_0 in Eq. (1.24). We would reject the null hypothesis H_0 if either

$$Z_0 > Z_{\alpha/2}$$

or

$$Z_0 < -Z_{\alpha/2}$$

and we should fail to reject H_0 if

$$-Z_{\alpha/2} \leq Z_0 \leq Z_{\alpha/2}$$

Figure 1.17a shows the location of the critical regions and the critical values for this test.

Sometimes we are interested in rejecting the null hypothesis only if the mean is greater than μ_0. The hypotheses are

$$\begin{aligned} H_0 &: \mu = \mu_0 \\ H_1 &: \mu > \mu_0 \end{aligned} \tag{1.25}$$

This is called a one-sided or upper-tailed alternative hypothesis. In this situation, only values of the sample mean $\overline{y}$ that are greater than μ_0 are consistent with the Alternative hypothesis, so only positive values of the test statistic should lead to rejection of H_0. As a result, we should reject only if

$$Z_0 > Z_\alpha.$$

Refer to Figure 1.17b. Finally, consider the other one-sided test

$$\begin{aligned} H_0 &: \mu = \mu_0 \\ H_1 &: \mu < \mu_0 \end{aligned} \tag{1.26}$$

This is a one-sided lower-tailed alternative hypothesis. In this situation, only values of the sample mean $\overline{y}$ that are less than μ_0 are consistent with the Alternative hypothesis, so only negative values of the test statistic should lead to rejection of H_0. As a result, we should reject only if

$$Z_0 < Z_\alpha.$$

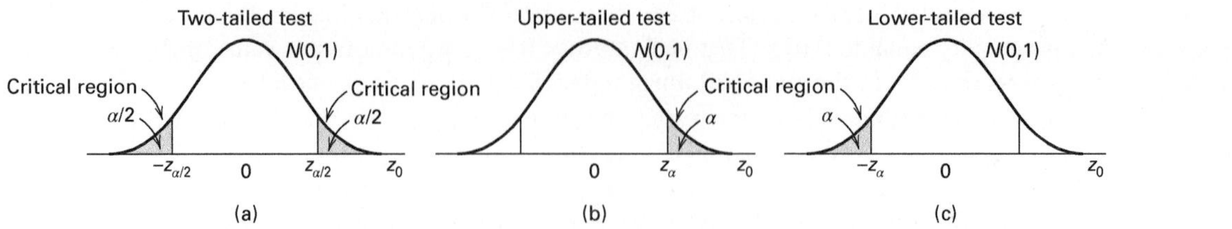

FIGURE 1.17 The distribution of Z_0 when $H_0 : \mu = \mu_0$ is true, with critical region for (a) The two-sided alternative $H_1 : \mu \neq \mu_0$. (b) The one-sided alternative $H_1 : \mu > \mu_0$. (c) The one-sided alternative $H_1 : \mu < \mu_0$

EXAMPLE 1.1 Golf Balls

Recall the golf ball overall distance data introduced earlier. Ten balls were selected at random from the balls produced by a golf equipment manufacturer and tested using a mechanical robot to hit the balls. The distances in yards traveled by the balls (carry plus roll) are as follows: 292, 270, 271, 278, 279, 268, 284, 271, 262, and 283. Assume that the standard deviation of overall distance is known to be $\sigma = 10$ yards. Can the manufacturer support the claim that the mean overall distance for this type of golf ball is at least 270 yards? Use $\alpha = 0.05$.

The appropriate hypotheses for this problem are

$$H_0 : \mu = 270$$
$$H_1 : \mu > 270$$

The sample average is $\bar{y} = 275.8$, so the value of the test statistic is

$$Z_0 = \frac{\bar{y} - \mu_0}{\sigma/\sqrt{n}} = \frac{275.8 - 270}{10/\sqrt{10}} = 1.83$$

Because we have chosen $\alpha = 0.05$, we will reject the null hypothesis if the test statistic value Z_0 is greater than $Z_{\alpha/2} = 1.645$. Since $Z_0 = 1.83 > Z_{\alpha/2} = 1.645$, we reject H_0 and conclude that the mean overall distance for this type of golf ball exceeds 270 yards.

1.6.3.2 The Use of *P*-values in Hypothesis Testing

The method that we have described to report the results of a hypothesis test essentially states that the null hypothesis was or was not rejected at a specified α-value or level of significance. This is called **fixed significance level** testing. The fixed significance level approach to hypothesis testing is useful in that it leads directly to methods for determining the appropriate sample sizes to use in hypothesis testing. However, the fixed significance level approach does have some disadvantages.

For example, in the golf ball problem above, we can say that H_0: $\mu = 270$ was rejected at the 0.05 level of significance. This statement of conclusions may be often inadequate because it gives the decision-maker no idea about whether the computed value of the test statistic was just barely in the rejection region or whether it was very far into this region. Furthermore, stating the results this way imposes the predefined level of significance on other users of the information. This approach may be unsatisfactory because some decision-makers might be uncomfortable with the risks implied by $\alpha = 0.05$.

To avoid these difficulties, the *P*-value approach has been adopted widely in practice. Roughly speaking, the *P*-value is the probability that the test statistic will take on a value that is at least as extreme as the observed value of the statistic when the null hypothesis H_0 is true. Thus, a *P*-value directly measures the weight of evidence against H_0; the smaller the *P*-value, the stronger the evidence against H_0. Therefore, a decision-maker can draw a conclusion at any specified level of significance. We now give a formal definition:

> The *P*-value is the smallest level of significance that would lead to rejection of the null hypothesis H_0 with the given data.

Another way to think of this is that the *P*-value is the probability that you are wrong if you reject the null hypothesis.

Let us consider the case

$$H_0 : \mu = \mu_0$$
$$H_1 : \mu \neq \mu_0$$

To test the null hypothesis using the *P*-value approach, we would find the probability of observing a value of the sample mean that is at least as extreme as the observed value $\bar{y}$, given that the null hypothesis is true. The standard normal *Z*-value that corresponds to $\bar{y}$ is found from the test statistic:

$$Z_0 = \frac{\bar{y} - \mu_0}{\sigma/\sqrt{n}}$$

In terms of the standard normal cumulative distribution, the probability we are seeking is $1 - \Phi(|Z_0|)$. The reason that the argument of the standard normal cumulative distribution is $|Z_0|$ is that the value of Z_0 could be either positive or negative, depending on the observed sample mean.

Because this is a two-tailed test, this is only one-half of the *P*-value. Therefore, for the two-sided alternative hypothesis we are considering, the *P*-value is

$$P = 2[1 - \Phi(|Z_0|)] \tag{1.27}$$

Figure 1.18a illustrates the computation of the *P*-value.

Now let us consider the one-sided alternatives. Suppose that we are testing

$$H_0 : \mu = \mu_0$$
$$H_1 : \mu > \mu_0$$

In this situation, only values of the sample mean $\bar{y}$ that are greater than μ_0 are consistent with the alternative hypothesis, so only positive values of the test statistic should lead to rejection of H_0. Therefore, the *P*-value would be the probability that the standard normal random variable is greater than the observed value of the test statistic Z_0. This *P*-value is computed as

$$P = 1 - \Phi(Z_0) \tag{1.28}$$

Refer to Figure 1.18b. Finally, consider the other one-sided test:

$$H_0 : \mu = \mu_0$$
$$H_1 : \mu < \mu_0$$

In this situation, only values of the sample mean $\bar{y}$ that are less than μ_0 are consistent with the alternative hypothesis, so only negative values of the test statistic should lead to rejection of H_0.

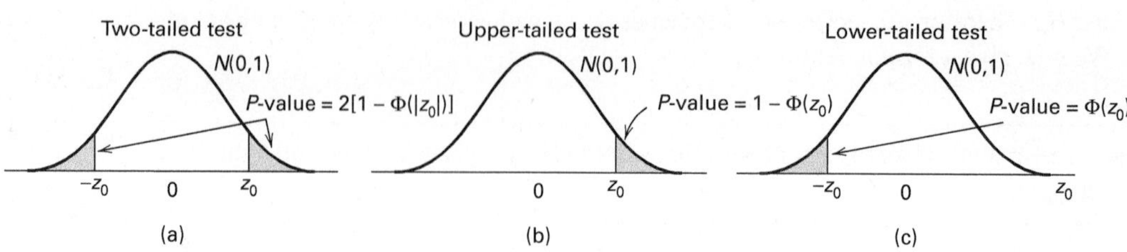

FIGURE 1.18 The *P*-value for a *z*-test. (a) The two-sided alternative $H_1 : \mu \neq \mu_0$. (b) The one-sided alternative $H_1 : \mu > \mu_0$. (c) The one-sided alternative $H_1 : \mu < \mu_0$

Therefore, the *P*-value would be the probability that the standard normal random variable is less than the observed value of the test statistic Z_0. This *P*-value is computed as

$$P = \Phi(Z_0) \tag{1.29}$$

Refer to Figure 1.18c.

EXAMPLE 1.2 Golf Balls

Reconsider the golf ball test data from Example 1.1. Recall that the hypotheses for this problem are

$$H_0 : \mu = 270$$
$$H_1 : \mu > 270$$

The sample average is $\overline{y} = 275.8$, so the value of the test statistic is $Z_0 = 1.83$. Because this is an upper-tailed one-sided test, the *P*-value is computed from Eq. (1.28):

$$P = 1 - \Phi(Z_0) = 1 - \Phi(1.83) = 0.0336$$

This is a small *P*-value, smaller than the "default" level of significance specified in the original problem statement, so the decision-maker should reject the null hypothesis. More formally, we can say that H_0 is rejected at the 0.0336 level of significance.

1.6.3.3 Confidence Interval on the Mean

In many situations, a point estimate does not provide enough information about a parameter. For example, in the golf ball problem, we have rejected the null hypothesis H_0: $\mu = 270$ and our estimate of the mean overall distance is $\overline{y} = 275.8$ yards. This is called a **point estimate**. It is never the case that the true mean overall distance is exactly equal to the point estimate 275.8 yards. In many cases, it would be preferable to have an **interval** in which we would expect to find the true mean distance. One way to accomplish this is with an interval estimate called a **confidence interval** (CI).

An interval estimate of the unknown parameter, such as μ, is an interval of the form

$$L \le \mu \le U$$

where the endpoints L and U depend on the numerical value of the sample mean for a particular sample. Because different samples will produce different values of the sample mean and, consequently, different values of the endpoints L and U, these endpoints are values of random variables. From the sampling distribution of the sample mean, we will be able to determine the values of L and U such that the following probability statement is true:

$$P(L \le \mu \le U) = 1 - \alpha \tag{1.30}$$

where $0 < \alpha < 1$. Thus, we have a probability of $1 - \alpha$ of selecting a sample that will produce an interval containing the true value of μ. The resulting interval

$$L \le \mu \le U \tag{1.31}$$

is the $100(1 - \alpha)\%$ CI for the mean μ. The quantities L and U are called the lower- and upper-confidence limits, respectively, and $1 - \alpha$ is called the **confidence coefficient**.

A strict interpretation of a CI is that, if an infinite number of random samples are collected and a $100(1-\alpha)\%$ CI for μ is computed from each sample, $100(1-\alpha)\%$ of these intervals will contain the true value of μ.

In practice, we obtain only one random sample and calculate one CI. Because this interval either will or will not contain the true value of μ, it is not reasonable to attach a probability level to this specific event. The appropriate statement is that the observed interval $[L, U]$ brackets the true value of μ with confidence $100(1-\alpha)$. This statement has a frequency interpretation; that is, we do not know whether the statement is true for this specific sample, but the method used to obtain the interval $[L, U]$ yields correct statements $100(1-\alpha)\%$ of the time.

The length $U - L$ of the observed two-sided CI is an important measure of the quality of the information obtained from the sample. The half-interval length $\mu - L$ or $U - \mu$ is called the **precision** of the estimator. The longer the CI, the more confident we are that the interval actually contains the true value of μ. On the other hand, the longer the interval is, the less information we have about the true value of μ. In an ideal situation, we obtain a relatively short interval with high confidence.

It is very easy to find the quantities L and U that define the two-sided CI for μ. We know that the sampling distribution of $\overline{X}$ is normal with mean μ and variance σ^2/n. Therefore, the distribution of the statistic

$$Z = \frac{\overline{X} - \mu}{\sigma/\sqrt{n}}$$

is a standard normal distribution.

Therefore

$$P\{-Z_{\alpha/2} \le Z \le Z_{\alpha/2}\} = 1 - \alpha$$

so that

$$P\left\{-Z_{\alpha/2} \le \frac{\overline{X} - \mu}{\sigma/\sqrt{n}} \le Z_{\alpha/2}\right\} = 1 - \alpha$$

This can be rearranged as

$$P\left\{\overline{X} - \frac{Z_{\alpha/2}\sigma}{\sqrt{n}} \le \mu \le \overline{X} + \frac{Z_{\alpha/2}\sigma}{\sqrt{n}}\right\} = 1 - \alpha$$

From the consideration of Eq. (1.30), the lower and upper limits of the inequalities in the last equation are the lower- and upper-confidence limits L and U, respectively. This leads to the following definition:

The $100(1-\alpha)\%$ CI on the mean μ with known standard deviation is

$$\overline{y} - Z_{\alpha/2}\frac{\sigma}{\sqrt{n}} \le \mu \le \overline{y} + Z_{\alpha/2}\frac{\sigma}{\sqrt{n}} \tag{1.32}$$

where $Z_{\alpha/2}$ is the upper $100\alpha/2$ percentage point and $-Z_{\alpha/2}$ is the corresponding lower $100\alpha/2$ percentage point of the standard normal distribution.

To illustrate the construction of a CI, reconsider the golf ball distance data. Suppose that we want a 95% CI. Then $Z_{\alpha/2} = Z_{0.025} = 1.96$ and the 95% CI is

$$\overline{y} - Z_{\alpha/2}\frac{\sigma}{\sqrt{n}} \le \mu \le \overline{y} + Z_{\alpha/2}\frac{\sigma}{\sqrt{n}}$$

$$275.8 - 1.96\frac{10}{\sqrt{10}} \leq \mu \leq 275.8 + 1.96\frac{10}{\sqrt{10}}$$
$$275.8 - 1.96(3.16) \leq \mu \leq 275.8 + 1.96(3.16)$$
$$269.6 \leq \mu \leq 282.0$$

Remember how to interpret the CI; this specific interval either contains the true mean or it does not (and we do not know which), but because of the procedure that we used to construct the CI, in repeated sampling 95% of the intervals that we would compute will contain the true value of μ. It is reasonable to think of this interval as a range of highly likely values for the mean.

Both hypothesis testing and CI construction can be done using software. The output from JMP for the golf ball data is shown in Figure 1.19. JMP presents a normal quantile plot, which is useful in determining if the distribution of overall distance is normal, and since the observations are reasonably well approximated by a straight line, it is reasonable to assume that overall distance is at least approximately normally distributed. The quantiles are shown next, followed by the statistical test of the hypothesis that the mean overall distance equals 270 yards. We supplied the standard deviation as input to the software, so JMP performed the Z-test described above. The test statistic agrees closely with the value we computed manually. JMP reports a two-tailed P-value (0.0666) and both one-sided P-values. Because the alternative hypothesis was an upper-tailed one-sided test, the appropriate P-value is 0.0333. This agrees with our previous results.

1.6.3.4 The Experimental Design Problem

The experimental design problem in the case of statistical inference (hypothesis tests, CIs) on the mean of a population with known variance is fairly simple; it consists mainly of determining the appropriate sample size to use. Determination of sample size requires discussion of a type II error of a statistical test. Recall that a type I error is made if the null hypothesis is rejected when it is actually true. You control the probability of making a type I error when you select the significance level (α) for the test. The type II error is defined as failing to reject the null hypothesis when it is false. The probability of the type II error is denoted by β. Sometimes, it is more convenient to work in terms of the power of the test, where power = $1 - \beta$. Therefore, power is the probability of rejecting the null hypothesis when it is in fact false. High power (or small β) in a statistical test is desirable.

The type II error probability depends on four parameters: the type I error probability, the sample size, the population standard deviation, and the true value of the mean. Fortunately, the type I error probability is controlled and the standard deviation is known. So to determine the appropriate sample size, all we need to do is to define a value of the true population mean, μ, different from the hypothesized value μ_0, such that if the true mean really is this value μ we want to reject the null hypothesis with high probability. Let the true value of the mean be

$$\mu = \mu_0 + \delta$$

Then the relationship between β, δ, α, and the sample size n for the two-sided alternative hypothesis is given by

$$\beta = \Phi\left(Z_{\alpha/2} - \frac{\delta\sqrt{n}}{\sigma}\right) - \Phi\left(-Z_{\alpha/2} - \frac{\delta\sqrt{n}}{\sigma}\right) \tag{1.33}$$

See Montgomery and Runger (2018) for the development of this equation. This equation can be solved for the sample size n, yielding

$$n \cong \frac{(Z_{\alpha/2} + Z_\beta)^2\sigma^2}{\delta^2} \tag{1.34}$$

Distance

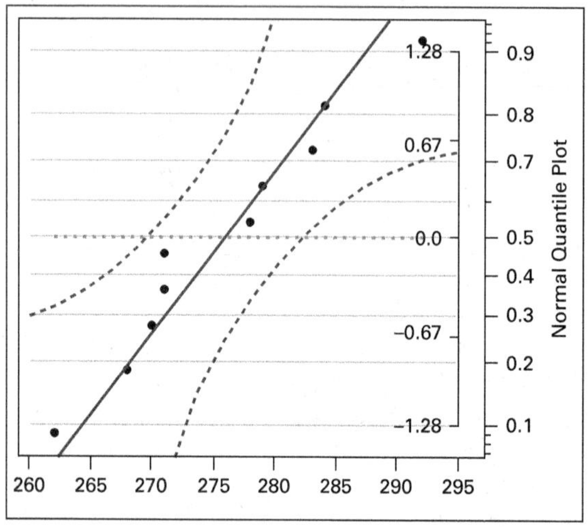

Quantiles

100.0%	maximum	292
99.5%		292
97.5%		292
90.0%		291.2
75.0%	quartile	283.25
50.0%	median	274.5
25.0%	quartile	269.5
10.0%		262.6
2.5%		262
0.5%		262
0.0%	minimum	262

Test Mean

Hypothesized Value	270
Actual Estimate	275.8
DF	9
Std Dev	8.99135
Sigma given	10

	z Test
Test Statistic	1.8341
Prob > |z|	0.0666
Prob > z	0.0333*
Prob < z	0.9667

FIGURE 1.19 JMP Output for the Golf Ball Distance Data

For the one-sided alternatives, the analogous equation is

$$n = \frac{(Z_\alpha + Z_\beta)^2 \sigma^2}{\delta^2} \tag{1.35}$$

When using Eqs. (1.34) and (1.35), round up if the sample size is not an integer.

EXAMPLE 1.3 Golf Balls

Recall that in the golf ball overall distance problem we are testing

$$H_0 : \mu = 270$$
$$H_1 : \mu > 270$$

with $\alpha = 0.05$. We want to reject the null hypothesis with a high probability, say power of 0.90, if the mean is as large as 275 yards. What sample size is required?

Because $\alpha = 0.05$ and $\beta = 1 - \text{power} = 1 - 0.90 = 0.10$, we find $Z_\alpha = Z_{0.05} = 1.645$ and $Z_\beta = Z_{0.10} = 1.28$. We also know that $\sigma = 10$ and since the true value of the mean that we want to "detect" is $\mu = 275, \delta = \mu - \mu_0 = 275 - 270 = 5$. Using Eq. (1.35) the required sample size is

$$n = \frac{(Z_\alpha + Z_\beta)^2\sigma^2}{\delta^2} = \frac{(1.645 + 1.28)^2(10)^2}{5^2} = 34.22 \cong 35$$

1.6.3.5 Comparing a Single Mean to a Specified Value, Variance Unknown

We now consider the case of hypothesis testing and CI construction on a single mean where the population variance is unknown. If the sample size is large, say at least 30 or 40, the procedures presented previously for the known variance case can be used. In other words, they are **large-sample methods**. However, there are many situations where the sample size is small, so the methods we will describe are extremely useful. When the variance of the population is unknown, we must make the additional assumption that the population is normally distributed, although moderate departures from normality will not seriously affect the results.

Suppose that $y_1, y_2, \ldots, y_n$ is a random sample of size n from a population that has a normal distribution with mean μ and variance σ^2, and let $\bar{y}$ and S^2 be the sample average and sample variance of the observations, respectively. Then as noted in Eq. (1.19), the ratio

$$t = \frac{\bar{y} - \mu}{S/\sqrt{n}}$$

has a t-distribution with $n - 1$ degrees of freedom. Let $t_{\alpha/2,n-1}$ be the upper $\alpha/2$ percentage point of the t-distribution with $n - 1$ degrees of freedom; that is, the probability beyond $t_{\alpha/2,n-1}$ is $\alpha/2$. Because the t-distribution is symmetric about zero, the lower-tail percentage point that has probability $\alpha/2$ below it is $-t_{\alpha/2,n-1}$.

Consider testing the two-sided hypotheses

$$H_0 : \mu = \mu_0$$
$$H_1 : \mu \neq \mu_0 \tag{1.36}$$

The **test statistic** is

$$t_0 = \frac{\bar{y} - \mu_0}{S/\sqrt{n}} \tag{1.37}$$

Notice that the test statistic is just the difference between the sample and hypothesized values of the mean divided by the estimated standard error. The decision procedure is very similar to the Z-test, except that the t-distribution is used to find the critical values instead of the standard normal. If the null hypothesis is true, t_0 has a t-distribution with $n - 1$ degrees of freedom, and we would reject H_0 if $t_0 > t_{\alpha/2,n-1}$ or if $t_0 < -t_{\alpha/2,n-1}$. If the alternative hypothesis is one-sided, say

$$H_0 : \mu = \mu_0$$
$$H_1 : \mu > \mu_0$$

we would reject H_0 if $t_0 > t_{\alpha,n-1}$, and if the hypotheses are

$$H_0 : \mu = \mu_0$$
$$H_1 : \mu < \mu_0$$

we would reject H_0 if $t_0 < -t_{\alpha,n-1}$.

EXAMPLE 1.4 Cloud Seeding

Cloud seeding has been studied for many decades as a weather modification procedure (for an interesting study of this subject, see the article in *Technometrics*, "A Bayesian Analysis of a Multiplicative Treatment Effect in Weather Modification," Vol. 17, 1975, pp. 161–166). The rainfall in acre-feet from 20 clouds that were selected at random and seeded with silver nitrate is as follows: 18.0, 30.7, 19.8, 27.1, 22.3, 18.8, 31.8, 23.4, 21.2, 27.9, 31.9, 27.1, 25.0, 24.7, 26.9, 21.8, 29.2, 34.8, 26.7, and 31.6. Can you support the claim that mean rainfall from seeded clouds exceeds 25 acre-feet? Use $\alpha = 0.05$.

From the sample data, we find that $\bar{y} = 26.035$ and $S = 4.785$. The hypotheses we are going to test are

$$H_0 : \mu = 25$$
$$H_1 : \mu > 25$$

The value of the test statistic is

$$t_0 = \frac{\bar{y} - \mu_0}{S/\sqrt{n}} = \frac{26.035 - 25}{4.785/\sqrt{20}} = 0.9674$$

Since this is an upper-tailed one-sided test, we would reject the null hypothesis if the value of the test statistic $t_0 > t_{\alpha,n-1} = t_{0.05,19}$. Using JSL commands from the appendix, we find that $t_{0.05,19} = 1.729$, so we cannot reject the null hypothesis. There is little evidence in the data to support the claim that the mean rainfall from seeded clouds exceeds 25 acre-feet.

1.6.3.6 *P*-values in the *t*-test

We can use P-values in the t-test in a manner similar to their usage in the Z-test described earlier. Statistics software packages calculate and display P-values. However, in working problems by hand, it is useful to be able to find the P-value for a t-test. Fortunately, it is easy to find lower and upper bounds on the P-value by using JMP scripting functions described in the appendix. It is also helpful to refer to Figure 1.20.

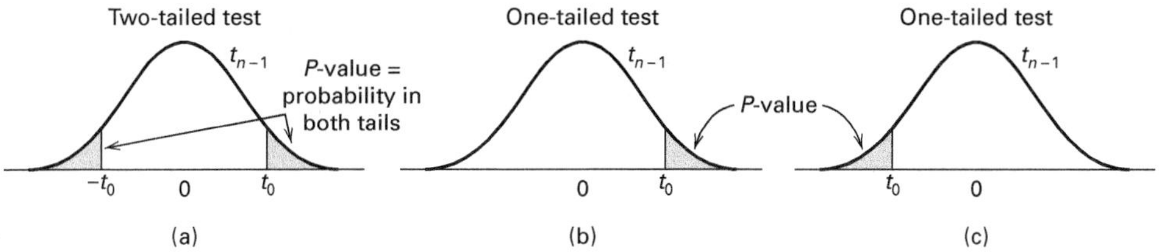

FIGURE 1.20 Calculating the P-value for a t-test: (a) $H_1: \mu \neq \mu_0$; (b) $H_1: \mu > \mu_0$; (c) $H_1: \mu < \mu_0$

To illustrate, suppose that we are conducting an upper-tailed t-test (so H_1: $\mu > \mu_0$) with 14 degrees of freedom. The relevant critical values from the appendix are as follows:

Critical value:	0.258	0.692	1.345	1.761	2.145	2.624	2.977	3.326	3.787	4.140
Tail area:	0.40	0.25	0.10	0.05	0.025	0.01	0.005	0.0025	0.001	0.0005

After calculating the test statistic, we find that $t_0 = 2.8$. Now, $t_0 = 2.8$ is between two tabulated values 2.624 and 2.977. Therefore, the P-value must be between 0.01 and 0.005. These are effectively the upper and lower bounds on the P-value.

This illustrates the procedure for an upper-tailed test. If the test is lower tailed, just change the sign on the lower and upper bounds for t_0 and proceed as above. Remember that for a two-tailed test, the level of significance associated with a particular critical value is twice the corresponding tail area in the column heading. This consideration must be taken into account when we compute the bound on the P-value. For example, suppose that $t_0 = 2.8$ for a two-tailed alternative based on 14 degrees of freedom. The value of the test statistic $t_0 > 2.624$ (corresponding to $\alpha = 2 \times 0.01 = 0.02$) and $t_0 < 2.977$ (corresponding to $\alpha = 2 \times 0.005 = 0.01$), so the lower and upper bounds on the P-value would be $0.01 < P < 0.02$ for this case.

We can also find CIs on the mean μ of a normal distribution where the variance is unknown. The CI equation is as follows:

The $100(1 - \alpha)\%$ CI on the mean μ of a normal distribution with unknown standard deviation is

$$\bar{y} - t_{\alpha/2,n-1}\frac{S}{\sqrt{n}} \leq \mu \leq \bar{y} + t_{\alpha/2,n-1}\frac{S}{\sqrt{n}} \tag{1.38}$$

where $t_{\alpha/2,n-1}$ is the upper $100\alpha/2$ percentage point and $-t_{\alpha/2,n-1}$ is the corresponding lower $100\alpha/2$ percentage point of the t-distribution with $n - 1$ degrees of freedom in the appendix.

The JMP output for the cloud seeding problem is shown in Figure 1.21. Notice that the P-value reported is 0.1728. You should be able to verify that $0.10 < P < 0.25$, so we cannot reject the null hypothesis. There is little evidence to indicate that the mean rainfall from seeded clouds exceeds 25 acre-feet. JMP also provides the 95% CI from Eq. (1.38).

We can also determine sample sizes for the t-test. Unfortunately, there are no closed-form equations for this as there were in the Z-test because the probability calculations involve the noncentral t-distribution, which has a very messy probability density function. When the standard deviation is unknown, we must work in terms of the difference between the true mean and the hypothesized mean that we want to detect as a ratio, say δ/σ. If we are interested in detecting small differences between the means, we should select a small ratio, say $0 < \delta/\sigma \leq 1$. For detecting moderate differences, we could select $1 < \delta/\sigma \leq 2$, and for detecting large differences we could

Rainfall

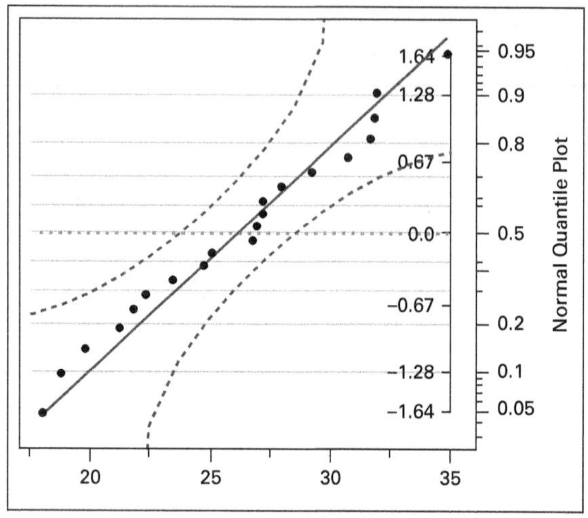

Quantiles

100.0%	maximum	34.8
99.5%		34.8
97.5%		34.8
90.0%		31.89
75.0%	quartile	30.325
50.0%	median	26.8
25.0%	quartile	21.925
10.0%		18.9
2.5%		18
0.5%		18
0.0%	minimum	18

Moments

Mean	26.035
Std Dev	4.7847647
Std Err Mean	1.0699059
Upper 95% Mean	28.274339
Lower 95% Mean	23.795661
N	20

Test Mean

Hypothesized Value	25
Actual Estimate	26.035
DF	19
Std Dev	4.78476

	t Test
Test Statistic	0.9674
Prob > |t|	0.3455
Prob > t	0.1728
Prob < t	0.8272

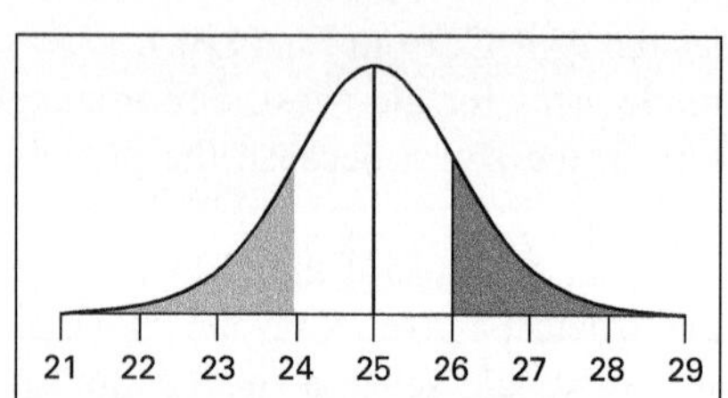

FIGURE 1.21 JMP Output for the Rainfall Data

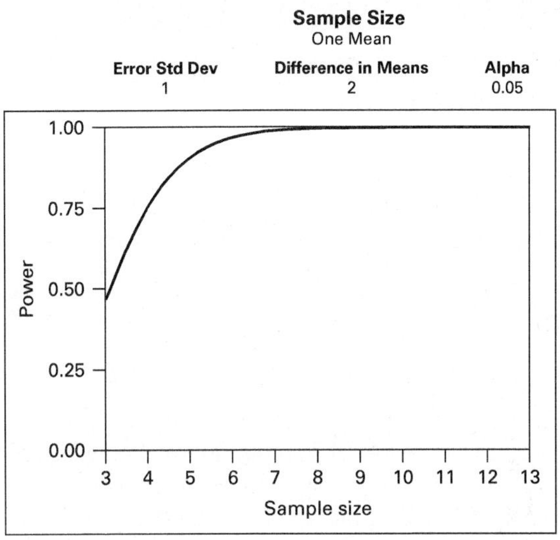

FIGURE 1.22 Power Versus Sample Size Curve for the Cloud Seeding Data

select $\delta/\sigma > 2$. Software can be used to determine the appropriate sample size for many situations. Suppose that in the cloud seeding problem we wanted to detect a moderately large difference between the true mean and the hypothesized value of 25. Figure 1.22 is an output from the JMP power and sample size calculator for $\delta/\sigma = 2$. If the sample size is at least $n = 8$, the statistical test will have very high power, exceeding 0.99.

2 Simple Comparative Experiments

2.1 Introduction

In this chapter, we consider experiments to compare two **levels** of a **single factor**. In the experimental design literature, these levels are often called **treatments**. These types of experiments are usually called **simple comparative experiments**. We concentrate on the case where the factor is **qualitative**, meaning that there is no natural numerical scale that can be used to define the levels. For example, the factor could be machines, and the two levels could be different types or brands of the machines; or the factor could be the suppliers of a particular raw material, and the levels would correspond to two different suppliers. We begin with an example of an experiment performed to determine whether two different types of a component used in a laser printer differ regarding an important characteristic.

A team of engineers is developing a new type of laser printer that has increased printing speed. They have designed a paper handling system that uses a newly designed gear wheel that makes paper handling faster. This gear can be fabricated from two different types of plastic, and because the gear is subject to impact forces, the team is interested in determining if there is any difference in impact strength between gears made of the two plastics. They want to plan and conduct an experiment to specifically investigate whether there is any difference in mean impact strength between gears manufactured with the two different types of plastic. This is a single-factor experiment (the factor is material) with two levels, plastic type 1 and plastic type 2. The experimental design questions are (1) how many observations should be taken (or what sample size to use) and (2) how should the total number of observations be allocated to the two types of plastic (or how many gears made of each type of plastic should be tested)?

Before we can properly address the experimental design questions, we must discuss how the data from the experiment will be analyzed, because the design of an experiment and the analysis of the experimental data are linked. You cannot do a good job of design without understanding the analysis method, and how you do the analysis depends on the experimenter's choice of design. The two topics are closely related.

We will use **parameter estimation** and statistical **hypothesis testing** to compare the mean impact strengths of these two types of gears. Because the factor has two levels, the procedure we will use is an extension of the single-sample methods reviewed in Chapter 1. We will now describe the procedures to be used.

2.2 Statistical Methods for Comparing Two Population Means

2.2.1 Parameter Estimation and Confidence Intervals

Suppose that we have two independent populations with unknown means μ_1 and μ_2 and known variances σ_1^2 and σ_2^2. We are interested in knowing whether the two means differ. Let $\delta = \mu_1 - \mu_2$

be the true difference in means. Clearly, if $\delta = 0$ then $\mu_1 = \mu_2$, so estimating the difference in means δ is a simple way to answer the question about equality of the two means. Suppose that we have a random sample of n_1 observations from the first population, say $y_{11}, y_{12}, \ldots, y_{1,n_1}$ and a random sample of n_2 observations from the second population, say $y_{21}, y_{22}, \ldots, y_{2,n_2}$, and let $\overline{y}_1$ and $\overline{y}_2$ be the two sample means. Now $\overline{y}_1$ is an unbiased estimator of μ_1 and $\overline{y}_2$ is an unbiased estimator of μ_2. So

$$\hat{\delta} = \overline{y}_1 - \overline{y}_2 \tag{2.1}$$

is an unbiased estimator of δ That is,

$$\begin{aligned} E(\hat{\delta}) &= E(\overline{y}_1 - \overline{y}_2) \\ &= E(\overline{y}_1) - E(\overline{y}_2) \\ &= \mu_1 - \mu_2 \\ &= \delta \end{aligned}$$

Therefore, finding a **point estimate** of the difference in means is easy; all we have to do is to calculate the observed difference in the two sample averages. If $\hat{\delta}$ is close to zero, then we might logically conclude that the two means are equal; if $\hat{\delta}$ is fairly different from zero, we might conclude that the two means are not equal.

Now suppose that we want an interval estimate of the difference in means. **Confidence intervals (CIs)**, introduced in Chapter 1, provide such an interval estimate. Because the two populations are independent and the samples are assumed to be random, the variance of the difference in the two sample means is

$$V(\hat{\delta}) = V(\overline{y}_1 - \overline{y}_2) = \frac{\sigma_1^2}{n_1} + \frac{\sigma_2^2}{n_2} \tag{2.2}$$

and the standard error of the difference in the two means is

$$\sigma_{\hat{\delta}} = \sqrt{V(\hat{\delta})} = \sqrt{V(\overline{y}_1 - \overline{y}_2)} = \sqrt{\frac{\sigma_1^2}{n_1} + \frac{\sigma_2^2}{n_2}} \tag{2.3}$$

Let us assume that either the two populations are normally distributed or if they are not normal that the conditions of the central limit theorem apply. Then, $\hat{\delta}$ is normally distributed with mean $\delta = \mu_1 - \mu_2$ and variance as given by Eq. (2.2). Then the quantity

$$Z = \frac{\hat{\delta} - \delta}{\sqrt{\dfrac{\sigma_1^2}{n_1} + \dfrac{\sigma_2^2}{n_2}}}$$

has a standard normal distribution, and the following probability statement is true:

$$P\left(-Z_{\alpha/2} \le \frac{\hat{\delta} - \delta}{\sqrt{\dfrac{\sigma_1^2}{n_1} + \dfrac{\sigma_2^2}{n_2}}} \le Z_{\alpha/2}\right) = 1 - \alpha$$

This can be rearranged to

$$P\left(\hat{\delta} - Z_{\alpha/2}\sqrt{\frac{\sigma_1^2}{n_1} + \frac{\sigma_2^2}{n_2}} \le \delta \le \hat{\delta} + Z_{\alpha/2}\sqrt{\frac{\sigma_1^2}{n_1} + \frac{\sigma_2^2}{n_2}}\right) = 1 - \alpha \tag{2.4}$$

Equation (2.4) is the definition of a $100(1-\alpha)\%$ CI on the true difference in means δ. The equation for the $100(1-\alpha)\%$ CI on the difference in the two means is

$$\bar{y}_1 - \bar{y}_2 - Z_{\alpha/2}\sqrt{\frac{\sigma_1^2}{n_1} + \frac{\sigma_2^2}{n_2}} \leq \mu_1 - \mu_2 \leq \bar{y}_1 - \bar{y}_2 + Z_{\alpha/2}\sqrt{\frac{\sigma_1^2}{n_1} + \frac{\sigma_2^2}{n_2}} \tag{2.5}$$

The **experimental design problem** here is very interesting: Given that we are going to take a total of $N = n_1 + n_2$ observations, how should we allocate these resources? In other words, how do we choose n_1 and n_2? It is convenient to think of the difference in population means $\delta = \mu_1 - \mu_2$ as a **model** with a single parameter δ. It seems reasonable to want to estimate this parameter precisely. Suppose that we choose the two sample sizes n_1 and n_2 to minimize the variance of $\hat{\delta}$ or, equivalently, its standard error. This is a logical criterion, because it will

1. provide the most precise estimate possible of the true difference in means; and
2. ensure that the CI on the difference in means is as short as possible for a given level of confidence.

It is easy to do this. The variance of the difference in means is given in Eq. (2.2), and we want to choose n_1 and n_2 to minimize this quantity subject to the constraint that $N = n_1 + n_2$. To do this, first write the variance of the difference in means in terms of only n_1 by using the constraint $n_2 = N - n_1$:

$$V(\bar{y}_1 - \bar{y}_2) = \frac{\sigma_1^2}{n_1} + \frac{\sigma_2^2}{N - n_1}$$

Now differentiate this objective function with respect to n_1, which results in

$$\frac{dV(\bar{y}_1 - \bar{y}_2)}{dn_1} = \frac{-\sigma_1^2}{n_1^2} + \frac{\sigma_2^2}{(N - n_1)^2}$$

Finally, replace $N - n_1$ by n_2, equate the derivative to zero, and solve. This results in

$$\begin{aligned} \frac{-\sigma_1^2}{n_1^2} + \frac{\sigma_2^2}{(N - n_1)^2} &= 0 \\ \frac{-\sigma_1^2}{n_1^2} + \frac{\sigma_2^2}{n_2^2} &= 0 \\ \frac{n_2}{n_1} &= \frac{\sigma_2}{\sigma_1} \end{aligned} \tag{2.6}$$

The intuitively pleasing result is that the samples should be allocated to the two populations in a ratio that equals the ratio of the population standard deviations. This is a special case of the ***D*-optimal** design optimality criterion that we will introduce more formally later and make great use of throughout the book. Loosely speaking, a *D*-optimal design minimizes the variances of the estimated model parameters. Here, we have only one model parameter, δ, and we have designed the experiment to minimize the variance of its estimate $\bar{y}_1 - \bar{y}_2$.

EXAMPLE 2.1 Plastic Gears

Recall the team of engineers who are developing a new laser printer that has increased printing speed. Their new paper handling system uses a newly designed gear wheel that increases paper handling speed. This gear can be fabricated from two different types of plastic, and because the gear is subject to impact forces, the team is interested in determining if there is any difference in impact strength between gears made of the two plastics. The Izod impact strength test (see ASTM D256) is used. In this test, a pendulum swings on a track and strikes a notched, cantilevered plastic sample that is approximately 1/8 in. thick. The response variable is the energy lost in ft-lb/in. required for the pendulum to break the sample.

The engineering team has considerable experience with the plastic formulation that is currently used in the gear wheel, and they are comfortable with the assumption that the standard deviation of impact strength of gear wheels made from this material is around 0.20 ft-lb/in. The new material should result in gear wheels that have a much smaller standard deviation of impact strength, probably around 0.10 ft-lb/in. They want to estimate the difference in mean impact strength for gears manufactured from the two materials.

The team can afford to fabricate and test a total of 30 gear wheels. Because the two standard deviations are $\sigma_1 = 0.10$ and $\sigma_2 = 0.20$, from Eq. (2.6), we would allocate the $N = 30$ total observations to the two gear wheel types as follows:

$$\frac{n_2}{n_1} = \frac{\sigma_2}{\sigma_1} = \frac{0.20}{0.10} = 2$$

$$n_2 = 2n_1$$

Consequently, we should use $n_1 = 10$ for the new plastic and $n_2 = 20$ for the current material.

Figure 2.1 shows output from the JMP Profiler displaying how the standard error of the difference in means behaves in this problem as a function of the two population variances and the sample sizes. With the optimal allocation, JMP reports the standard error as 0.054772. The profiler is interactive, and any of the three parameters σ_1, σ_2, and n_1 with $n_2 = 30 - n_1$ can be changed. Figure 2.2 shows the profiler with equal allocation of the observations to the two populations, or $n_1 = n_2 = 15$. Now the standard error is 0.057735, so clearly the optimal sample allocation has resulted in a smaller value of the standard error. By adjusting the sample sizes in the profiler, it is easy to see that the optimal sample allocation minimizes the standard error.

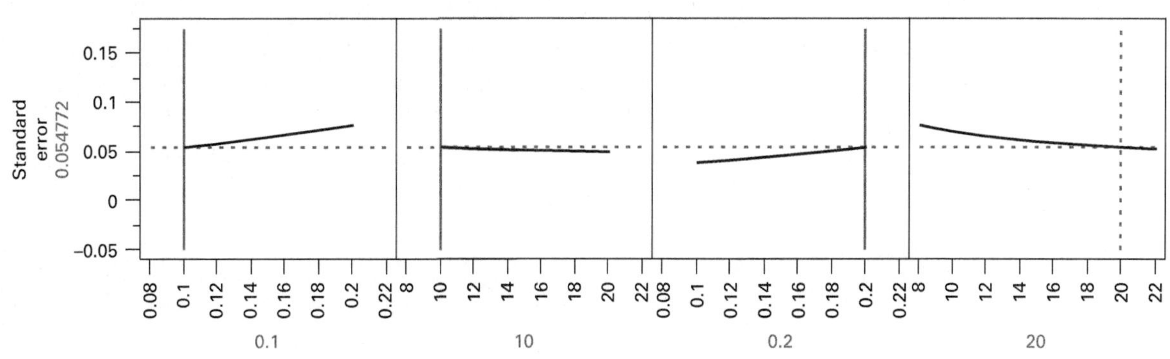

FIGURE 2.1 Prediction Profiler from JMP for the Standard Error in Example 2.1 as a Function of σ_1, σ_2, n_1, and n_2 with $n_1 = 10$ and $n_2 = 20$

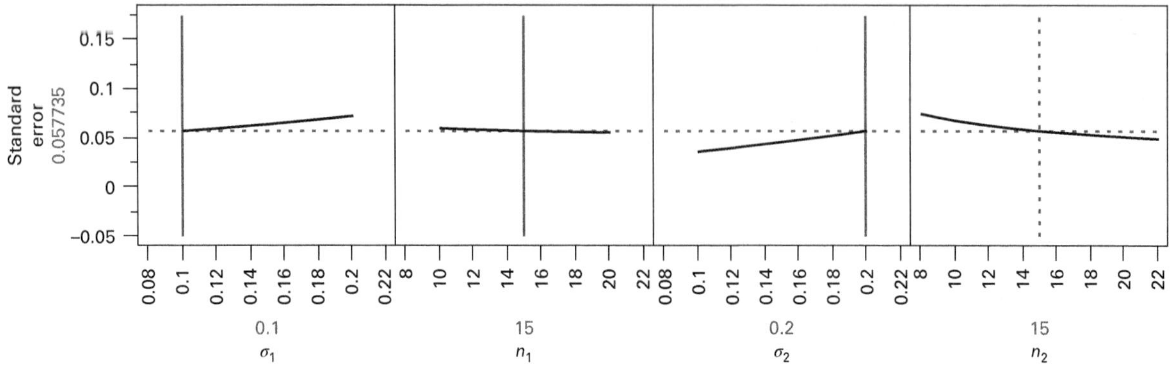

FIGURE 2.2 Prediction Profiler from JMP for the Standard Error in Example 2.1 as a Function of σ_1, σ_2, n_1, and n_2 with $n_1 = n_2 = 15$

The two samples of gear wheels are fabricated and tested, and Table 2.1 shows the resulting impact strength measurements. These observations are assumed to be random samples from the two populations of gear wheels that could possibly be fabricated using these two plastic materials. This is an example of sampling from a **conceptual** as opposed to a **physical** or **actual** population. Many engineering experiments involve sampling from conceptual populations, and the sample objects are often prototypes of the actual objects that will be produced in the future. One must take great care to ensure that the prototypes in such samples are really representative

Table 2.1 Gear Wheel Impact Strength (ft-lb/in.)

Sample	Plastic 1	Plastic 2
1	3.8	3.3
2	4.1	3.9
3	3.7	3.3
4	3.9	3.3
5	4.1	3.9
6	3.8	3.0
7	4.0	3.2
8	3.9	3.7
9	4.0	3.3
10	4.1	3.5
11		3.4
12		3.4
13		3.3
14		3.5
15		3.4
16		3.8
17		3.4
18		3.5
19		3.3
20		3.7

of the objects that will be produced later. Often prototypes are not representative of future production; they are "too good," because they are produced by a group of expert technicians and use equipment that is not comparable with actual production tools. Overreliance on such prototypes has resulted in significant problems in **confirmation**, that is, experimental results from engineering development that do not hold up in production.

The sample averages are for plastic 1, $\overline{y}_1 = 3.94$ and for plastic 2, $\overline{y}_2 = 3.455$. The point estimate of δ is $\hat{\delta} = 3.94 - 3.455 = 0.4850$. The 95% CI on the difference in population means is found from Eq. (2.5) as

$$\overline{y}_1 - \overline{y}_2 - Z_{\alpha/2}\sqrt{\frac{\sigma_1^2}{n_1} + \frac{\sigma_2^2}{n_2}} \leq \mu_1 - \mu_2 \leq \overline{y}_1 - \overline{y}_2 + Z_{\alpha/2}\sqrt{\frac{\sigma_1^2}{n_1} + \frac{\sigma_2^2}{n_2}}$$

$$3.94 - 3.455 - 1.96\sqrt{\frac{(0.10)^2}{10} + \frac{(0.20)^2}{20}} \leq \mu_1 - \mu_2 \leq 3.94 - 3.455 + 1.96\sqrt{\frac{(0.10)^2}{10} + \frac{(0.20)^2}{20}}$$

$$0.3776 \leq \mu_1 - \mu_2 \leq 0.5924$$

The length of the CI is determined by the quantity $Z_{\alpha/2}\sqrt{\frac{\sigma_1^2}{n_1} + \frac{\sigma_2^2}{n_2}} = 0.1074$. We have chosen the two sample sizes to minimize this quantity for any confidence level, so we have obtained an estimate of δ that is as precise as possible for a total sample size of $N = 30$.

The CI implies that a range of likely values for δ is between 0.3776 and 0.5924 ft-lb/h. Notice that the 95% CI does not include the value $\delta = 0$. Therefore, it is unlikely that the true difference in means is zero.

2.2.2 Statistical Hypothesis Testing on the Difference in Means

We can also use formal statistical hypothesis testing to investigate the difference in means. Often after estimating the parameters of a statistical model, formal hypothesis testing is useful to help solidify our conclusions. Suppose that we wish to test the hypothesis that the two means are equal against an alternative hypothesis that the means are not equal, say

$$\begin{aligned} H_0 &: \mu_1 = \mu_2 \\ H_1 &: \mu_1 \neq \mu_2 \end{aligned} \tag{2.7}$$

A slightly more general form of these hypotheses is

$$\begin{aligned} H_0 &: \mu_1 - \mu_2 = \delta_0 \\ H_1 &: \mu_1 - \mu_2 \neq \delta_0 \end{aligned} \tag{2.8}$$

In Eq. (2.8), we are testing that the difference in the two means is equal to a constant, δ_0. Obviously, if $\delta_0 = 0$ we are testing the equality of the two means, and Eqs. (2.7) and (2.8) are equivalent statements of the hypotheses.

We will continue to let $y_{11}, y_{12}, \ldots, y_{1,n_1}$ be a random sample of n_1 observations from the first population and $y_{21}, y_{22}, \ldots, y_{2,n_2}$ be a random sample of n_2 observations from the second population and let $\overline{y}_1$ and $\overline{y}_2$ be the two sample means. If $\hat{\delta} = \overline{y}_1 - \overline{y}_2$ is close to δ_0, this is evidence that supports the null hypothesis, while if $\hat{\delta} = \overline{y}_1 - \overline{y}_2$ is substantially different from δ_0, this is evidence against H_0. Therefore, it seems reasonable to use $\overline{y}_1 - \overline{y}_2$ as the basis of a statistical test of hypothesis.

The appropriate **test statistic** for the hypotheses in Eq. (2.8) is

$$Z_0 = \frac{\bar{y}_1 - \bar{y}_2 - \delta_0}{\sqrt{\dfrac{\sigma_1^2}{n_1} + \dfrac{\sigma_2^2}{n_2}}} \tag{2.9}$$

If the null hypothesis is true, the test statistic in Eq. (2.9) has a standard normal distribution. Therefore, the decision rule for a fixed significance level test is identical to the one-sample Z-test discussed in Chapter 1. We should reject H_0 if the value of the test statistic $Z_0 > Z_{\alpha/2}$ or if $Z_0 < -Z_{\alpha/2}$, where $Z_{\alpha/2}$ is the upper $\alpha/2$ percentage point of the standard normal distribution, and fail to reject otherwise. A P-value approach can also be used. The P-value for this two-sample Z-test is

$$P = 2[1 - \Phi(|Z_0|)] \tag{2.10}$$

The one-sided alternatives in the two-sample case are treated just as they were in the single-sample case. The hypotheses are

$$\begin{aligned} H_0 &: \mu_1 = \mu_2 \\ H_1 &: \mu_1 > \mu_2 \end{aligned} \tag{2.11}$$

We should reject H_0 if the value of the test statistic $Z_0 > Z_\alpha$, where Z_α is the upper α percentage point of the standard normal distribution, and fail to reject otherwise. The P-value for the upper-tail one-sided test is computed as

$$P = 1 - \Phi(Z_0) \tag{2.12}$$

The other one-sided hypotheses are

$$\begin{aligned} H_0 &: \mu_1 = \mu_2 \\ H_1 &: \mu_1 < \mu_2 \end{aligned} \tag{2.13}$$

We should reject H_0 if the value of the test statistic $Z_0 < -Z_\alpha$, where Z_α is the upper α percentage point of the standard normal distribution, and fail to reject otherwise. The P-value for the lower-tail one-sided test is computed as

$$P = \Phi(Z_0) \tag{2.14}$$

Figure 1.16 provides a graphical view of how the P-values are determined for a Z-test. For convenience, we have reproduced this as Figure 2.3.

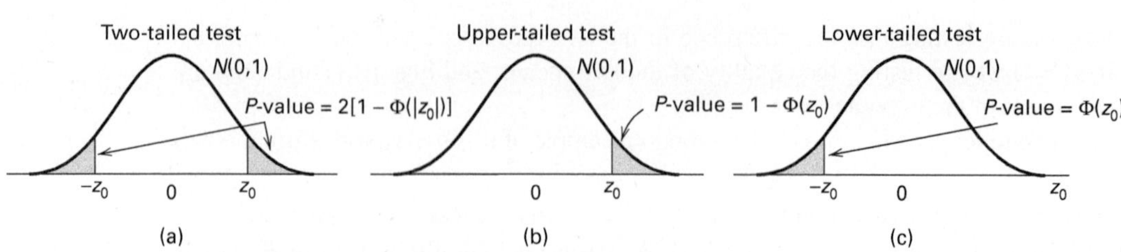

FIGURE 2.3 Determining P-values in the Z-test. (a) The Two-sided Alternative $H_0 : \mu \neq \mu_0$ (b) The One-sided Alternative $H_0 : \mu > \mu_0$ (c) The One-sided Alternative $H_0 : \mu < \mu_0$

Suppose that we use the **optimal design** for this hypothesis testing problem; that is, we allocate the observations to the two populations as in Eq. (2.6). This will maximize the power of the statistical test or equivalently minimize the probability of making a type II error for the difference in the two means.

We can also calculate the type II error probability for the sample sizes that we have selected. If δ is the true difference in means and we are testing the two-sided alternative hypothesis, then the type II error probability β is

$$\beta = \Phi\left(Z_{\alpha/2} - \frac{\delta - \delta_0}{\sqrt{\frac{\sigma_1^2}{n_1} + \frac{\sigma_2^2}{n_2}}}\right) - \Phi\left(-Z_{\alpha/2} - \frac{\delta - \delta_0}{\sqrt{\frac{\sigma_1^2}{n_1} + \frac{\sigma_2^2}{n_2}}}\right) \tag{2.15}$$

For more details about the development of this equation for β, refer to Montgomery and Runger (2018).

EXAMPLE 2.2 Plastic Gears (Continued)

Recall the team of engineers who are developing a new laser printer that has increased printing speed. The team wants a formal test of whether the mean impact strengths for gears manufactured from the two materials are equal, so the hypotheses are

$$H_0 : \mu_1 = \mu_2$$
$$H_1 : \mu_1 \neq \mu_2$$

In this problem, $\delta_0 = 0$. The value of the test statistic is

$$Z_0 = \frac{\bar{y}_1 - \bar{y}_2}{\sqrt{\frac{\sigma_1^2}{n_1} + \frac{\sigma_2^2}{n_2}}} = \frac{3.94 - 3.455}{\sqrt{\frac{(0.10)^2}{10} + \frac{(0.20)^2}{20}}} = \frac{0.4850}{0.0548} = 8.85$$

Suppose that we are testing at the 0.05 level of significance. Then, this value of the test statistic would be compared to $Z_{0.025} = 1.96$ and $-Z_{0.025} = -1.96$, and clearly we would reject the null hypothesis. There is strong evidence of a difference in means between the two plastics. Wheel gears made of plastic 1 have statistically significantly higher impact strength than gear wheels made of the older material. The *P*-value for this test is computed from Eq. (2.6):

$$P = 2[1 - \Phi(|Z_0|)] = 2[1 - \Phi(8.85)] \cong 0$$

because $\Phi(8.85)$ is approximately unity (a pocket calculator was used to find $1 - \Phi(8.85) \cong 4.38E - 19$, which is essentially zero). Once again, a small *P*-value leads to rejection of H_0. Some computer software reports *P*-values of zero in cases like this where the test statistic is very large (or unity when it is almost zero), but that cannot be correct because the standard normal distribution is continuous and there will always be some nonzero probability beyond or below the value of the *Z*-test statistic.

In this experiment, an optimal allocation of observations between the two populations was used. It is interesting to observe how this impacted the standard error of the difference in the two sample means. If we had used equal allocation of the 30 observations, then $n_1 = n_2 = 15$, and the standard error would have been

$$\sqrt{\frac{\sigma_1^2}{n_1} + \frac{\sigma_2^2}{n_2}} = \sqrt{\frac{(0.10)^2}{15} + \frac{(0.20)^2}{15}} = 0.0580$$

which is about 5.8% greater than the actual standard error of 0.0548, so the 95% CI would have been longer (these values agree with those reported in the JMP Profiler in Figures 2.1 and 2.2 after rounding). Clearly, the optimal allocation of sample observations has improved the precision of estimation in this problem.

Suppose that we are interested in detecting a difference in mean impact strength of $\delta = 0.2$ ft-lb/in. in a statistical test, where α, the probability of type I error, is specified to be $\alpha = 0.05$. From Eq. (2.15), the probability of type II error for the two-sided alternative hypothesis is

$$\begin{aligned}
\beta &= \Phi\left(Z_{\alpha/2} - \frac{\delta - \delta_0}{\sqrt{\frac{\sigma_1^2}{n_1} + \frac{\sigma_2^2}{n_2}}}\right) - \Phi\left(-Z_{\alpha/2} - \frac{\delta - \delta_0}{\sqrt{\frac{\sigma_1^2}{n_1} + \frac{\sigma_2^2}{n_2}}}\right) \\
&= \Phi\left(1.96 - \frac{0.20 - 0}{\sqrt{\frac{(0.10)^2}{10} + \frac{(0.20)^2}{20}}}\right) - \Phi\left(-1.96 - \frac{0.20 - 0}{\sqrt{\frac{(0.10)^2}{10} + \frac{(0.20)^2}{20}}}\right) \\
&= \Phi(1.96 - 3.6496) - \Phi(-1.96 - 3.6496) \\
&= \Phi(-1.6896) - \Phi(-5.6096) \\
&= 0.0456 - 0.0000 \\
&\cong 0.0456
\end{aligned}$$

The power of the test is about 0.95. That is, if the true means differ by as much as 0.2 ft-lb/in., we have about a 95% chance of detecting it with a sample of 30 observations allocated optimally. If we used equal allocation with $n_1 = n_2 = 15$, the type II error probability would be

$$\begin{aligned}
\beta &= \Phi\left(Z_{\alpha/2} - \frac{\delta - \delta_0}{\sqrt{\frac{\sigma_1^2}{n_1} + \frac{\sigma_2^2}{n_2}}}\right) - \Phi\left(-Z_{\alpha/2} - \frac{\delta - \delta_0}{\sqrt{\frac{\sigma_1^2}{n_1} + \frac{\sigma_2^2}{n_2}}}\right) \\
&= \Phi\left(1.96 - \frac{0.20 - 0}{\sqrt{\frac{(0.10)^2}{15} + \frac{(0.20)^2}{15}}}\right) - \Phi\left(-1.96 - \frac{0.20 - 0}{\sqrt{\frac{(0.10)^2}{15} + \frac{(0.20)^2}{15}}}\right) \\
&= \Phi(1.96 - 3.4483) - \Phi(-1.96 - 3.4483) \\
&= \Phi(-1.4883) - \Phi(-5.4083) \\
&= 0.0683 - 0.0000 \\
&\cong 0.0683
\end{aligned}$$

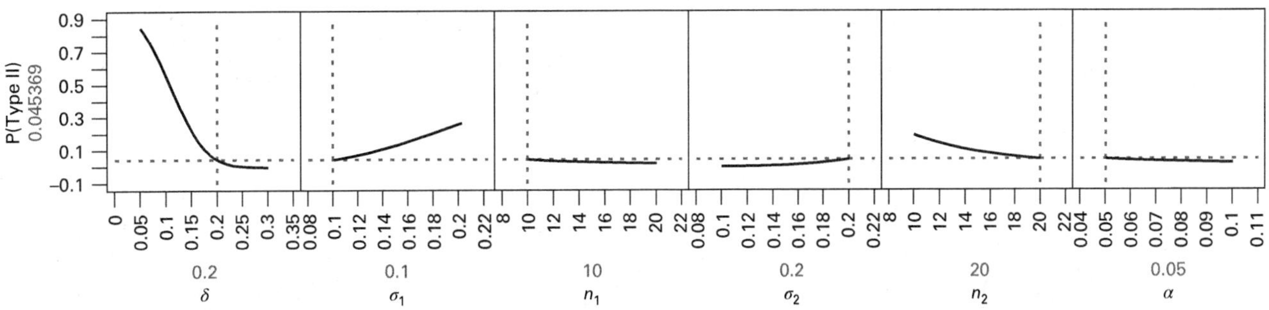

FIGURE 2.4 Prediction Profiler from JMP for the Probability of Type II Error in Example 2.1 as a Function of δ, σ_1, σ_2, n_1, n_2, and α with $\delta = 0.2$, $n_1 = 10$, $n_2 = N - n_1 = 30 - 10 = 20$ and $\alpha = 0.05$

so the power is smaller, about 0.93. The optimal experimental design produces a more powerful test.

Figure 2.4 is output from the JMP Profiler for the probability of type II error for Example 2.1 with $\delta = 0.2$, $n_1 = 10$, $n_2 = N - n_1 = 30 - 10 = 20$, and $\alpha = 0.05$. The probability of type II error is reported as 0.045369, which agrees closely with the value calculated above manually. The profiler gives some insight into the relative effect of changing the parameters of this system on the probability of type II error. For example, notice how steep the line is on the plots of type II error probability versus δ. This illustrates that as the difference in means that the experimenter is interested in detecting δ gets smaller, the probability of type II error increases fairly quickly. So if smaller differences than $\delta = 0.20$ are of interest, larger sample sizes need to be used to maintain adequate power of the test. Also, the line is very flat on the plot of probability of type II error versus α. This indicates that changing the significance level from $\alpha = 0.05$ to $\alpha = 0.10$ (say) would have little effect on the probability of type II error. The sample size n_2 has a larger effect; reducing n_2 would substantially increase the probability of type II error.

Using an optimal design in this problem has resulted in improved precision of estimation, both in terms of point estimation of the difference in means and the 95% CI, and higher power of the statistical test. The improvement was fairly modest, and using equal allocation of the 30 observations would not have been a bad design, but if the difference in population standard deviations had been larger or if a smaller total sample size had been required, more dramatic differences between the optimal design and just naively using equal sample sizes could have resulted.

2.3 Comparison of Two Means, Variances Unknown

We now consider the problem of comparing the means μ_1 and μ_2 of two independent populations where the variances σ_1^2 and σ_2^2 are unknown and not necessarily equal. This is a more general and much more realistic version of the problem of the previous section. The formal development of the procedures that we will describe requires that the two populations are normally distributed. Later on, we will discuss the sensitivity of the procedures to the normality assumption.

Suppose that we have a random sample of n_1 observations from the first normal population, say $y_{11}, y_{12}, \ldots, y_{1,n_1}$ and a random sample of n_2 observations from the second normal population, say $y_{21}, y_{22}, \ldots, y_{2,n_2}$. Let $\bar{y}_1$ and $\bar{y}_2$ be the two sample means and S_1^2 and S_2^2 be the two sample variances. First, we show how to obtain a CI on the difference in the two population means $\delta = \mu_1 - \mu_2$ and then we discuss statistical hypothesis testing.

2.3.1 Confidence Intervals on the Difference in Means of Two Normal Distributions, Variances Unknown

As in the previous section, let $\delta = \mu_1 - \mu_2$ be the true difference in means and $\hat{\delta} = \bar{y}_1 - \bar{y}_2$ be the point estimate. In Section 2.2.1, we used the ratio

$$Z = \frac{\hat{\delta} - \delta}{\sqrt{\dfrac{\sigma_1^2}{n_1} + \dfrac{\sigma_2^2}{n_2}}}$$

to establish the CI on the difference in means $\mu_1 - \mu_2$. When the variances are known, Z has a standard normal distribution. When the variances are unknown, it seems logical to replace the population variances σ_1^2 and σ_2^2 by the sample variances S_1^2 and S_2^2, which results in

$$t = \frac{\hat{\delta} - \delta}{\sqrt{\dfrac{S_1^2}{n_1} + \dfrac{S_2^2}{n_2}}} = \frac{\bar{y}_1 - \bar{y}_2 - (\mu_1 - \mu_2)}{\sqrt{\dfrac{S_1^2}{n_1} + \dfrac{S_2^2}{n_2}}} \tag{2.16}$$

If $\bar{y}_1, \bar{y}_2, S_1^2$, and S_2^2 are the sample means and variances of random samples from two independent normal populations and the two population means are equal, then the ratio t in Eq. (2.16) has (approximately) the t-distribution with degrees of freedom

$$v = \frac{\left(\dfrac{S_1^2}{n_1} + \dfrac{S_2^2}{n_2}\right)^2}{\dfrac{(S_1^2/n_1)^2}{n_1 - 1} + \dfrac{(S_2^2/n_2)^2}{n_2 - 1}} \tag{2.17}$$

Let $t_{\alpha,v}$ be the upper α percentage point of the t-distribution with v degrees of freedom defined such that the $P(t_v \geq t_{\alpha,v}) = \alpha$. Because the t-distribution is symmetric around zero, the corresponding lower α percentage point is $-t_{\alpha,v}$. The appendix provides JMP scripting commands to compute these percentage points. The following probability statement is true:

$$P\left(-t_{\alpha/2,v} \leq \frac{\bar{y}_1 - \bar{y}_2 - (\mu_1 - \mu_2)}{\sqrt{\dfrac{S_1^2}{n_1} + \dfrac{S_2^2}{n_2}}} \leq t_{\alpha/2,v}\right) \cong 1 - \alpha \tag{2.18}$$

This can be rearranged to

$$P\left(\bar{y}_1 - \bar{y}_2 - t_{\alpha/2,v}\sqrt{\frac{S_1^2}{n_1} + \frac{S_2^2}{n_2}} \leq \mu_1 - \mu_2 \leq \bar{y}_1 - \bar{y}_2 + t_{\alpha/2,v}\sqrt{\frac{S_1^2}{n_1} + \frac{S_2^2}{n_2}}\right) \cong 1 - \alpha$$

which is the definition of an approximate $100(1 - \alpha)\%$ CI on the difference in means $\mu_1 - \mu_2$. The CI is computed from

$$\bar{y}_1 - \bar{y}_2 - t_{\alpha/2,v}\sqrt{\frac{S_1^2}{n_1} + \frac{S_2^2}{n_2}} \leq \mu_1 - \mu_2 \leq \bar{y}_1 - \bar{y}_2 + t_{\alpha/2,v}\sqrt{\frac{S_1^2}{n_1} + \frac{S_2^2}{n_2}} \tag{2.19}$$

EXAMPLE 2.3 Plastic Gears (Continued)

Reconsider the plastic gear experiment from Example 2.1. The engineers are interested in estimating the difference in mean impact strength in printer gears formulated from two different plastics. Suppose that the team is comfortable with the assumption that impact strength is normally distributed, but they do not know the variances of the two populations. Because the variances are unknown, the approach to designing the experiment used in Example 2.1 cannot be used here. That is, because we do not know the variances, we cannot use Eq. (2.6) to allocate the total sample size of $N = 30$ to the two populations. However, suppose that we are willing to make an assumption about the magnitude of the ratio of the standard deviations in Eq. (2.6), say $d = \sigma_1/\sigma_2$. Then, we can use d to determine a sample allocation. Suppose that the team thinks that $d = 2$ is very likely. This would lead to the same allocation of observations used in Example 2.1 ($n_1 = 10$ and $n_2 = 20$) and the same data that is in Table 2.1. We will use this data to construct the t CI in Eq. (2.19).

Figure 2.5 presents the summary information about the two samples. This figure contains normal probability plots of the two samples. These plots do not indicate any problems with the normality assumption for either sample. The slope of the straight line on a normality plot is proportional to the sample standard deviation, and since these two plots have lines with somewhat different slopes, this indicates that the two populations have different variances. The values $\bar{y}_1 = 3.94$, $S_1 = 0.14298$, $\bar{y}_2 = 3.455$, and $S_2 = 0.23725$ will be used to construct a 95% CI on the difference in the two population means. From Eq. (2.17), the number of degrees of freedom for t is

$$v = \frac{\left(\frac{S_1^2}{n_1} + \frac{S_2^2}{n_2}\right)^2}{\frac{(S_1^2/n_1)^2}{n_1 - 1} + \frac{(S_2^2/n_2)^2}{n_2 - 1}} = \frac{\left(\frac{(0.14298)^2}{10} + \frac{(0.23725)^2}{20}\right)^2}{\frac{[(0.14298)^2/10]^2}{9} + \frac{[(0.23725)^2/20]^2}{19}} = 26.7$$

This number, as is usually the case, is not an integer, and rounding down to 26 degrees of freedom is a conservative approach. From JMP's t-distribution quantile function, we find that $t_{0.025,26} = 2.056$ (computer programs can actually work with fractional numbers of degrees of freedom). The 95% CI on the difference in the two means is computed from Eq. (2.19):

$$\bar{y}_1 - \bar{y}_2 - t_{\alpha/2,v}\sqrt{\frac{S_1^2}{n_1} + \frac{S_2^2}{n_2}} \leq \mu_1 - \mu_2 \leq \bar{y}_1 - \bar{y}_2 + t_{\alpha/2,v}\sqrt{\frac{S_1^2}{n_1} + \frac{S_2^2}{n_2}}$$

$$3.94 - 3.455 - 2.056\sqrt{\frac{(0.14298)^2}{10} + \frac{(0.23725)^2}{20}} \leq \mu_1 - \mu_2 \leq 3.94 - 3.455$$

$$+2.056\sqrt{\frac{(0.14298)^2}{10} + \frac{(0.23725)^2}{20}}$$

$$0.3417 \leq \mu_1 - \mu_2 \leq 0.6283$$

If we compare the length of this t CI with the one we computed in Example 2.1, when we assumed that the variances were known, we find that the t CI is longer (0.2866 versus 0.2184).

This is a reflection of the additional uncertainty introduced into the situation when we cannot make the assumption of known variances.

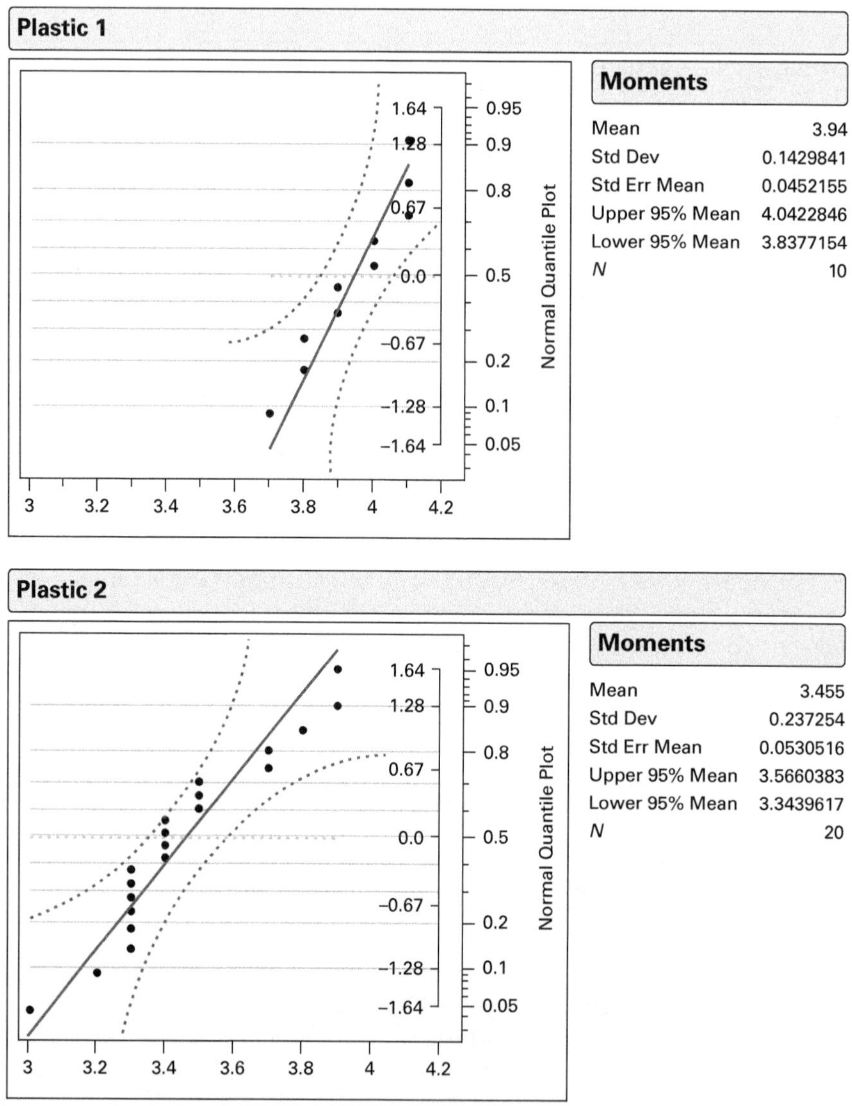

FIGURE 2.5 Summary Information About the Two Samples of Gear Impact Strength from Table 2.1

2.3.2 Hypothesis Testing on the Difference in Means of Two Normal Distributions with Unknown Variances

We can also use formal statistical hypothesis testing to investigate the difference in means of two normal distributions with unknown variances. Suppose that we wish to test the hypothesis that the difference in the two means is equal to a quantity δ_0 against an alternative hypothesis that the difference in the two means is not equal to δ_0, say

$$\begin{aligned} H_0 &: \mu_1 - \mu_2 = \delta_0 \\ H_1 &: \mu_1 - \mu_2 \neq \delta_0 \end{aligned} \tag{2.20}$$

Obviously, if $\delta_0 = 0$ we are testing the equality of the two means.

Suppose that $y_{11}, y_{12}, \ldots, y_{1,n_1}$ is a random sample of n_1 observations from the first population and $y_{21}, y_{22}, \ldots, y_{2,n_2}$ is a random sample of n_2 observations from the second population. The appropriate **test statistic** for the hypotheses in Eq. (2.20) is

$$t_0 = \frac{\bar{y}_1 - \bar{y}_2 - \delta_0}{\sqrt{\dfrac{S_1^2}{n_1} + \dfrac{S_2^2}{n_2}}} \tag{2.21}$$

If the null hypothesis is true, the test statistic in Eq. (2.21) has a t-distribution with degrees of freedom computed from Eq. (2.16). The procedure is usually called the **two-sample t-test**. It is also sometimes called **Welch's test**. The decision rule for a fixed significance level test is similar to the one-sample t-test discussed in Chapter 1. We should reject H_0 if the value of the test statistic $t_0 > t_{\alpha/2,v}$ or if $t_0 < -t_{\alpha/2,v}$, where $t_{\alpha/2,v}$ is the upper $\alpha/2$ percentage point of the t-distribution with v degrees of freedom and fail to reject otherwise. A P-value approach can also be used. The P-value for this two-sample t-test is

$$P = P(\text{above } |t_0|) + P(\text{below } - |t_0|) \tag{2.22}$$

The one-sided alternatives in the two-sample case are treated just as they were in the single-sample case. If the hypotheses are

$$\begin{aligned} H_0 &: \mu_1 = \mu_2 \\ H_1 &: \mu_1 > \mu_2 \end{aligned} \tag{2.23}$$

We should reject H_0 if the value of the test statistic $t_0 > t_{\alpha,v}$, where $t_{\alpha,v}$ is the upper α percentage point of the t-distribution with v degrees of freedom, and fail to reject otherwise. The P-value for the upper-tail one-sided two-sample t-test is computed as

$$P = P(\text{above } |t_0|) \tag{2.24}$$

The other one-sided hypotheses are

$$\begin{aligned} H_0 &: \mu_1 = \mu_2 \\ H_1 &: \mu_1 < \mu_2 \end{aligned} \tag{2.25}$$

We should reject H_0 if the value of the test statistic $t_0 < -t_{\alpha,v}$, where $t_{\alpha,v}$ is the upper α percentage point of the t-distribution with v degrees of freedom, and fail to reject otherwise. The P-value for the lower-tail one-sided t-test is computed as

$$P = P(\text{below } - |t_0|) \tag{2.26}$$

If the P-value approach is used for decision-making, the JMP procedure described in the appendix can be used to compute the P-value as explained in Chapter 1.

EXAMPLE 2.4 Plastic Gears (Continued)

We will use the two-sample *t*-test to test the equality of means in the plastic gear problem. The hypotheses are

$$H_0 : \mu_1 = \mu_2$$
$$H_1 : \mu_1 \neq \mu_2$$

The value of the test statistic is

$$t_0 = \frac{\bar{y}_1 - \bar{y}_2}{\sqrt{\frac{S_1^2}{n_1} + \frac{S_2^2}{n_2}}} = \frac{3.94 - 3.455}{\sqrt{\frac{(0.14298)^2}{10} + \frac{(0.23725)^2}{20}}} = \frac{0.4850}{0.0697} = 6.96$$

Suppose that we are testing at the 0.05 level of significance. Then, this value of the test statistic would be compared to $t_{0.025,\,26} = 2.056$ and $-t_{0.025,\,26} = -2.056$, and clearly we would reject the null hypothesis. Wheel gears made of plastic 1 have statistically significantly higher impact strength than gear wheels made of the older material. An approximate *P*-value for this test can be found by looking at Appendix Table II and noting that for 26 degrees of freedom the largest tabulated percentage point of $t_{0005,\,26} = 3.707$. Therefore, because $t_0 = 6.96 > 3.707$, the *P*-value is less than 2(0.0005) = 0.0010 (remember the test is two sided). There is strong evidence of a difference in means between the two plastics.

JMP will perform the two-sample *t*-test. Simply use the Fit Y by X analysis with y as continuous and x categorical. Figure 2.6 contains the output. Both the *t*-test statistic and the degrees of freedom are very close to the values we calculated manually, and the conclusions match the ones from Example 2.4. JMP also reports an *F*-statistic. This is associated with the analysis of variance, a technique we will introduce in the next chapter.

Welch's Test

Welch Anova Testing Means Equal, Allowing Std Devs Not Equal

F Ratio	DFNum	DFDen	Prob > *F*
48.4110	1	26.788	<.0001*

t Test
6.9578

FIGURE 2.6 JMP Two-sample *t*-test for Example 2.4

2.3.3 Comparison of Means of Two Normal Distributions with Variances Unknown but Assumed Equal

We now consider the case of comparing the means μ_1 and μ_2 of two independent populations where the variances are unknown but can be assumed to be equal, that is, $\sigma_1^2 = \sigma_2^2 = \sigma^2$. Since the variances are assumed to be equal and we can estimate the variance with the sample variance from each of the two samples, it makes sense to combine the two individual sample variances. This results in a **pooled estimate of the common variance**:

$$S_p^2 = \frac{(n_1 - 1)S_1^2 + (n_2 - 1)S_2^2}{n_1 + n_2 - 2} \tag{2.27}$$

The pooled variance estimate is a weighted average of the two individual sample variances. If the sample sizes are equal, then S_p^2 is just the average of the two individual sample variances. The number of degrees of freedom in the equal variance case will be $v = n_1 + n_2 - 2$. So the equation for the $100(1 - \alpha)\%$ CI on the difference in means becomes

$$\bar{y}_1 - \bar{y}_2 - t_{\alpha/2,n_1+n_2-2}S_p\sqrt{\frac{1}{n_1} + \frac{1}{n_2}} \le \mu_1 - \mu_2 \le \bar{y}_1 - \bar{y}_2 + t_{\alpha/2,n_1+n_2-2}S_p\sqrt{\frac{1}{n_1} + \frac{1}{n_2}} \tag{2.28}$$

The **test statistic** for the hypothesis testing problem becomes

$$t_0 = \frac{\bar{y}_1 - \bar{y}_2 - \delta_0}{S_p\sqrt{\frac{1}{n_1} + \frac{1}{n_2}}} \tag{2.29}$$

and the number of degrees of freedom in the statistical test would be $v = n_1 + n_2 - 2$. This version of the two-sample *t*-test is usually called the **pooled *t*-test**.

2.3.4 Power and Sample Size Calculations

The type II error probability for the *t*-test depends on the distribution of the test statistic when the null hypothesis is false. When the true value of the difference in means is not the hypothesized value, the distribution of the test statistic is the noncentral *t*-distribution. Finding the type II error probability for the *t*-test involves finding probabilities associated with the noncentral *t*-distribution. Because the noncentral *t* random variable has a messy density function, this calculation must be done numerically.

Fortunately, this difficult task has already been done and implemented in software. JMP has a power and sample size calculator. Suppose we have a two-sample *t*-test and we want to detect a difference in means of about one standard deviation (we assume both populations have the same variance). Simply enter the same number into the JMP calculator for the standard deviation and the difference in means that you want to detect, and JMP will produce a power curve, that is, a plot of power versus sample size. Figure 2.7 is the plot obtained from JMP for the two-sample *t*-test when we are interested in detecting a difference in means of one standard deviation. The sample size on the graph is the total sample size, N. Figure 2.7 indicates that a total sample size of about $N = 40$ would provide a power of about 0.80, or, in other words, it would give us about an 80% chance to detect a true difference in means of about one standard deviation. If you want to determine the sample size for detecting a two standard deviation difference in means, just make the difference in means that you enter twice the value of the standard deviation.

2.3.5 The Normality Assumption

The formal development of the *t*-test and the *t* CIs in this section assume that the two random samples are taken from independent normal populations. There is some confusion among

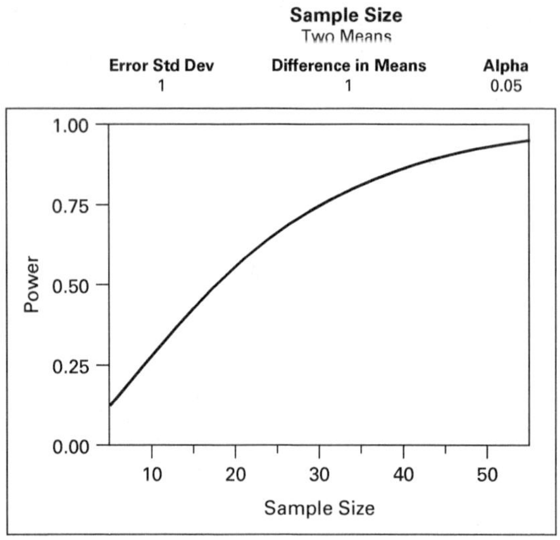

FIGURE 2.7 Power Curve from JMP for Detecting a Difference in Means of One Standard Deviation

practitioners about the importance of this assumption. It is true that if the normality assumption holds and both population variances are equal, the pooled t-test and the CI described in this section are exact procedures. That is, the reference distribution for the pooled t-test is the t-distribution with $n_1 + n_2 - 2$ degrees of freedom. However, the t-test is very insensitive to this assumption. Only very large departures from normality would cause concern. The most important assumptions are independence of the two populations and random sampling. Violation of these assumptions is serious and could lead to erroneous conclusions.

A simple but very effective way to check the normality assumption is to construct a normality probability plot of the sample data. If the data fall approximately along a straight line on this plot, there should be no concerns about the normality assumption. If the normal probability plot is problematic, we recommend transformation of the data. Often working on a log scale (for example) results in data that is approximately normally distributed. We will illustrate transformations later in the book. Some authors recommend using nonparametric methods when the data is not normally distributed. We are always very cautious about this, because while nonparametric methods do not require assumptions about the form of the distribution, they do have other assumptions that may not be met in practice, such as assuming that the populations have symmetric distributions. We feel that transformations are a better approach.

Experiments with a Single Categorical Factor: Design Issues and the Analysis of Variance

3

In Chapter 2, we showed how to compare the means of two treatments. For example, the plastic gears experiment in Example 2.1 involved gears made from two different types of plastic. You could describe this experiment as a one-factor experiment where the factor is gear (or plastic) type and the two levels are plastic type 1 and 2. A natural way to extend this kind of comparison is to consider more than two levels of the factor. This chapter deals with the design and analysis of experiments with one categorical factor having more than two levels.

3.1 Motivating Example

A cement manufacturer wants to find a mixing technique that will maximize the tensile strength of its concrete. Engineers at the company have developed four different economical methods of mixing cement. The question is, "Does one of these four methods result in concrete with demonstrably higher tensile strength?" To answer this question, the engineering team decides to mix four batches of cement using each of the four mixing methods.

This is an example of a one-factor experiment with $k = 4$ **levels** of the factor and $r = 4$ **replicates**. Since there is the same number of replicates for each level, the total number of test runs is the product of the number of levels and the number of replicates, so there are $N = 16$ tests or runs.

It might seem more organized to do all the runs for a given mixture method in sequence. That is, the team could do all the runs for a given method in a row and then proceed to the next method and so on. However, one of the fundamental principles of designed experimentation is **randomization**. It is good practice to run these mixing tests in random order so that it is unpredictable which mixing method is next in the sequence of runs. Why is this? Remember the discussion of this topic from Chapter 1. Perhaps the eventual tensile strength is related to the humidity in the room while the cement is being mixed. It is likely that in the course of a day of test the humidity will change. If the mixing tests are run in a systematic order, then the effect of the humidity change will bias the effects of the different tests. This bias could result in two possible errors. First, the team might conclude that there was no effect of the mixing technique because the effect of humidity canceled out the effect of changing technique. Another possibility is that mixing technique has a negligible effect, but the effect of humidity makes one or more mixing technique look different than the others.

Table 3.1 shows the result of sorting a column of random numbers to create a random ordering of test runs of the mixing technique.

The second column of the table shows the mixing technique in order. The third column contains random numbers from the Random Integer function in JMP. The fourth through sixth columns show the result of sorting the first three columns by the third column. The actual tests should be run in the order shown in the fifth column labeled "Randomized mixing technique."

Suppose the engineering team then runs the experiment in the order given by the fifth column. Table 3.2 shows observations they obtain for all the mixing techniques.

Table 3.1 Creating a Randomized Run Order for an Experiment

Test sequence	Ordered mixing technique	Random numbers	Original sequence	Randomized mixing technique	Random numbers
1	1	18868	6	2	13106
2	1	31552	13	4	17375
3	1	55831	1	1	18868
4	1	65797	2	1	31552
5	2	37168	7	2	32821
6	2	13106	9	3	32821
7	2	32821	8	2	36923
8	2	36923	5	2	37168
9	3	32821	15	4	40501
10	3	60074	12	3	44373
11	3	48477	11	3	48477
12	3	44373	3	1	55831
13	4	17375	16	4	57693
14	4	80196	10	3	60074
15	4	40501	4	1	65797
16	4	57693	14	4	80196

Table 3.2 Tensile Strength Data from the Cement Mixing Experiment

Mixing technique	rep 1	rep 2	rep 3	rep 4	Totals	Averages
1	3129	3000	2865	2890	11884	2971.00
2	3300	2975	3150	3200	12625	3156.25
3	2800	3050	2985	2900	11735	2933.75
4	2600	2600	2765	2700	10665	2666.25

While numerical summaries such as those shown in Table 3.2 are useful, **graphical** display of the data makes it much easier to understand what is happening. Thus, the first rule of data analysis is "**Plot** your data, then *look* at the plot." Figure 3.1a shows a scatter plot of the tensile strength for each of the four mixing techniques. Notice that mixing technique 4 appears to only have three values. That is because two replicates (1 and 2) have the same value, 2600. Figure 3.1b shows **box plots** of the four samples. Both graphs provide visual evidence that mixing technique 2 results in generally higher tensile strength.

Before the engineering team goes to management with a recommendation, it is important to know whether the results of the experiment could actually be due to the random variation in the mixing process rather than the inherent superiority of mixing technique 2. After all, both plots also show substantial variation in the replicate observations for each technique. Also, the box plots for techniques 1 and 3 overlap the box plot for technique 2.

To judge whether experimental results are due to the random variation or the effect of changing the level of the categorical factor, the team first needs to choose a statistical model for the data. Ideally then, the design of the experiment and the analysis of the data are appropriate to the chosen model.

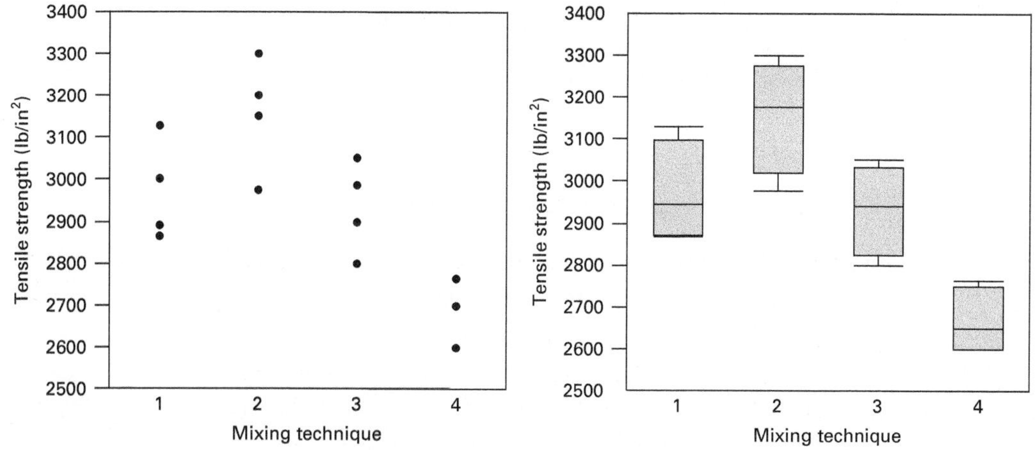

(a) Scatter plot (b) Comparative box plots

FIGURE 3.1 Plots of the Data in Table 3.2: (a) Scatter Plot (b) Comparative Box Plots

3.2 Statistical Model for the Data

For the sake of generality, let us say that we have k **treatments** or **levels** of a single **categorical** factor. Table 3.3 is a generic version of Table 3.2 for k treatments.

Notice that n_i is the number of tests at the ith level of the factor. Specifically, n_1 is the number of tests of the first mixing technique and n_2 is the number of tests of the second mixing technique. The total number of tests, N, is the sum of the number of tests at each level, $N = \sum_{i=1}^{k} n_i$.

The underpinning for any statistical analysis is a mathematical model. In our case, the **means model** is convenient. Equation (3.1) shows the means model symbolically.

$$y_{ij} = \mu_i + \varepsilon_{ij} \begin{cases} i = 1,2,\ldots,k \\ j = 1,2,\ldots,n_i \end{cases} \tag{3.1}$$

In Eq. (3.1), y_{ij} is the jth test (or replicate) of the ith treatment or level, μ_i is the mean of the ith level, ε_{ij} is a **random error** component. Each ε_{ij} is the result of all variations in the system that are not due to the change in the level of the factor. This includes variation in the measurement system, variation from differences in the test material (experimental units), and variation from all the uncontrolled factors affecting the system. The standard assumption about each ε_{ij} is that it is normally distributed, has an expected value of zero, variance, σ^2, and that it is independent of all the other random error components. The assumption that each random error component is

Table 3.3 Generic View of Data for an Experiment with a Single Categorical Factor

Treatment (level)	Data				Totals	Averages
1	y_{11}	y_{12}	…	y_{1n_1}	$y_{1.}$	$\bar{y}_{1.}$
2	y_{21}	y_{22}	…	y_{2n_2}	$y_{2.}$	$\bar{y}_{2.}$
…	…	…	…	…	…	…
k	y_{k1}	y_{k2}	…	y_{kn_μ}	$y_{k.}$	$\bar{y}_{k.}$

independent of the others is the most critical, and the purpose of randomizing the order of the tests is to provide assurance that this assumption holds.

These assumptions imply that y_{ij} is distributed as $N(\mu_i, \sigma^2)$ and that the observations are mutually independent.

3.3 Design Considerations

In Chapter 1, we learned that the variance of the mean of a sample is $\frac{\sigma^2}{N}$, where N is the number of observations. By assumption, the variance of y_{ij} is σ^2. Since the number of tests of the ith level is n_i, the variance of the ith treatment mean is $\frac{\sigma^2}{n_i}$.

We could minimize this variance by doing all the tests at the ith level. This would provide the greatest possible information about the ith level but at the expense of providing no information about any of the other levels. Clearly, this is undesirable.

Rather than minimizing the variance of any individual mean, consider minimizing the average variance of all the means. Symbolically, we have

$$\overline{\text{var}(\bar{y}_i)} = \frac{\sigma^2}{k} \sum_{i=1}^{k} \frac{1}{n_i} \tag{3.2}$$

In our example, there are four mixing methods, so k is 4. Thus, both σ^2 and k are fixed quantities in Eq. (3.2). The variance, σ^2, is fixed by assumption and the number of levels, k, is fixed by the structure of the problem. Often, the total number of runs, N, is fixed by budgetary constraints. If this is true, then the only way to minimize the average variance of the means is by a crafty allocation of the N tests over the k levels of the categorical factor.

The engineering team allocated four runs to each of the four levels. Suppose that $\sigma^2 = 1$, then we can calculate the average variance of the means as $\frac{1}{4}\left(\frac{1}{4} + \frac{1}{4} + \frac{1}{4} + \frac{1}{4}\right) = \frac{1}{4}1$. What if instead of an equal allocation of tests for each level, the team had done three tests for the first two mixing methods and five tests for the other two? Repeating the above calculation, we have $\frac{1}{4}\left(\frac{1}{3} + \frac{1}{3} + \frac{1}{5} + \frac{1}{5}\right) = \frac{1}{4}\left(\frac{5}{15} + \frac{5}{15} + \frac{3}{15} + \frac{3}{15}\right) = \frac{1}{4}\frac{16}{15}$. We can see that the average variance of the means for this allocation of tests to mixing methods is greater because $\frac{16}{15} > 1$. So, an equal allocation of tests over the levels is better.

In general, it is true that if the variance, σ^2, is constant, then an equal allocation of tests to treatments minimizes the average variance of the means. It turns out that a design that minimizes the average variance of the parameter estimates is an A-optimal design. So, for this model, the team made the best possible allocation of tests for each mixing method.

What if it were possible to perform a test of the last mixing method faster than the other three so that in the allotted time for the experiment, the team could do five tests instead of four? Should the team perform the extra test? If they perform the extra test, there will no longer be an equal allocation of tests to treatments. Once more, calculating the effect of doing the extra run on the average variance of the means, we have $\frac{1}{4}\left(\frac{1}{4} + \frac{1}{4} + \frac{1}{4} + \frac{1}{5}\right) = \frac{1}{4}\left(\frac{5}{20} + \frac{5}{20} + \frac{5}{20} + \frac{4}{20}\right) = \frac{1}{4}\frac{19}{20}$. So, in this case performing the extra run has a beneficial effect on the average variance of the means.

3.4 Statistical Analysis of the Data

In Chapter 2, we learned about the t-test for comparing the means of two treatments. To determine whether the difference in the means of the four mixing methods is due to something other than chance, what we need is something like a t-test but more generally applicable to an arbitrary number of treatments. Happily there is a statistical procedure called the **analysis of variance** that allows for testing for the equality of any number of means.

To get started, we first need to develop some notation. Remember that we use $y_{i.}$ for the sum of the observed responses for tests using the *i*th level of the categorical factor. We denote the average of all these values by $\bar{y}_{i.}$. We also need expressions for the sum of all the responses, $y_{..}$, and their grand average, $\bar{y}_{..}$. We now have

$$\begin{aligned} y_{i.} &= \sum_{j=1}^{n_i} y_{ij} \qquad \bar{y}_{i.} = y_{i.}/n_i \quad i = 1,2,\ldots,k \\ y_{..} &= \sum_{i=1}^{k}\sum_{j=1}^{n_i} y_{ij} \qquad \bar{y}_{..} = y_{..}/N \end{aligned} \tag{3.3}$$

The "dot" subscript is an abbreviated way to express a summation over the subscript that it replaces.

The team wants to test whether all the treatment means are equal. The appropriate null and alternative hypotheses are

$$\begin{aligned} H_0 &: \mu_1 = \mu_2 = \cdots = \mu_k \\ H_1 &: \mu_i \neq \mu_j \quad \text{for at least one } (i,j) \end{aligned} \tag{3.4}$$

The analysis of variance (ANOVA) is a statistical procedure for testing the null hypothesis above. In practice, we let statistical software do all the computations but we think that it is useful to understand the derivation of the various parts of the standard ANOVA output.

3.4.1 Partitioning the Variance of the Response

In the ANOVA, for one categorical factor the variation in the observed response has two parts. One part is due to the difference in the average values for each level of the factor. The other part is due to the random effect of all the other uncontrolled sources of variability. For the sake of simplicity of calculation, let the number of tests for each level be the same so that $n_i = n$ for all *i*. We call the total corrected sum of squares the sum of the squared differences between each observed response and the grand average of all the responses. Symbolically, we have $SS_T = \sum_{i=1}^{k}\sum_{j=1}^{n}(y_{ij} - \bar{y}_{..})^2$. We now break this quantity into two parts:

$$\sum_{i=1}^{k}\sum_{j=1}^{n}(y_{ij} - \bar{y}_{..})^2 = n\sum_{i=1}^{k}(\bar{y}_{i.} - \bar{y}_{..})^2 + \sum_{i=1}^{k}\sum_{j=1}^{n}(y_{ij} - \bar{y}_{i.})^2 \tag{3.5}$$

The first term on the right-hand side of Eq. (3.5) is the sum of the squared differences between the treatment averages and the grand average. The second term is the sum of the squared differences of observed values within a treatment and the treatment average. We sometimes call these terms the **between** sum of squares and the **within** sum of squares, respectively. The between sum of squares has $k - 1$ **degrees of freedom**, because while there are k terms in the sum, only $k - 1$ of them are independent. To see this, note that $n\sum_{i=1}^{k}(\bar{y}_{i.} - \bar{y}_{..}) = 0$. That is, the sum of deviations of numbers from their own average is always zero. This means that $k - 1$ of the individual averages $\bar{y}_{i.}$ can be chosen arbitrarily, but the last one must be chosen as their sum because the total must always equal zero. In general, the number of degrees of freedom of a sum of squares equals the number of independent terms used to compute it. A similar argument to the preceding one reveals that the within sum of squares has $k(r-1)$ degrees of freedom.

3.4.2 The ANOVA

If there is no difference in the treatment means, it turns out that both the between and within sums of squares, when divided by their respective degrees of freedom, estimate the variance of the experimental error, σ^2. If the treatment means differ, the value of the between sum of squares is expected to be larger than σ^2. Define the ratio

$$F_0 = \frac{SS_{\text{Between}}/(k-1)}{SS_{\text{Within}}/k(r-1)} \tag{3.6}$$

Values of this ratio that are close to unity do not convey any evidence against the claim that the treatment means are all equal. However, values of this ratio that are greater than 1 are evidence against this claim. The larger the value of $F_{0,}$ the greater the weight of evidence against the claim of equal treatment means. Stated formally, F_0 can be used as **test statistic** for the hypotheses

$$\begin{aligned} H_0 &: \mu_1 = \mu_2 = \cdots = \mu_k \\ H_1 &: \text{Above not true for at least one treatment mean} \end{aligned} \tag{3.7}$$

To operationalize this test statistic, we need a way to decide if the observed value of the test statistic F_0 is large enough to reject the null hypothesis, that is, the claim of equal treatment means. Fortunately, there is an easy way to do this. If the assumptions made earlier that the observations are normally and independently distributed with constant variance, and the null hypothesis in Eq. (3.7) of equal treatment means is true, the ratio F_0 has an F-distribution with $k-1$ numerator degrees of freedom and $k(r-1)$ denominator degrees of freedom (refer to Chapter 1 for a review of the F-distribution). Let $F_{\alpha,k-1,k(r-1)}$ be the upper α percentage point of the F-distribution with $k-1$ numerator degrees of freedom and $k(r-1)$ denominator degrees of freedom; that is, the area or probability above $F_{\alpha,k-1,k(r-1)}$ is α. Therefore, we would reject the null hypothesis in Eq. (3.7) at the α level of significance if

$$F_0 > F_{\alpha,k-1,k(r-1)}$$

and fail to reject otherwise. It is customary to call the between sum of squares the **treatment** sum of squares and the within sum of squares the **error** sum of squares and to arrange the ANOVA procedure in a table, such as shown in Table 3.4. This display is called an **ANOVA table**.

Table 3.4 The ANOVA Table for a Single-factor Experiment

Source of variation	Sum of squares	Degrees of freedom	Mean square	F_0
Between treatments	$SS_{\text{Treatments}} = n\sum_{i=1}^{k}(\bar{y}_{i.}-\bar{y}_{..})^2$	$k-1$	$MS_{\text{Treatments}}$	$F_0 = \frac{MS_{\text{Treatments}}}{MS_E}$
Error (within treatments)	$SS_E = SS_T - SS_{\text{Treatments}}$	$N-k$	MS_E	
Total	$SS_T = \sum_{i=1}^{k}\sum_{j=1}^{r}(y_{ij}-\bar{y}_{..})^2$	$N-1$		

The ANOVA is implemented in many software packages, so one rarely needs to resort to manual computing of Table 3.4. For completeness, we give convenient equations for computing the sums of squares in Table 3.4:

$$SS_T = \sum_{i=1}^{k}\sum_{j=1}^{r} y_{ij}^2 - \frac{y_{..}^2}{N}$$

$$SS_{\text{Treatments}} = \frac{1}{r}\sum_{i=1}^{k} y_{i.}^2 - \frac{y_{..}^2}{N} \tag{3.8}$$

$$SS_E = SS_T - SS_{\text{Treatments}}$$

The JMP ANOVA for the concrete strength data is shown in Table 3.5.

The computed value of the test statistic is $F_0 = 12.7381$, and the upper 0.05 percentage point of the F-distribution with $k - 1 = 4 - 1 = 3$ numerator degrees of freedom and $k(r - 1) = 4(4 - 1) = 12$ denominator degrees of freedom is $F_{0.05,3,12} = 3.49$. Since the computed value of the test statistic exceeds this critical value, we conclude that mixing technique affects the mean strength of the resulting concrete. That is, the mean strengths of the concrete are different depending on the mixing technique used. JMP also reports a P-value of 0.0005 for the F-statistic, strong evidence that there is a difference in treatment means.

3.4.3 Post-ANOVA Comparison of Treatment Means

When the analysis leads to rejection of the null hypothesis in Eq. (3.7), we can conclude that there are differences between the treatment means. However, we do not know specifically which means differ. Put another way, we do not know which treatments are responsible for the differences among the means. Determining this is usually a necessary follow-up to the formal ANOVA so that we may draw practical conclusions from the experiment. Sometimes graphical displays such as box plots and scatter diagrams are adequate for this, but often more formal **multiple comparison** methods are helpful. This section presents some analysis techniques for post-ANOVA comparison of treatment means. For discussion of other methods, see Montgomery (2020).

In many practical situations, we will wish to compare only **pairs of means**. Frequently, we can determine which means differ just by testing the differences between *all* pairs of treatment means. Thus, we are interested in hypotheses of the form $H_0 : \mu_1 = \mu_j$ for all $i \neq j$. The Fisher least significant difference (LSD) method for comparing all pairs of means controls the type I error rate α for each individual pairwise comparison but does not control the **experimentwise** or **family** error rate. The procedure makes use of the familiar two-sample t-statistic for testing the equality of two means:

$$t_0 = \frac{\overline{y}_{i.} - \overline{y}_{j.}}{\sqrt{MS_E\left(\frac{1}{n_i} + \frac{1}{n_j}\right)}} \tag{3.9}$$

Assuming a two-sided alternative, the pair of means μ_i and μ_j would be declared significantly different if $|\overline{y}_{i.} - \overline{y}_{j.}| > t_{\alpha/2,N-k}\sqrt{MS_E(1/r_i + 1/r_j)}$. The quantity

$$LSD = t_{\alpha/2,N-k}\sqrt{MS_E\left(\frac{1}{r_i} + \frac{1}{r_j}\right)} \tag{3.10}$$

Table 3.5 JMP ANOVA for the Concrete Strength Data in Table 3.2

Oneway Anova

Summary of Fit

Rsquare	0.760881
Adj Rsquare	0.701102
Root Mean Square Error	113.2506
Mean of Response	2931.813
Observations (or Sum Wgts)	16

Analysis of Variance

Source	DF	Sum of Squares	Mean Square	*F* Ratio	Prob > *F*
Technique	3	489740.19	163247	12.7281	0.0005*
Error	12	153908.25	12826		
C. Total	15	643648.44			

Means for Oneway Anova

Level	Number	Mean	Std Error	Lower 95%	Upper 95%
1	4	2971.00	56.625	2847.6	3094.4
2	4	3156.25	56.625	3032.9	3279.6
3	4	2933.75	56.625	2810.4	3057.1
4	4	2666.25	56.625	2542.9	2789.6

Std Error uses a pooled estimate of error variance

Means Comparisons

Comparisons for each pair using Student's *t*

t	Alpha
2.17881	0.05

Abs(Dif)-LSD

	2	1	3	4
2	−174.48	10.77	48.02	315.52
1	10.77	−174.48	−137.23	130.27
3	48.02	−137.23	−174.48	93.02
4	315.52	130.27	93.02	−174.48

Positive values show pairs of means that are significantly different.

Level		Mean
2	A	3156.2500
1	B	2971.0000
3	B	2933.7500
4	C	2666.2500

Levels not connected by same letter are significantly different.

(continued)

Table 3.5 **(Continued)**

Means Comparisons

Level	– Level	Difference	Std Err Dif	Lower CL	Upper CL	p-Value
2	4	490.0000	80.08023	315.520	664.4798	<.0001*
1	4	304.7500	80.08023	130.270	479.2298	0.0025*
3	4	267.5000	80.08023	93.020	441.9798	0.0059*
2	3	222.5000	80.08023	48.020	396.9798	0.0167*
2	1	185.2500	80.08023	10.770	359.7298	0.0392*
1	3	37.2500	80.08023	−137.230	211.7298	0.6501

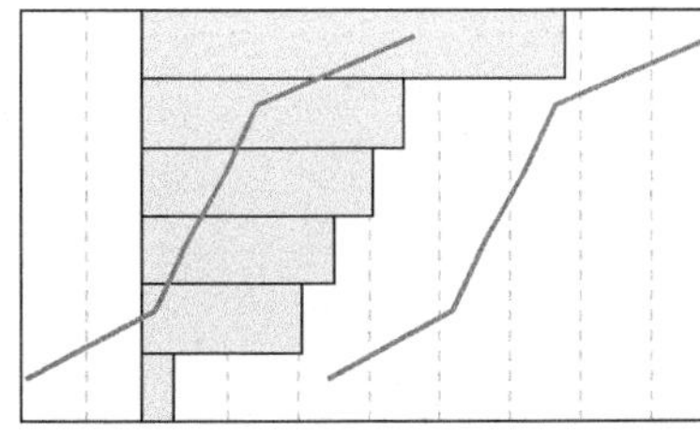

is called the **least significant difference**. If the design is balanced, $r_1 = r_2 = \cdots = r_a = r$, and

$$LSD = t_{\alpha/2,N-k}\sqrt{\frac{2MS_E}{r}} \tag{3.11}$$

To use the Fisher LSD procedure, we simply compare the observed difference between each pair of averages to the corresponding *LSD*. If $|\bar{y}_{i.} - \bar{y}_{j.}| > LSD$, we conclude that the population means μ_i and μ_j differ. The *t*-statistic in Eq. (3.9) could also be used.

Another way to implement the Fisher LSD procedure is to use the test statistic Eq. (3.9) directly for each pair of means and either compare the computed values to the *t*-distribution percentage point $t_{\alpha/2,N-k}$ or to construct a 100(1 − α)% confidence interval for each pair of means using

$$-LSD \le \mu_i - \mu_j \le LSD \tag{3.12}$$

Every interval that includes zero indicates a pair of means that are not significantly different from zero.

JMP implements the Fisher LSD procedure. The lower section of Table 3.5 titled "Means Comparisons" displays the JMP output for the Fisher LSD. The confidence intervals for each pair of means are shown. Mixing techniques 1 and 3 result in the same mean strength. All other pairs of means are different. Mixing technique 2 results in the highest mean strength and mixing technique 4 the lowest mean strength.

Note that the overall type I error rate may be considerably inflated using the Fisher LSD method. Specifically, as the number of treatments *k* gets larger, the experimentwise or family type I error rate (the ratio of the number of experiments in which at least one type I error is made to the total number of experiments) becomes large. This is considered a disadvantage of the procedure.

Tukey (1953) proposed a procedure for testing hypotheses for which the overall significance level is exactly α when the sample sizes are equal and at most α when the sample sizes are unequal. His procedure can also be used to construct confidence intervals on the differences in all pairs of means. For these intervals, the simultaneous confidence level is 100(1 − α)% when the sample sizes are equal and at least 100(1 − α)% when sample sizes are unequal. In other words, the Tukey procedure controls the experimentwise or "family" error rate at the selected level α. It is often called the Tukey honestly significantly different (HSD) method.

Tukey's procedure uses the distribution of the **studentized range statistic**

$$q = \frac{\bar{y}_{\max} - \bar{y}_{\min}}{\sqrt{MS_E/n}}$$

where $\bar{y}_{max}$ and $\bar{y}_{min}$ are the largest and smallest sample means, respectively, out of a group of p sample means. For equal sample sizes, Tukey's test declares two means significantly different if the absolute value of their sample differences exceeds

$$T_\alpha = q_\alpha(a,f)\sqrt{\frac{MS_E}{n}} \tag{3.13}$$

Equivalently, we could construct a set of 100(1 – α)% confidence intervals for all pairs of means as follows:

$$\begin{aligned}\bar{y}_{i.} - \bar{y}_{j.} - q_\alpha(a,f)\sqrt{\frac{MS_E}{n}} &\le \mu_i - \mu_j \\ &\le \bar{y}_{i.} - \bar{y}_{j.} + q_\alpha(a,f)\sqrt{\frac{MS_E}{n}}, i \neq j\end{aligned} \tag{3.14}$$

When sample sizes are not equal, Eqs. (3.13) and (3.14) become

$$T_\alpha = \frac{q_\alpha(a,f)}{\sqrt{2}}\sqrt{MS_E\left(\frac{1}{n_i}+\frac{1}{n_j}\right)} \tag{3.15}$$

and

$$\begin{aligned}\bar{y}_{i.} - \bar{y}_{j.} - \frac{q_\alpha(a,f)}{\sqrt{2}}\sqrt{MS_E\left(\frac{1}{n_i}+\frac{1}{n_j}\right)} &\le \mu_i - \mu_j \\ &\le \bar{y}_{i.} - \bar{y}_{j.} + \frac{q_\alpha(a,f)}{\sqrt{2}}\sqrt{MS_E\left(\frac{1}{n_i}+\frac{1}{n_j}\right)}, i \neq j\end{aligned} \tag{3.16}$$

respectively. The unequal sample size version is sometimes called the **Tukey–Kramer procedure**.

JMP implements the Tukey HSD method. The output is shown in Table 3.6. This procedure indicates that mixing techniques 1, 2, and 3 all result in the same mean strength and that mixing technique 4 produces concrete with significantly lower mean strength.

When using any procedure for pairwise testing of means, we occasionally find that the overall F-test from the ANOVA is significant, but the pairwise comparison of means fails to reveal any significant differences. This situation occurs because the F-test is simultaneously considering all possible contrasts involving the treatment means, not just pairwise comparisons. That is, in the data at hand, the significant contrasts may not be of the form $\mu_i - \mu_j$.

A logical question which one of these procedures should one use? Unfortunately, there is no clear-cut answer to this question, and professional statisticians often disagree over the utility of the various procedures. Carmer and Swanson (1973) have conducted Monte Carlo simulation studies of a number of multiple comparison procedures, including others not discussed here. They report that the least significant difference method is a very effective test for detecting true differences in means if it is applied *only after* the F-test in the ANOVA is significant at 5%. However, this method does not control the experimentwise error rate. Because the Tukey method does control the overall error rate, many statisticians prefer this procedure.

3.4.4 Comparing Treatment Means with a Control

In many experiments, one of the treatments is a **control**, and the analyst is interested in comparing each of the other $a - 1$ treatment means with the control. Thus, only $a - 1$ comparisons are to

Table 3.6 JMP Output for the Concrete Experiment, Tukey HSD Method

Means Comparisons

Comparisons for all pairs using Tukey-Kramer HSD

q*	Alpha
2.96880	0.05

Abs(Dif)-HSD

	2	1	3	4
2	−237.74	−52.49	−15.24	252.26
1	−52.49	−237.74	−200.49	67.01
3	−15.24	−200.49	−237.74	29.76
4	252.26	67.01	29.76	−237.74

Positive values show pairs of means that are significantly different.

Level		Mean
2	A	3156.2500
1	A	2971.0000
3	A	2933.7500
4	B	2666.2500

Levels not connected by same letter are significantly different.

Level	− Level	Difference	Std Err Dif	Lower CL	Upper CL	p-Value
2	4	490.0000	80.08023	252.257	727.7425	0.0003*
1	4	304.7500	80.08023	67.007	542.4925	0.0116*
3	4	267.5000	80.08023	29.757	505.2425	0.0262*
2	3	222.5000	80.08023	−15.243	460.2425	0.0693
2	1	185.2500	80.08023	−52.493	422.9925	0.1494
1	3	37.2500	80.08023	−200.493	274.9925	0.9653

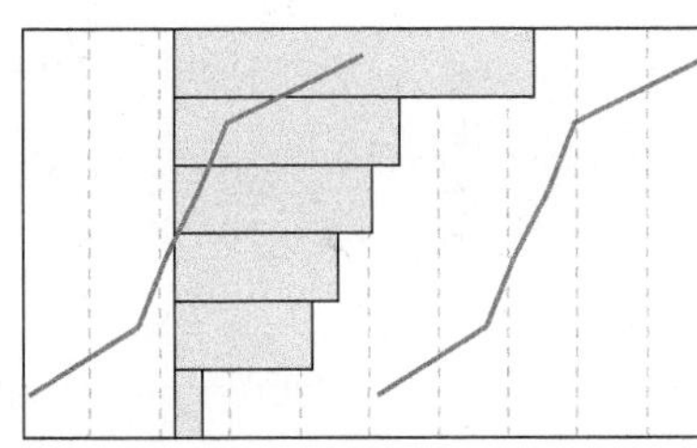

be made. A procedure for making these comparisons has been developed by Dunnett (1964). Suppose that treatment a is the control and we wish to test the hypotheses

$$
\begin{aligned}
H_0 &: \mu_i = \mu_a \\
H_1 &: \mu_i \neq \mu_a
\end{aligned}
$$

for $i = 1, 2, \ldots, k-1$. Dunnett's procedure is a modification of the usual t-test. For each hypothesis, we compute the observed differences in the sample means.

$$|\bar{y}_{i.} - \bar{y}_{k.}| \quad i = 1,2,\ldots,k-1$$

The null hypothesis $H_0 : \mu_i = \mu_a$ is rejected using a type I error rate α if

$$|\bar{y}_{i.} - \bar{y}_{k.}| > d_\alpha(k-1,f)\sqrt{MS_E\left(\frac{1}{n_i} + \frac{1}{n_k}\right)} \tag{3.17}$$

Table 3.7 Dunnett's Test for the Concrete Experiment with Method 4 Assumed to Be the Control

Means Comparisons

Comparisons with a control using Dunnett's method

Control Group = 4

|d|	Alpha
2.68292	0.05

Level	Abs(Dif)-LSD	p-Value
2	275.2	0.0001*
1	89.9	0.0066*
3	52.65	0.0153*
4	−215	1.0000

Positive values show pairs of means that are significantly different.

Note that α is the **joint significance level** associated with all $k - 1$ tests.

JMP implements Dunnett's test. Table 3.7 is the JMP output for Dunnett's test for the concrete experiment, assuming that method 4 is the control. All of the mixing techniques 1–3 result in concrete with a mean strength that exceeds the mean strength from the control method 4.

3.4.5 The Effects Model

In Section 3.2, we introduced the **means model**

$$y_{ij} = \mu_i + \varepsilon_{ij} \begin{cases} i = 1,2, \dots, k \\ j = 1,2, \dots, n_i \end{cases} \tag{3.18}$$

for the ANOVA. Many authors and textbooks use a slightly different parameterization of the ANOVA model called the **effects model**:

$$y_{ij} = \mu + \tau_i + \varepsilon_{ij} \begin{cases} i = 1,2, \dots, k \\ j = 1,2, \dots, n_i \end{cases} \tag{3.19}$$

In the effects model, μ is an **overall mean**, whose value is common to all treatments, and τ_i is a parameter that is unique to the ith treatment, usually called the ith **treatment effect**. There is a very simple connection between the means model and the effects model; namely, the mean of the ith treatment is just the sum of the overall mean plus the ith treatment effect:

$$\mu_i = \mu + \tau_i$$

Clearly, if the treatment effects are all equal to zero, then all of the treatment means are equal. Sometimes the ANOVA hypotheses are stated in terms of the treatment effects:

$$\begin{aligned} H_0 &: \tau_1 = \tau_2 = \cdots = \tau_k = 0 \\ H_1 &: \text{At least one } \tau_i \neq 0 \end{aligned}$$

This is equivalent to the hypotheses written in terms of the treatment means given previously as Eq. (3.4).

It is straightforward to estimate the parameters in the effects model. The overall mean μ is estimated by the grand average; that is, $\hat{\mu} = \bar{y}_{..}$, and each individual treatment mean is estimated by the corresponding treatment average $\hat{\mu}_i = \bar{y}_{i.}$. Consequently, the treatment effects are estimated as

$$\begin{aligned}\hat{\tau}_i &= \hat{\mu}_i - \hat{\mu} \\ &= \bar{y}_{i.} - \bar{y}_{..}, \quad i = 1,2,\ldots,k\end{aligned}$$

3.5 Model Adequacy Checking

The decomposition of the variability in the observations through an analysis of variance identity (Eq. (3.5)) is a purely algebraic relationship. However, the use of the partitioning to test formally for no differences in treatment means requires that certain assumptions be satisfied. Specifically, these assumptions are that the observations are adequately described by the (effects) model

$$y_{ij} = \mu + \tau_i + \varepsilon_{ij}$$

and that the errors are normally and independently distributed with mean zero and constant but unknown variance σ^2. If these assumptions are valid, the analysis of variance procedure is an exact test of the hypothesis of no difference in treatment means.

In practice, however, these assumptions will usually not hold exactly. Consequently, it is usually unwise to rely on the analysis of variance until the validity of these assumptions has been checked. Violations of the basic assumptions and model adequacy can be easily investigated by the examination of **residuals**. We define the residual for observation j in treatment i as

$$e_{ij} = y_{ij} - \hat{y}_{ij} \tag{3.20}$$

where $\hat{y}_{ij}$ is an estimate of the corresponding observation y_{ij} obtained as follows:

$$\begin{aligned}\hat{y}_{ij} &= \hat{\mu} + \hat{\tau}_i \\ &= \bar{y}_{..} + (\bar{y}_{i.} - \bar{y}_{..}) \\ &= \bar{y}_{i.}\end{aligned} \tag{3.21}$$

Equation (3.21) gives the intuitively appealing result that the estimate of any observation in the ith treatment is just the corresponding treatment average.

Examination of the residuals should be an automatic part of any analysis of variance. The experimenter should be cautious about adopting the conclusions from the ANOVA until model adequacy has been checked. If the model is adequate, the residuals should be **structureless**; that is, they should contain no obvious patterns. Through analysis of residuals, many types of model inadequacies and violations of the underlying assumptions can be discovered. In this section, we discuss how model diagnostic checking can be done easily by graphical analysis of residuals and how to deal with several commonly occurring abnormalities.

3.5.1 Checking the Normality Assumption

A check of the normality assumption could be made by plotting a histogram of the residuals. Conceptually, if the normality assumption on the errors is satisfied, this plot should look like a sample from a normal distribution centered at zero. Unfortunately, histograms are large sample tools, and most designed experiments utilize relatively small samples. With small samples, considerable fluctuation in the shape of a histogram often occurs, so the histogram is not a very reliable diagnostic. A moderate departure from normality does not necessarily imply a serious

Table 3.8 Residuals from the Concrete Mixing Experiment

Mixing technique	Residuals			
1	158	29	−106	−81
2	143.54	−181.25	−6.25	43.75
3	−133.75	116.25	51.25	−33.75
4	−66.25	−66.25	98.75	33.75

violation of the assumptions. Gross deviations from normality are potentially serious and require further analysis.

A useful procedure is to construct a **normal probability plot** of the residuals. Recall from Chapter 2 that we used a normal probability plot of the raw data to check the assumption of normality when using the t-test. In the analysis of variance, it is usually more effective (and straightforward) to do this with the **residuals.** If the underlying error distribution is normal, this plot will resemble a straight line. In visualizing the straight line, more emphasis should be placed on the central values of the plot than on the extremes.

Table 3.8 contains the residuals from the concrete mixing experiment. Figure 3.2 shows the normal probability plot of these residuals. This plot does not reveal any potential problem with the normality assumption. As noted earlier, moderate departures from normality are of little concern in the ANOVA. An error distribution that has considerably thicker or thinner tails than the normal is of more concern than a skewed distribution. Because the F-test is only slightly affected, we say that the analysis of variance (and related procedures such as multiple comparison methods) is **robust** to the normality assumption. Departures from normality usually cause both the true significance level and the power to differ slightly from the advertised values, with the power generally being lower.

A very common defect that often shows up on normal probability plots is one residual that is very much larger than any of the others. Such a residual is often called an **outlier**. The presence of one or more outliers can seriously distort the analysis of variance, so when a potential outlier is located, careful investigation is called for. Frequently, the cause of outliers is a mistake in calculations or a data coding or copying error. If this is not the cause, the experimental circumstances surrounding this run must be carefully studied. If the outlying response is a particularly desirable value (high strength, low cost, etc.), the outlier may be more informative than the rest of the data.

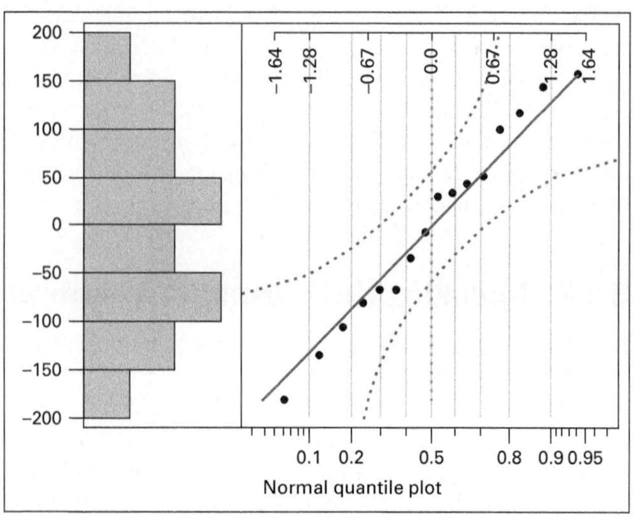

FIGURE 3.2 Normal Probability Plot of the Residuals from the Concrete Mixing Experiment

We should be careful not to reject or discard an outlying observation unless we have reasonable nonstatistical grounds for doing so. At worst, you may end up with two analyses: one with the outlier and one without.

Several formal statistical procedures may be used for detecting outliers (e.g., see Stefansky (1972), John and Prescott (1975), and Barnett and Lewis (1994)). Some statistical software packages report the results of a statistical test for normality (such as the Anderson–Darling test) on the normal probability plot of residuals. This should be viewed with caution as those tests usually assume that the data to which they are applied are independent and residuals are not independent.

A rough check for outliers may be made by examining the **standardized residuals**:

$$d_{ij} = \frac{e_{ij}}{\sqrt{MS_E}} \tag{3.22}$$

If the errors ε_{ij} are $N(0, \sigma^2)$, the standardized residuals should be approximately normal with mean zero and unit variance. Thus, about 68% of the standardized residuals should fall within the limits ±1, about 95% of them should fall within ±2, and virtually all of them should fall within ±3. A residual bigger than 3 or 4 standard deviations from zero is a potential outlier.

The plot that we find most useful is a plot of **externally studentized** residuals versus run order. JMP puts this in the effect screening personality of Fit Model. This plot is shown in Figure 3.3. This plot is preferable to plotting either the standardized residuals or the ordinary residuals.

3.5.2 Checking for Nonconstant Variance

If the model is correct and the assumptions are satisfied, the residuals should be structureless; in particular, they should be unrelated to any other variable including the predicted response. A simple check is to plot the residuals versus the fitted values. For the single-factor experiment model, remember that the fitted values for the observations in treatment i are just the ith treatment average. This plot should not reveal any obvious pattern. It is also useful to plot residuals versus the levels of each treatment. In the single-factor experiment, this is identical to plotting the residuals versus the fitted values. Figure 3.4 plots the residuals versus the levels of mixing technique for the concrete strength experiment. No unusual structure is apparent.

A defect that occasionally shows up on this plot is **nonconstant variance**. Sometimes, the variance of the observations increases as the magnitude of the observation increases. This would be the case if the error or background noise in the experiment was a constant percentage of the size of the observation. (This commonly happens with many measuring instruments—error is a percentage of the scale reading.) If this were the case, the residuals would get larger as y_{ij} gets larger, and the plot of residuals versus would look like an outward-opening funnel or megaphone.

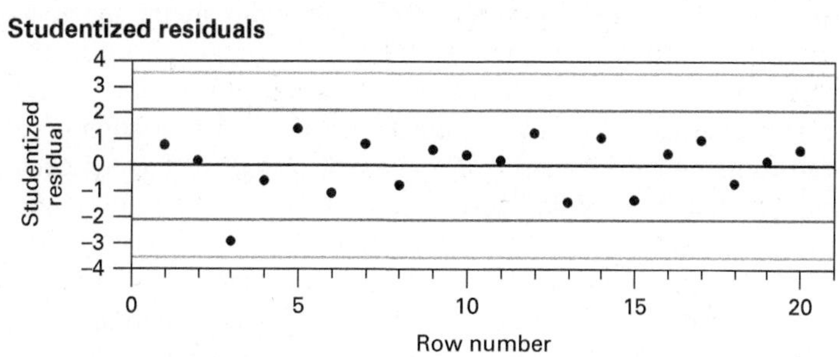

FIGURE 3.3 Externally Studentized Residuals with 95% Simultaneous Limits (Bonferroni) in Light Gray, Individual Limits in Dark Gray

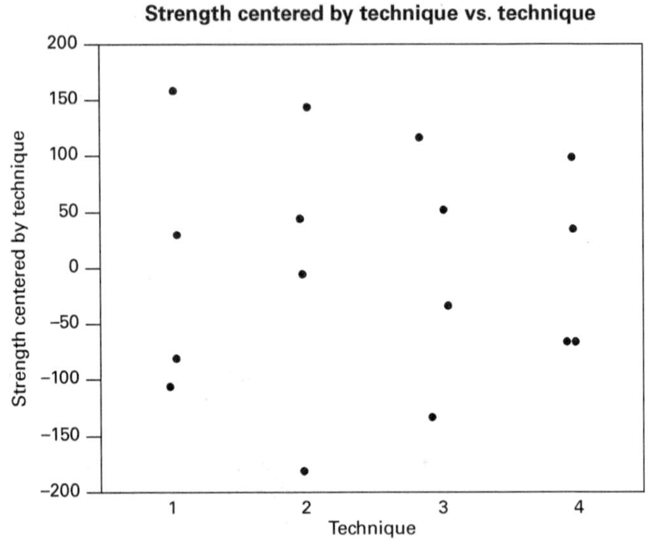

FIGURE 3.4 Residuals Plotted Versus Mixing Technique for the Concrete Experiment

Nonconstant variance also arises in cases where the data follow a nonnormal, skewed distribution because in skewed distributions the variance tends to be a function of the mean.

If the assumption of homogeneity of variances is violated, the F-test is only slightly affected in the balanced (equal sample sizes in all treatments) design case. However, in unbalanced designs or in cases where one variance is very much larger than the others, the problem is more serious. Specifically, if the factor levels having the larger variances also have the smaller sample sizes, the actual type I error rate is larger than anticipated (or confidence intervals have lower actual confidence levels than were specified). Conversely, if the factor levels with larger variances also have the larger sample sizes, the significance levels are smaller than anticipated (confidence levels are higher).

It is also a good idea to plot the residuals versus the run order in which the experiment was conducted. Inequality of variance also shows up occasionally on this plot. An outward-opening funnel pattern indicates that variability is increasing over time. This is also potentially problematic and it should be investigated. An increase in variability over time could result from operator/subject fatigue, accumulated stress on equipment, changes in material properties such as catalyst degradation, or tool wear, or any of a number of causes.

The usual approach to dealing with nonconstant variance when it occurs for the above reasons is to apply a **variance-stabilizing transformation** and then to run the analysis of variance on the transformed data. In this approach, one should note that the ANOVA conclusions apply to the *transformed* populations. The most widely used transformations are the square root, logarithmic, and reciprocal transformations. Often the form of the transformation is selected empirically; that is, try a particular transformation and observe the effect that it has on the residual plots. If the plots improve or if the variance seems to stabilize, then the transformation is successful. If not, select another transformation and try again. There are also analytical approaches to selecting the form of the transformation. The method developed by Box and Cox (1964) is available in many software packages and is widely used. We will illustrate the use of transformations later.

Transformations made for inequality of variance also affect the form of the error distribution. In most cases, the transformation brings the error distribution closer to normal. For more discussion of transformations, refer to Bartlett (1947), Dolby (1963), Box and Cox (1964), and Draper and Hunter (1969).

3.6 Power and Sample Size

In any experimental design problem, a critical decision is the choice of sample size—that is, determining the number of replicates to run. Generally, if the experimenter is interested in detecting small effects, more replicates are required than if the experimenter is interested in detecting large effects. In this section, we discuss a simple and useful approach to determining the sample size. Although our discussion focuses on a single-factor design, the approach can be extended to more complex experimental situations.

In order to determine the appropriate sample size for an experiment, we need to be able to determine the type II error probability β or equivalently the **power** $1 - \beta$ of the ANOVA. The type II error is the probability that a false null hypothesis will not be rejected, or the probability that a real difference in treatment means will not be detected. The probability statement is

$$\begin{aligned}\beta &= 1 - P\{\text{Reject } H_0 | H_0 \text{ is false}\}\\ &= 1 - P\{F_0 > F_{\alpha,k-1,k(r-1)} | H_0 \text{ is false}\}\end{aligned}$$

To evaluate this probability statement, we need to know the distribution of the ANOVA test statistic F_0 if the null hypothesis is false. It can be shown that, if H_0 is false, the statistic F_0 is distributed as a **noncentral *F*** random variable with $k - 1$ and $k(r - 1)$ degrees of freedom and a **noncentrality parameter** δ. The noncentrality parameter reflects how much the treatment means differ from each other relative to the variance of the experimental errors σ^2. If the treatment means are all equal, then $\delta = 0$, and the noncentral F-distribution becomes the usual (central) F-distribution. For a specified difference in the set of k treatment means, number of replicates r, and σ^2, it is possible to numerically evaluate the type II error probability β. Historically, experimenters could refer to a set of published **operating characteristic** (OC) curves that plotted β versus a function of the difference in treatment means relative to σ^2 for different sample sizes to determine the appropriate sample size for their specific experiment. Power curves plot $1 - \beta$ instead of β. For examples of these OC curves and details about how to use them, refer to Montgomery (2020).

Modern computer software can also perform power and sample size calculations making the use of printed OC curves unnecessary. We will illustrate the use of the power and sample size calculator in JMP. Consider an experiment similar to the concrete mixing example, with $k = 4$ treatments. Suppose we are interested in rejecting the null hypothesis of equal treatment means of one of the mixing techniques produces concrete with a mean strength that is one standard deviation different from the others. Notice that this does not require us to actually know either the value of the experimental error σ^2 or the actual values of the treatment means. Figure 3.5 is the JMP output for this situation. JMP plots power versus the total sample size, so to determine the sample size for each treatment divide the chosen sample size on the horizontal axis by $k = 4$. For example, suppose that the experimenter wants the power to be at least 0.70. The power curve indicates that a total sample size of 60 would result in a power of about 0.70, so this means that $r = 60/4$ or 15 replicates would result in an experiment with the desired power.

To perform the calculations in Figure 3.5, it is necessary to define the difference in means that the experimenter wants to detect in terms of the treatment effects, that is, the τ_i from the effects model. Since the treatment effects are customarily defined so that they sum to zero, a one standard deviation difference in one treatment mean (say treatment one) would imply that the treatment effects are $\tau_1 = 0.75$ and $\tau_2 = \tau_3 = \tau_4 = -0.25$. The difference in means reported by JMP in Figure 3.5 is

$$D = \sqrt{\frac{1}{\sigma^2}\left[\sum_{i=1}^{4} \tau_i^2\right]} = \sqrt{\frac{1}{1}[(0.75)^2 + 3(-0.25)^2]} = \sqrt{\frac{3}{4}} = 0.866$$

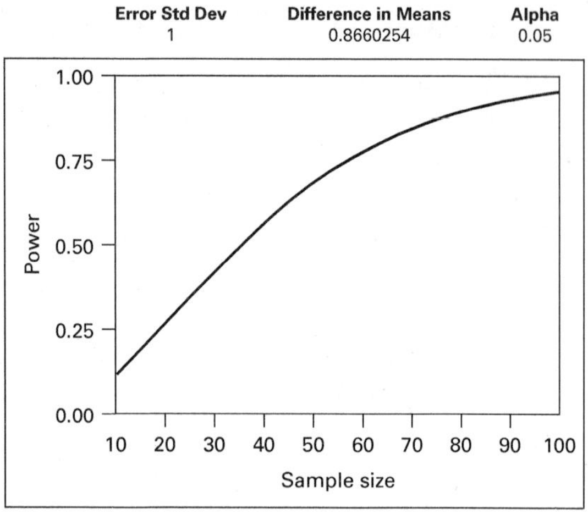

FIGURE 3.5 JMP Power and Sample Size Results

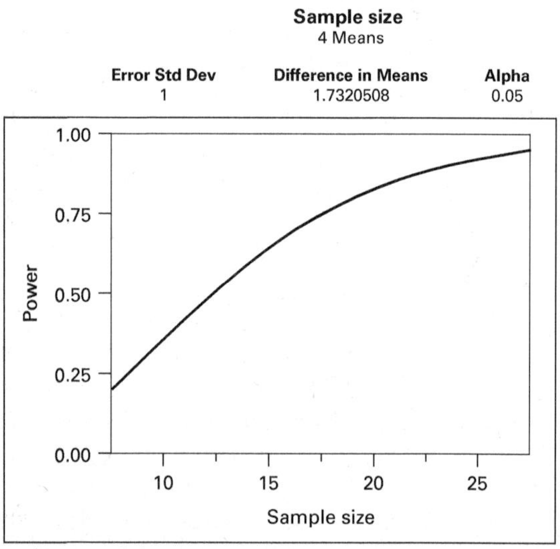

FIGURE 3.6 JMP Power and Sample Size Results

The sample size in this example is relatively large because the difference in treatment means that the experimenter is interested in detecting, a magnitude of one standard deviation in one of the $k = 4$ treatment means, is relatively small. Let us suppose that the difference that the experimenter wants to detect in one treatment mean is two standard deviations. Figure 3.6 is the JMP power and sample size output for this situation. Now the required sample size to detect this difference in means with power of at least 0.70 is $r = 18/4 = 4.5$ or about 5 replicates.

These two examples illustrate that larger differences in treatment means are easier to detect with a given sample size than are small ones. As a rough rule of thumb, we recommend designing experiments with power of at least 0.7–0.8 or higher, and with the capability of detecting differences in treatment means of about two standard deviations. There may be situations where smaller differences in means have economic or operational importance, but in many cases differences smaller than two standard deviations are unlikely to produce changes in the response variable that are much different than the background noise in the system.

Experiments with a Single Continuous Factor: Design Issues and the Regression Analysis

4

In Chapters 2 and 3, we showed how to design and analyze the data for a single categorical factor. For example, the concrete mixing experiment in Example 3.1 involved a study to evaluate the relative tensile strength of concrete prepared using four different mixing techniques. The goal of the study was to identify the mixing approach that demonstrated the highest strength concrete. If none of the mixing techniques had been superior, then choosing the technique that was simplest or least expensive would make sense. You could describe this experiment as a one-factor experiment where the factor is mixing technique and the mixing techniques are labeled 1–4. Studies involving a continuous factor are different for two reasons. First, continuous factors vary over a range of practical interest on which the engineering team needs to agree. Second, since continuous factors can, in theory, take any value between the low and high values chosen, the engineering team needs to decide which values to choose and how to allocate the budgeted number of runs to these values. This chapter deals with considerations involved in the design of experiments for one continuous factor. It also shows how to analyze data from such experiments.

4.1 Motivating Example

The nuclear power industry requires valves that can withstand high temperatures while retaining adequate strength to function reliably. Engineers have collected data on the strength of 11 valves tested at different temperatures from 21.1°C to 537.8 Celsius as shown in Table 4.1.

A graph of the data (Figure 4.1) clearly shows that as the temperature increases, the strength of the valves drops.

This data is a designed experiment with one continuous factor, Temperature. There are 11 different values for temperature that are roughly equally spaced.

While it does not require statistical expertise to see that the strength of valves decreases with increasing temperature, the engineering team would like to use the data to approximate the functional relationship between temperature and strength. The engineering team could then use this relationship to predict the strength when valves are subjected to a temperature of 400°C. Note that no valves were tested at this temperature.

Two questions might occur to you. First, what mathematical models should the engineers consider? Second, could the engineers have chosen better values of Temperature to estimate these models?

4.2 Statistical Models for the Data

The graph of the data looks roughly linear, so a **linear regression model** would be a natural candidate for consideration.

Table 4.1 Test Data for Valve Strength at 11 Different Temperatures

Temperature	Strength
21.1	326.9
37.8	317.9
93.3	287.6
148.9	252.4
204.4	257.2
260	231.7
315.6	214.5
371.1	212.4
426.7	196.5
482.2	191.7
537.8	179.3

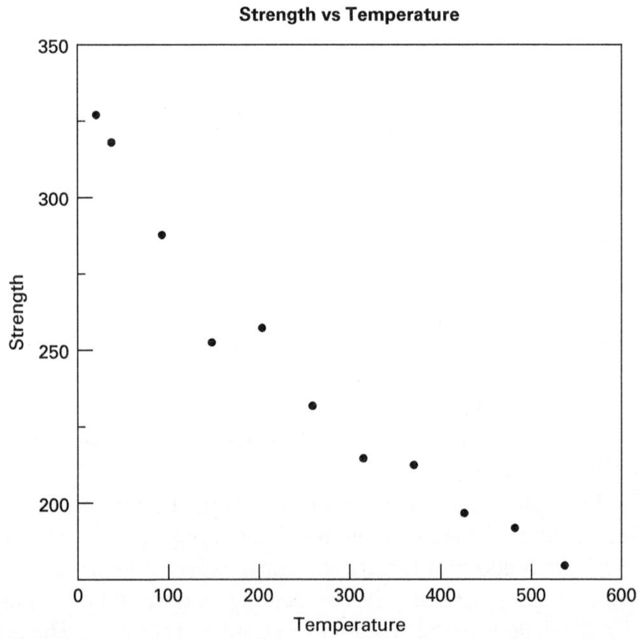

FIGURE 4.1 Plot of Strength Versus Temperature

Equation (4.1) shows the linear model symbolically.

$$y_i = \beta_0 + \beta_1 x_i + \varepsilon_i \tag{4.1}$$

In Eq. (4.1) y_i is the response value for the ith run, β_0 is the true value of y when x is zero, β_1 is true value of the change in y when x changes by 1 unit, also called the slope of line, and ε_i is the random error for run i. The coefficients β_0 and β_1 are called the **parameters** of the model.

Alternatively, you could add a quadratic term to the model as shown in Eq. (4.2).

$$y_i = \beta_0 + \beta_1 x_i + \beta_{11} x_i^2 + \varepsilon_i \tag{4.2}$$

In Eq. (4.2), β_{11} is the true value of the added coefficient of the quadratic term, x_i^2, which is the same as $x_i * x_i$. The effect of adding this term is to change the line in Eq. (4.1) to a curve.

To make predictions for the model in Eq. (4.1) requires estimates for the parameters, β_0 and β_1. For Eq. (4.2), we need to add an estimate for β_{11}.

4.3 Fitting a Statistical Model Using the Data

A more concise representation of Eqs. (4.1) and (4.2) uses matrices. The model for Eq. (4.1) in terms of the observed data may be written in matrix notation as

$$\boldsymbol{y} = \mathbf{X}\boldsymbol{\beta} + \boldsymbol{\varepsilon}$$

where $\boldsymbol{y} = \begin{bmatrix} y_1 \\ y_2 \\ \vdots \\ y_n \end{bmatrix}$, $\mathbf{X} = \begin{bmatrix} 1 & x_1 \\ 1 & x_2 \\ \vdots & \vdots \\ 1 & x_n \end{bmatrix}$, $\boldsymbol{\beta} = \begin{bmatrix} \beta_0 \\ \beta_1 \end{bmatrix}$, $\varepsilon = \begin{bmatrix} \varepsilon_1 \\ \varepsilon_2 \\ \vdots \\ \varepsilon_n \end{bmatrix}$, and n is the number of observations, which is 11 in Table 4.1.

The matrix notation for Eq. (4.2) is the same except for the entries in $\mathbf{X}$ and $\boldsymbol{\beta}$. These become

$$\mathbf{X} = \begin{bmatrix} 1 & x_1 & x_1^2 \\ 1 & x_2 & x_2^2 \\ \vdots & \vdots & \vdots \\ 1 & x_n & x_n^2 \end{bmatrix} \text{ and } \boldsymbol{\beta} = \begin{bmatrix} \beta_0 \\ \beta_1 \\ \beta_{11} \end{bmatrix}$$

to accommodate the squared term in $\mathbf{X}$.

The parameters, $\boldsymbol{\beta}$, in the model are fixed but unknown constants. One way to denote an estimate of a parameter is to put a circumflex (or a hat) above it. For example, an estimate of $\boldsymbol{\beta}$ is $\widehat{\boldsymbol{\beta}}$. Another convention is to use Roman characters instead of Greek characters for estimates of parameters. Thus, $\boldsymbol{b} = \widehat{\boldsymbol{\beta}}$ and $\boldsymbol{e} = \widehat{\boldsymbol{\varepsilon}}$. A popular way to obtain these estimates is the method of least squares generally attributed to Gauss who invented this method around 1795. What Gauss proposed was to obtain $\boldsymbol{b}$ by minimizing the sum of the squared elements of $\boldsymbol{e}$, that is $\boldsymbol{e}'\boldsymbol{e} = (\boldsymbol{y} - \mathbf{X}\boldsymbol{b})'(\boldsymbol{y} - \mathbf{X}\boldsymbol{b})$. This minimization requires some differential calculus that is not important here but yields the following result in Eq. (4.3).

$$\boldsymbol{b} = (\mathbf{X}'\mathbf{X})^{-1}\mathbf{X}'\boldsymbol{y} \tag{4.3}$$

Once the estimates, $\boldsymbol{b}$, are available, it is possible to make predictions of the values in the vector, $\boldsymbol{y}$, by plugging in the estimates into the statistical model. This gives Eq. (4.4).

$$\widehat{\boldsymbol{y}} = \mathbf{X}\boldsymbol{b} \tag{4.4}$$

The predictions of the vector of random errors, $\boldsymbol{e}$, called **residuals** are shown in Eq. (4.5).

$$\widehat{\boldsymbol{\varepsilon}} = \boldsymbol{e} = \boldsymbol{y} - \widehat{\boldsymbol{y}} \tag{4.5}$$

A standard assumption employed in least squares is that the random errors, ε_i, are statistically independent and normally distributed with a mean of zero and a common variance, σ^2, which is also an unknown constant that needs to be estimated. Equation (4.6) provides the formula for an unbiased estimate of σ^2 given that the model is correct.

$$\widehat{\sigma}^2 = \boldsymbol{e}'\boldsymbol{e}/(n - p) \tag{4.6}$$

In Eq. (4.6), n is the number of runs and p is the number of model parameters being estimated excluding σ^2.

A numeric convenience is to express the values of the factor in **coded units** where the minimum value is coded as -1 and the maximum value is coded as 1. To do this requires calculating the midrange and the half range of the data.

$$midrange = mr = \frac{maximum + minimum}{2}$$

$$half\ range = hr = \frac{maximum - minimum}{2}$$

$$coded\ x = \frac{x - mr}{hr}$$

Table 4.2 shows the coded values of Temperature from the data in Table 4.1.

For the coded temperature data and the model in Eq. (4.1)

$$(\mathbf{X'X})^{-1} = \begin{bmatrix} 0.092 & 0.013 \\ 0.013 & 0.209 \end{bmatrix}, \boldsymbol{b} = \begin{bmatrix} 238.18 \\ -71.01 \end{bmatrix}, \text{ and } \hat{\sigma}^2 = 169.9.$$

Figure 4.2 shows the fitted line for strength and the coded values of temperature.

The first element of $\boldsymbol{b}$ is called the ***y* intercept**. It is the predicted value of strength at the zero value of the coded temperature.

Looking at the graph you can see that the plotted points for the smallest and largest two values of the coded temperature are above the best fit line. Also, three of the middle values of coded temperature have plotted points that are below the line. This suggests that the quadratic model of Eq. (4.2) might be a more appropriate description of the relationship between strength and temperature.

For the quadratic regression model

$$(\mathbf{X'X})^{-1} = \begin{bmatrix} 0.2202 & -0.0077 & -0.2959 \\ -0.0077 & 0.2123 & 0.04745 \\ -0.2959 & 0.04745 & 0.6811 \end{bmatrix}, \boldsymbol{b} = \begin{bmatrix} 226.40 \\ -69.13 \\ 27.10 \end{bmatrix}, \text{ and } \hat{\sigma}^2 = 56.3.$$

There are several things to note about these results. First, the estimate of the error variance, $\hat{\sigma}^2$, has dropped by almost a factor of three, which is a strong indication that the quadratic regression model is better than the linear regression model. Second, the first two elements of $\boldsymbol{b}$ (226.40

Table 4.2 Coded Temperature Rounded to 3 Decimal Places

Coded values
−1
−0.935
−0.721
−0.505
−0.290
−0.075
0.14
0.355
0.570
0.785
1

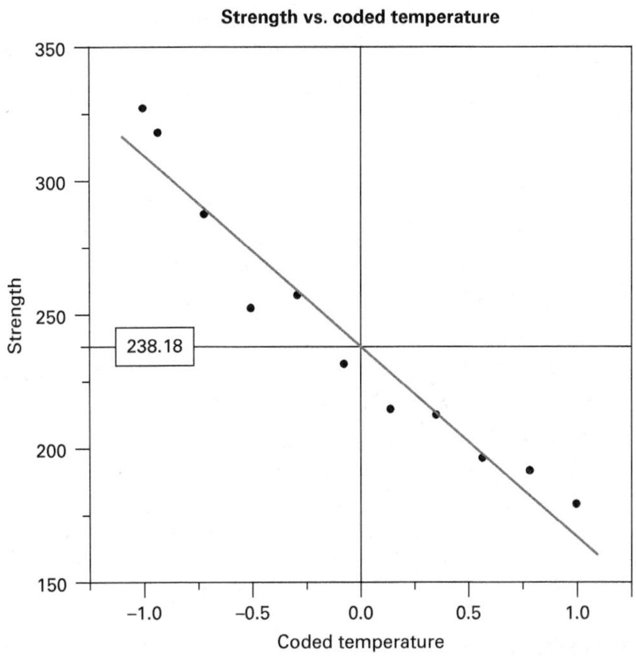

FIGURE 4.2 Plot of Linear Fit of Coded Temperature Versus Strength

and −69.13) for the quadratic regression model are different from the values of $\boldsymbol{b}$ (238.18 and −71.01) for the linear regression model. The statistical term for this effect is **bias**. Assuming the quadratic regression model is correct, the y intercept of the linear regression model is biased high by nearly 12 units. The estimate of the error variance of the linear regression model is also biased high. Compare 169.9 to 56.3. If we had assumed that the linear regression model was correct and neglected to fit the quadratic regression model, we would have concluded that the error variance was much greater than it is.

The **alias matrix** is a way to quantify the bias in the estimates of one model induced by a missing term or terms. If the fitted model is $\boldsymbol{y} = \mathbf{X}\boldsymbol{\beta} + \boldsymbol{\varepsilon}$, but the correct model is $\boldsymbol{y} = \mathbf{X}\boldsymbol{\beta} + \mathbf{Z}\boldsymbol{\gamma} + \boldsymbol{\varepsilon}$, then the expected value of $\boldsymbol{b}$ is $E(\boldsymbol{b}) = \boldsymbol{\beta} + \mathbf{A}\boldsymbol{\gamma}$, where $\mathbf{A}$ is the alias matrix. Equation (4.7) provides the formula for $\mathbf{A}$.

$$\mathbf{A} = (\mathbf{X}'\mathbf{X})^{-1}\mathbf{X}'\mathbf{Z} \tag{4.7}$$

For the strength versus coded temperature data, the matrix, $\mathbf{A}$, is shown in Table 4.3.

Assuming the true value of the parameter for the quadratic term is its estimated value, 27.10, the intercept is biased by 0.434*27.10 = 11.77, which is the difference between the two intercept estimates (i.e., 238.18 − 226.4). Similarly, the estimate of the linear term is biased by −0.69*27.10 = −1.89, which is the difference in the estimates of the two linear term parameters.

The graph in Figure 4.3 shows fitted relationship between the coded temperature and strength for the quadratic regression model. The points in this plot are substantially closer to the curve than those points near the line in Figure 4.2.

Table 4.3 **Aliasing of the Estimates of the Linear Regression Model by a Nonzero Quadratic Effect**

Effect	Coded temperature squared
Intercept	0.4344
Coded temperature	−0.0697

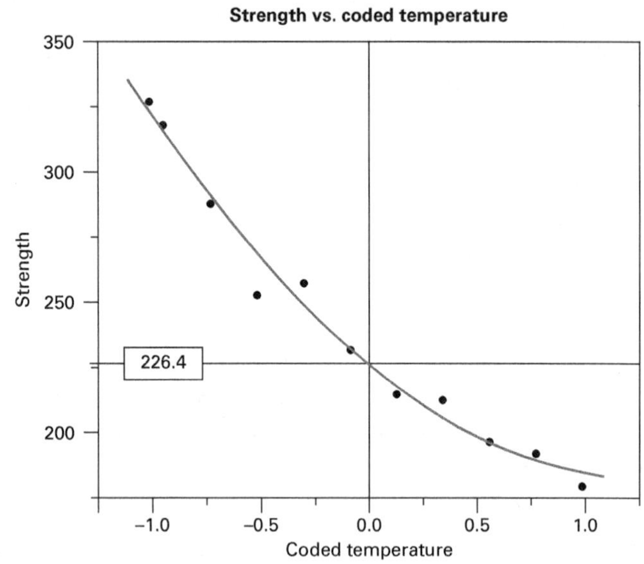

FIGURE 4.3 Fitted Quadratic Model. Notice the Curved Solid Line

4.4 Design Considerations

In this section, we consider whether the choice of temperature values was the most efficient use of resources.

Because the individual observations are subject to random variation, estimates of the parameters of the statistical model also have a variance. Equation (4.8) provides the covariance matrix of $\boldsymbol{b}$.

$$Cov(\boldsymbol{b}) = (\mathbf{X}'\mathbf{X})^{-1}\sigma^2 \tag{4.8}$$

Generally, the value of σ^2 is a property of the system and is not under the direct control of the experimenter. However, controlling the values of the matrix, $\mathbf{X}$, is the whole function of design in experimentation.

The diagonal elements of the covariance matrix of $\boldsymbol{b}$ are the variances of the parameter estimates, while the off-diagonal elements of this matrix are the covariances of pairs of parameter estimates. We want to choose the values of the matrix, $\mathbf{X}$, to make the covariance matrix of $\boldsymbol{b}$ small in some sense.

One way to do this is to choose $\mathbf{X}$ to maximize the determinant of $\mathbf{X}'\mathbf{X}$, or equivalently to minimize the determinant of $(\mathbf{X}'\mathbf{X})^{-1}$. Symbolically, the determinant of a square matrix, $\mathbf{M}$, is written $|\mathbf{M}|$. Designs that maximize $|\mathbf{X}'\mathbf{X}|$ are called D-optimal designs, which were the first type of optimal design for which computer code was available. Being first accounts to a great extent for the popularity of D-optimal designs.

The D-efficiency of a design is

$$D_{\text{eff}} = \left(\frac{|\mathbf{X}'\mathbf{X}|}{Max|\mathbf{X}'\mathbf{X}|}\right)^{1/p}$$

where the maximization in the denominator is taken over all possible designs in the same design region with n runs and p is the number of parameters in the model. In the cases where the D-optimal design is not known, it is standard practice to replace the expression in the denominator above with n^p. An $\mathbf{X}$ matrix with p orthogonal columns of plus and minus ones has a determinant of n^p. This makes the result of replacing the denominator above with n^p a lower bound on the D-efficiency. Generally, this lower bound is not tight and the true D-efficiency may be substantially higher than the reported lower bound.

To compare the relative D-efficiency of design 1 to design 2 with respect to a common model, one computes $(|\mathbf{X}_1'\mathbf{X}_1|/|\mathbf{X}_2'\mathbf{X}_2|)^{(1/p)}$. It is possible for design 2 to be less efficient than design 1 even if design 1 has fewer runs. Interestingly, for the linear regression model, replicating the minimum temperature five times and the maximum value five times to create a 10-run design is D-optimal. The relative D-efficiency of this design to the design in Table 4.1 is 1.379. Thus, our 10-run D-optimal design is 37.9% more D-efficient than the design in Table 4.1 and requires one fewer run.

Another way to make the covariance matrix of $\boldsymbol{b}$ small is to minimize the average of the diagonal elements of $(\mathbf{X}'\mathbf{X})^{-1}$, that is, the average variance of the parameter estimates. Designs that do this are called A-optimal designs. Both D-optimal and A-optimal designs are concerned with parameter estimation. Both design criteria assume that the statistical model is correct, so that the focus is on variance rather than bias.

The D-optimal design for the quadratic regression model puts points at the minimum value, the maximum value, and the midrange value. Ideally, the number of runs at each of these values would be the same. However, since 11 is not divisible by 3, the 11-run D-optimal design cannot have the same number of runs at each point. Any design that puts four runs at two of the three levels and three runs at the remaining level is D-optimal. The D-efficiency of any of these designs relative to the design in Table 4.1 is 1.355. Thus, for the quadratic regression model, any of the first three designs in Table 4.4 has the same D-efficiency that is 35.5% better than the design in Table 4.1.

Note that the A-optimal design, the last column in Table 4.4, has three runs at the maximum and minimum values and five runs at the midrange value. The variances of the estimates of the intercept and quadratic terms are smaller for the A-optimal design than for the D-optimal designs. However, the variance of the estimate of the linear term for all the D-optimal designs is smaller than that for the A-optimal design.

4.5 Design Comparison

Both the D-optimal and A-optimal designs were created by optimizing a single scalar quantity. For the D-optimal design, the software finds a design that has a larger D-efficiency than all other designs with the same number of runs and with respect to the same model. For the A-optimal design, the software finds a design where the sum of the variances of the parameters is smaller than any other design again with the same number of runs and with respect to the same model.

Table 4.4 **Three D-Optimal Designs and the A-Optimal Design**

Run	D-optimal design 1	D-optimal design 2	D-optimal design 3	A-optimal design
1	1	1	1	1
2	1	1	1	1
3	1	1	1	1
4	1	1	0	0
5	0	0	0	0
6	0	0	0	0
7	0	0	0	0
8	−1	0	−1	0
9	−1	−1	−1	−1
10	−1	−1	−1	−1
11	−1	−1	−1	−1

Design Diagnostics

	Efficiency of A-optimal design Relative to D-optimal with 3 zeros	Efficiency of A-optimal design Relative to D-optimal with 4 zeros
D-efficiency	0.979	0.979
G-efficiency	1.000	1.000
A-efficiency	1.250	1.080
I-efficiency	1.250	1.080
Additional Run Size	0	0

FIGURE 4.4 Relative Efficiencies of the A-optimal Design to the D-optimal Designs

Figure 4.4 shows the relative efficiencies of the A-optimal design to the D-optimal design with three zeros and both D-optimal designs with four zeros, which are equivalent for all design diagnostics. Note that the A-optimal design is approximately 2% less D-efficient than the D-optimal designs. However, it is 25% more A-efficient than the D-optimal design. G-efficiency is a measure of the maximum variance of the predicted response over the range of the factor values. In this case, all the designs have the same maximum prediction variance of $1/3\sigma^2$. The I-efficiency is a measure of the average variance of the predicted response of the range of the factor values. The A-optimal design is 25% more I-efficient than the D-optimal design with three zeros and 8% more efficient than the D-optimal design with four zeros.

Figure 4.5 shows that power each design has for correctly labeling each parameter estimate as being statistically significant when the true value of each coefficient is twice as large as the **root mean squared error (RMSE)**. Note that the anticipated RMSE is 0.5 and the anticipated coefficients are all twice as large as the true value, that is, 1.0.

Note that the A-optimal design has a power of greater than 0.8 for all the coefficients, whereas both D-optimal designs do not meet this threshold value for the coefficient of the quadratic term.

Figure 4.6 shows overlaid fraction of the design plots for the three designs. The y-axis of the plot is the prediction variance relative to σ^2, which is generally unknown before the experiment is run. The x-axis is the fraction of the design space covered and goes from 0 (minimum variance) to 1 (maximum variance). The y value where each plotted curve crosses the vertical line shows that half of the relative prediction variances are smaller than this value (and half are larger). The line for the A-optimal design is substantially lower than the other two for most of the plot. This indicates that the A-optimal design performs better with respect to the variance of the predicted response over most of the range of the factor.

Power Analysis

Significance Level 0.05

Anticipated RMSE 0.5

Term	Anticipated Coefficient	A-optimal design Power	D-optimal with 3 zeros Power	D-optimal with 4 zeros Power
Intercept	1	0.973	0.857	0.937
X1	1	0.989	0.998	0.995
X1*X1	1	0.824	0.735	0.795

FIGURE 4.5 Comparative Power for the Case Where the True Coefficient Is Twice As Large As the True Value of σ

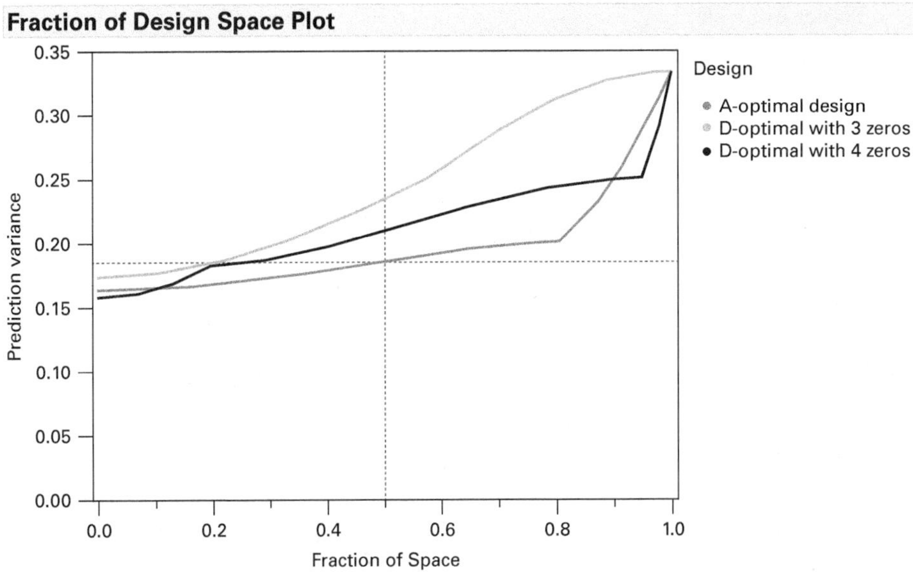

FIGURE 4.6
Overlaid Fraction of the Design Space Plots for the Three Designs

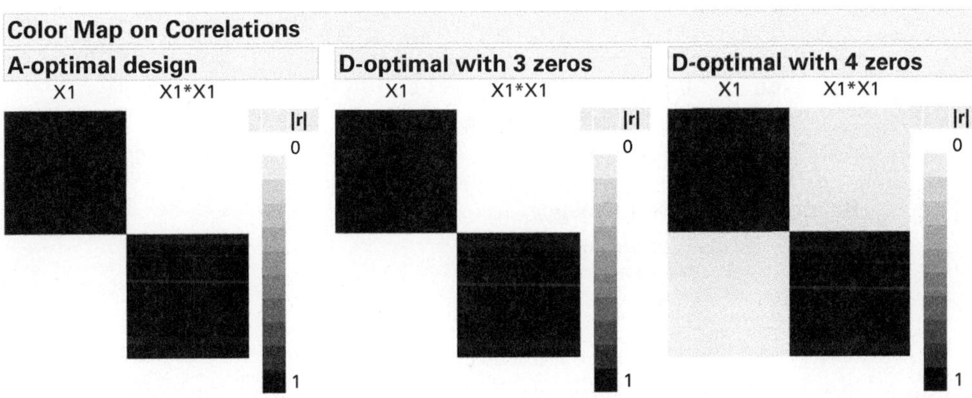

FIGURE 4.7
Comparative Correlation Cell Plots

Figure 4.7 shows the column correlations for X1 and X1*X1 for the three designs.

In this case, the column X1 and X1*X1 are uncorrelated for the A-optimal design and the D-optimal design with three zeros. The light gray cell in the right most plot shows that the X1 and X1*X1 columns are slightly correlated (0.0867). This correlation arises from the fact that the number of +1s and −1s are unequal.

All these diagnostic comparisons indicate that the A-optimal design would be preferable to the D-optimal designs in this case having one continuous factor and fitting a quadratic model. Figure 4.8 compares the A-optimal design to the coded values in Table 4.2. Note that coding the factors has no effect on design diagnostics.

The A-optimal design is substantially better than the evenly spaced design by every measure of design quality for fitting the quadratic model. Especially notable is the improved performance of the estimation of the quadratic effect.

The A-optimal design when fitting the quadratic term is (36% more efficient) than the evenly spaced design.

Power Analysis

Significance Level 0.05

Anticipated RMSE 0.5

Term	Anticipated Coefficient	A-optimal design Power	Engineer Design Power
Intercept	1	0.973	0.960
X1	1	0.989	0.966
X1*X1	1	0.824	0.567

Fraction of Design Space Plot

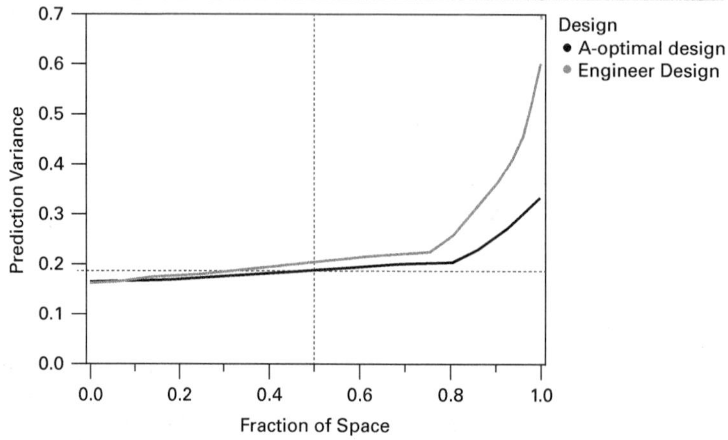

Relative Estimation Efficiency

Term	Efficiency of A-optimal design Relative to Engineer Design
Intercept	1.049
X1	1.129
X1*X1	1.363

Design Diagnostics

	Efficiency of A-optimal design Relative to Engineer Design
D-efficiency	1.326
G-efficiency	1.804
A-efficiency	1.519
I-efficiency	1.176
Additional Run Size	0

FIGURE 4.8 A-optimal Design Compared to the Evenly Spaced Design

Two-factor Factorial Experiments

5

5.1 Basic Concepts

Many experiments involve the study of the effects of two or more factors. In general, **factorial designs** are most efficient for this type of experiment. By a factorial experimental design, we mean that in each complete trial or replicate of the experiment, all possible combinations of the levels of the factors are investigated. For example, if there are a levels of factor A and b levels of factor B, each replicate contains all ab **treatment combinations**. When factors are arranged in a factorial design, they are often said to be **crossed**.

The effect of a factor is defined to be the change in response produced by a change in the level of the factor. This is frequently called a **main effect** because it refers to the primary factors of interest in the experiment. For example, consider the simple experiment in Figure 5.1. This is a two-factor factorial experiment with both design factors at two levels. We have called these levels "low" and "high" and denoted them "−" and "+," respectively. The main effect of factor A in this two-level design can be thought of as the difference between the average response at the low level of A and the average response at the high level of A. Numerically, this is

$$A = \frac{40 + 52}{2} - \frac{20 + 30}{2} = 21$$

That is, increasing factor A from the low level to the high level causes an **average response increase** of 21 units. Similarly, the main effect of B is

$$B = \frac{30 + 52}{2} - \frac{20 + 40}{2} = 11$$

If the factors appear at more than two levels, the above procedure must be modified because there are other ways to define the effect of a factor. The definition of effects in a factorial experiment when the factors have more than two levels is discussed more completely later.

In some experiments, the difference in response between the levels of one factor is not the same at all levels of the other factors. When this occurs, we say that the factors **interact** in their effect on the response. We call this an **interaction** effect. For example, consider the two-factor factorial experiment shown in Figure 5.2. At the low level of factor B (or B^-), the A effect is

$$A = 50–20 = 30$$

and at the high level of factor B (or B^+), the A effect is

$$A = 12–0 = -28$$

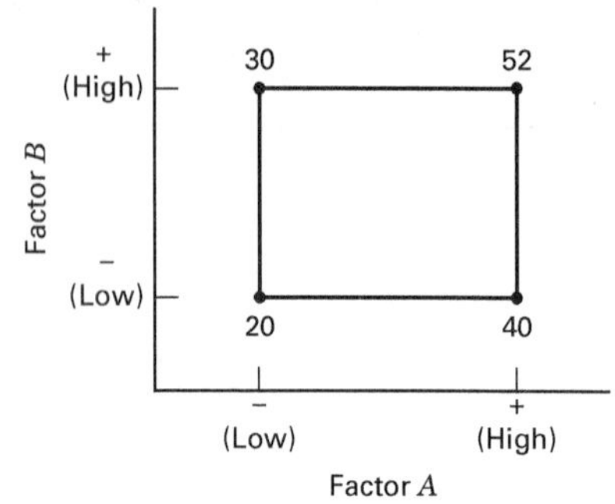

FIGURE 5.1 A Two-Factor Factorial Experiment, with the Response (y) shown at the Corners

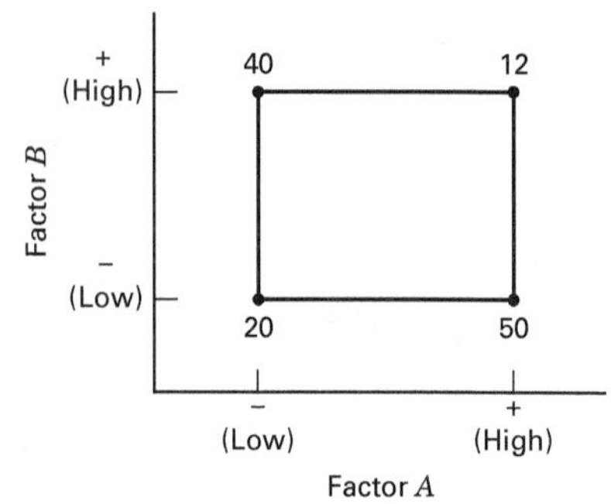

FIGURE 5.2 A Two-Factor Factorial Experiment with Interaction

Because the effect of A depends on the level chosen for factor B, we see that there is an interaction involving A and B. The magnitude of the interaction effect is the average *difference* in these two effects, or $AB = (-28 - 30)/2 = -29$. Clearly, the interaction is large in this experiment.

These ideas may be illustrated graphically. Figure 5.3 plots the response data in Figure 5.1 against factor A for both levels of factor B. Note that the B^- and B^+ lines are approximately parallel, indicating that the effect of A does not depend on the level of factor B, or a lack of interaction involving factors A and B. Similarly, Figure 5.4 plots the response data in Figure 5.2. Here we see that the B^- and B^+ lines are not parallel. This indicates an interaction involving factors A and B. Two-factor interaction graphs such as these are frequently very useful in interpreting significant interactions and in reporting results to nonstatistically trained personnel. However, they should not be utilized as the primary technique of data analysis because their interpretation is subjective and their appearance can be misleading.

This chapter focuses on the design and analysis of experiments with two factors. We consider three cases; first where both factors are categorical, second where one factor is continuous and the other factor is categorical, and finally where both factors are continuous. We demonstrate that the factorial design structure arises directly from a design optimality criterion.

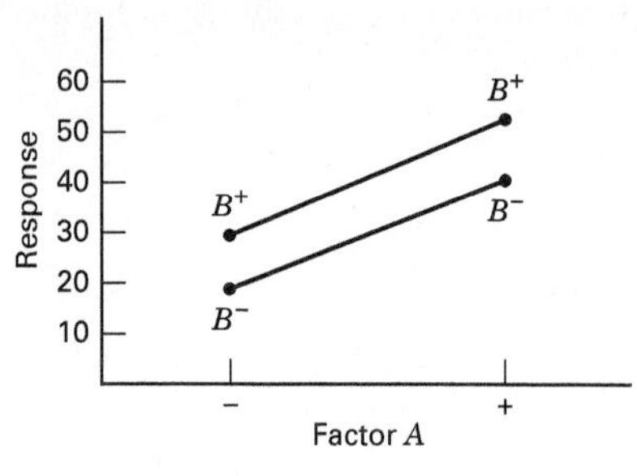

FIGURE 5.3 A Factorial Experiment without Interaction

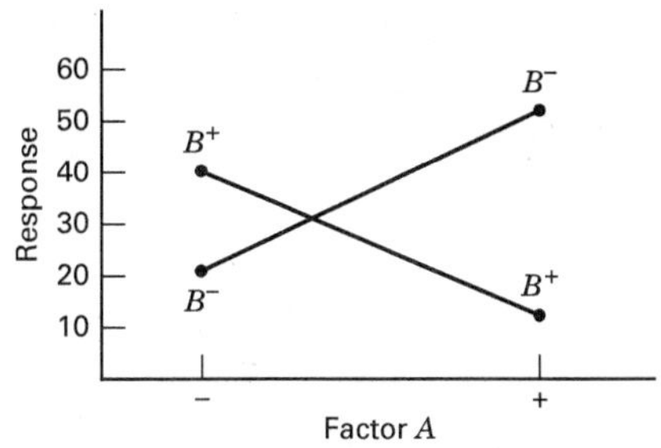

FIGURE 5.4 A Factorial Experiment with Interaction

5.2 Two Categorical Factors

Bone anchors are used by orthopedic surgeons in repairing torn rotator cuffs (a common shoulder tendon injury among baseball players). The bone anchor is a threaded insert that is screwed into a hole that has been drilled into the shoulder bone near the site of the torn tendon. The torn tendon is then sutured to the anchor. In a successful operation, the tendon is stabilized and reattaches itself to the bone. However, bone anchors can pull out if they are subjected to high loads.

A team of biomedical engineers investigating the development of a new bone anchor wanted to perform an experiment to study the force required to pull out the anchor for three anchor types and two different foam densities (the foam simulates the natural variability found in real bone). Since this experiment involves two categorical factors, a two-factor factorial design seems appropriate. It turns out that factorial designs can be constructed using the D-optimality criterion. The engineers used the software package JMP to construct a design for the bone anchor experiment. Figure 5.5 shows the output obtained from the JMP custom designer tool for this experiment. The recommended sample size for this experiment is 12 runs, and the D-optimal design is a 2×3 factorial with 2 replicates. In the design evaluation portion of the output, you see a display of the variance of the predicted response over the two factors Anchor Type (or X1) and Foam Density (or X2). Because these two factors are categorical, this prediction variance only exists at the six combination of the two factor levels, but it is the same at all six treatment combinations ($0.5\sigma^2$). Furthermore, the relative standard errors of the estimate of each single degree of freedom that defines the effects of the two factors are all the same, and they are all the minimum value ($1/\sqrt{12} = 0.289$). This happens because JMP uses the orthogonal coding for categorical variables. Recall that the D-optimality criterion tries to find a design that minimizes the variance (or standard error) of the estimates of the model parameters.

Finally, look at the color map on correlations toward the bottom of Figure 5.5. Notice that the correlations between all pairs of effects are zero. This means that the 2×3 factorial is an **orthogonal design**. In an orthogonal design, all effects are estimated independently of all other effects that could be in the model. Orthogonality is a property shared by most factorial designs. The engineers conducted the 12 runs from this experiment in random order, and the design and resulting data are shown in Table 5.1

The D-optimal design in Table 5.1 is an example of a **balanced design**; that is, it has the same number of runs in each treatment combination. Suppose that the experimenters had decided arbitrarily to replicate Anchor Type C three times at each Foam Type level and have only a single run at each treatment combination for Anchor Type C. This is not a D-optimal design. The boxed display on page 102 shows the estimation efficiency and design diagnostics for this design. Notice that the relative standard errors are not the minimum value and the D-efficiency is no longer 100%. The average variance of prediction has increased from 0.5 to 0.611. The design is no longer orthogonal.

Custom Design
Design

Run	Anchor Type	Foam Density
1	A	High
2	B	High
3	C	High
4	A	Low
5	B	Low
6	C	Low
7	A	High
8	B	Low
9	C	Low
10	A	Low
11	B	High
12	C	High

Design Evaluation
Prediction Variance Profile

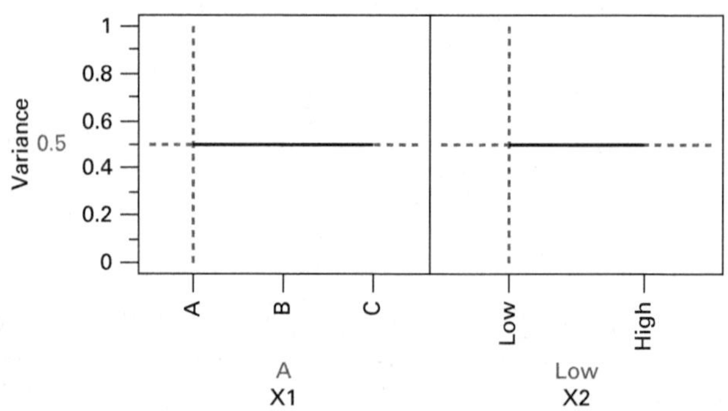

Estimation Efficiency

Term	Fractional Increase in CI Length	Relative Std Error of Estimate
Intercept	0	0.289
X1 1	0	0.289
X1 2	0	0.289
X2	0	0.289
X1*X2 1	0	0.289
X1*X2 2	0	0.289

FIGURE 5.5 JMP Custom Designer Output for the Bone Anchor Experiment

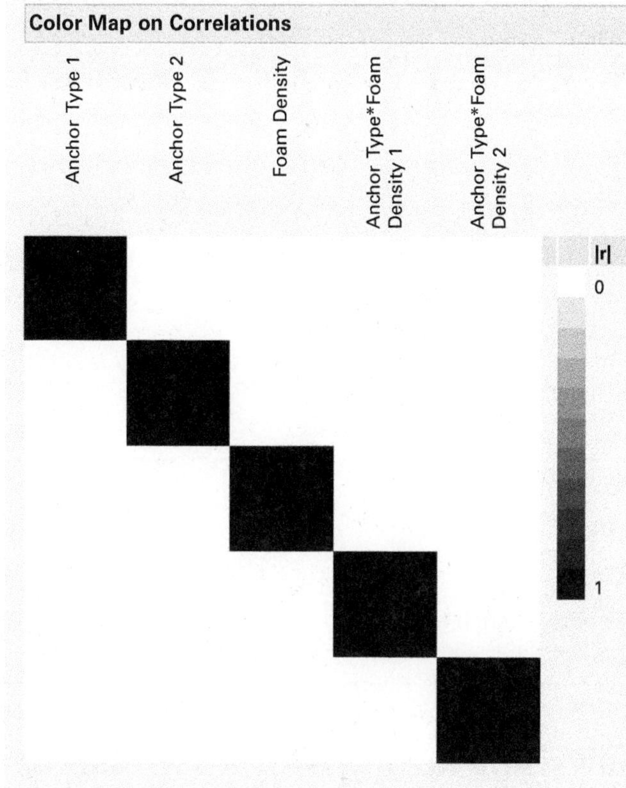

Design Diagnostics

D Optimal Design

D Efficiency 100

G Efficiency 100

A Efficiency 100

Average Variance of Prediction 0.5

Design Creation Time (seconds) 0.016667

FIGURE 5.5 (*Continued*)

Table 5.1 **The 2 × 3 Factorial Design for the Bone Anchor Experiment**

Anchor type	Foam density	
	Low	High
A	190, 200	241, 255
B	185, 190	230, 237
C	210, 205	256, 260

Estimation Efficiency

Term	Fractional Increase in CI Length	Relative Std Error of Estimate
Intercept	0.106	0.319
X1 1	0.155	0.333
X1 2	0.054	0.304
X2	0.106	0.319
X1*X2 1	0.155	0.333
X1*X2 2	0.054	0.304

Design Diagnostics

D Efficiency 90.8650

G Efficiency 70.7106

A Efficiency 81.8181

Average Variance of Prediction 0.61111

5.3 The Analysis of Variance for a Two-factor Factorial

The bone anchor experiment is a specific example of the general case of a two-factor factorial. To pass to the general case, let y_{ijk} be the observed response when factor A is at the ith level $(i = 1, 2, \dots, a)$ and factor B is at the jth level $(j = 1, 2, \dots, b)$ for the kth replicate $(k = 1, 2, \dots, n)$. In general, a two-factor factorial experiment will appear as in Table 5.2. The order in which the abn observations are taken is selected at random so that this design is a **completely randomized design**.

The observations in a factorial experiment can be described by a model. There are several ways to write the model for a factorial experiment. The **effects model** is

$$y_{ijk} = \mu + \tau_i + \beta_j + (\tau\beta)_{ij} + \varepsilon_{ijk} \begin{cases} i = 1,2,\dots,a \\ j = 1,2,\dots,b \\ k = 1,2,\dots,n \end{cases} \tag{5.1}$$

Table 5.2 General Arrangement for a Two-factor Factorial Design

		Factor B			
		1	2	...	b
Factor A	1	$y_{111}, y_{112}, \dots, y_{11n}$	$y_{121}, y_{122}, \dots, y_{12n}$		$y_{1b1}, y_{1b2}, \dots, y_{1bn}$
	2	$y_{211}, y_{212}, \dots, y_{21n}$	$y_{221}, y_{222}, \dots, y_{22n}$		$y_{2b1}, y_{2b2}, \dots, y_{2bn}$
	⋮				
	a	$y_{a11}, y_{a12}, \dots, y_{a1n}$	$y_{a21}, y_{a22}, \dots, y_{a2n}$		$y_{ab1}, y_{ab2}, \dots, y_{abn}$

where μ is the overall mean effect, τ_i is the effect of the ith level of the row factor A, β_j is the effect of the jth level of column factor B, $(\tau\beta)_{ij}$ is the effect of the interaction between τ_i and β_j, and ε_{ijk} is a random error component. Both factors are assumed to be **fixed**, and the treatment effects are defined as deviations from the overall mean, so $\sum_{i=1}^{a} \tau_i = 0$ and $\sum_{j=1}^{b} \beta_j = 0$. Similarly, the interaction effects are fixed and are defined such that $\sum_{i=1}^{a} (\tau\beta)_{ij} = \sum_{j=1}^{b} (\tau\beta)_{ij} = 0$. Because there are n replicates of the experiment, there are abn total observations.

By a **fixed factor,** we mean that the levels of the factor were chosen by the experimenter because they are of specific interest and that conclusions will be drawn about those specific levels. There are experiments where the factor levels are chosen randomly from a large population of levels and the conclusions extend to the population. These **random effects** experiments are discussed in a later chapter.

In the two-factor factorial, both row and column factors (or treatments), A and B, are of equal interest. Specifically, we are interested in **testing hypotheses** about the equality of row treatment effects, say

$$\begin{aligned} H_0 &: \tau_1 = \tau_2 = \cdots = \tau_a = 0 \\ H_1 &: \text{at least one } \tau_i \neq 0 \end{aligned} \tag{5.2a}$$

and the equality of column treatment effects, say

$$\begin{aligned} H_0 &: \beta_1 = \beta_2 = \cdots = \beta_b = 0 \\ H_1 &: \text{at least one } \beta_i \neq 0 \end{aligned} \tag{5.2b}$$

We are also interested in determining whether row and column treatments *interact*. Thus, we also wish to test

$$\begin{aligned} H_0 &: (\tau\beta)_{ij} = 0 \quad \text{for all } i,j \\ H_1 &: \text{at least one } (\tau\beta)_{ij} \neq 0 \end{aligned} \tag{5.2c}$$

We now discuss how these hypotheses are tested using a **two-factor analysis of variance.**

Let $y_{i..}$ denote the total of all observations under the ith level of factor A, $y_{.j.}$ denote the total of all observations under the jth level of factor B, $y_{ij.}$ denote the total of all observations in the ijth cell, and $y_{...}$ denote the grand total of all the observations. Define $\bar{y}_{i..}, \bar{y}_{.j.}, \bar{y}_{ij.}$, and $\bar{y}_{...}$ as the corresponding row, column, cell, and grand averages. Expressed mathematically,

$$\begin{aligned}
y_{i..} &= \sum_{b}^{j=1} \sum_{n}^{k=1} y_{ijk} & \bar{y}_{i..} &= \frac{y_{i..}}{bn} & i &= 1,2,\ldots,a \\
y_{.j.} &= \sum_{a}^{i=1} \sum_{n}^{k=1} y_{ijk} & \bar{y}_{.j.} &= \frac{y_{.j.}}{an} & j &= 1,2,\ldots,b \\
y_{ij.} &= \sum_{n}^{k=1} y_{ijk} & \bar{y}_{ij.} &= \frac{y_{ij.}}{n} & i &= 1,2,\ldots,a \quad j = 1,2,\ldots,b \\
y_{...} &= \sum_{a}^{i=1} \sum_{b}^{j=1} \sum_{n}^{k=1} y_{ijk} & \bar{y}_{...} &= \frac{y_{...}}{abn}
\end{aligned} \tag{5.3}$$

The **total corrected sum of squares** may be written as

$$\begin{aligned}\sum_{i=1}^{a}\sum_{j=1}^{b}\sum_{k=1}^{n}(y_{ijk}-\bar{y}_{...})^2 &= \sum_{i=1}^{a}\sum_{j=1}^{b}\sum_{k=1}^{n}[(\bar{y}_{i..}-\bar{y}_{...})+(\bar{y}_{.j.}-\bar{y}_{...})\\ &\quad +(\bar{y}_{ij.}-\bar{y}_{i..}-\bar{y}_{.j.}+\bar{y}_{...})+(y_{ijk}-\bar{y}_{ij.})]^2\\ &= bn\sum_{i=1}^{a}(\bar{y}_{i..}-\bar{y}_{...})^2+an\sum_{j=1}^{b}(\bar{y}_{.j.}-\bar{y}_{...})^2\\ &\quad +n\sum_{i=1}^{a}\sum_{j=1}^{b}(\bar{y}_{ij.}-\bar{y}_{i..}-\bar{y}_{.j.}+\bar{y}_{...})^2\\ &\quad +\sum_{i=1}^{a}\sum_{j=1}^{b}\sum_{k=1}^{n}(y_{ijk}-\bar{y}_{ij.})^2\end{aligned} \tag{5.4}$$

because the six cross products on the right-hand side are zero. Notice that the total sum of squares has been partitioned into a sum of squares due to "rows," or factor A, (SS_A); a sum of squares due to "columns," or factor B, (SS_B); a sum of squares due to the interaction between A and B, (SS_{AB}); and a sum of squares due to error, (SS_E). This is the fundamental ANOVA equation for the two-factor factorial. From the last component on the right-hand side of Eq. (5.4), we see that there must be at least two replicates ($n \geq 2$) to obtain an error sum of squares.

We may write Eq. (5.4) symbolically as

$$SS_T = SS_A + SS_B + SS_{AB} + SS_E \tag{5.5}$$

The number of degrees of freedom associated with each sum of squares is

Effect	Degrees of freedom
A	$a-1$
B	$b-1$
AB interaction	$(a-1)(b-1)$
Error	$ab(n-1)$
Total	$abn-1$

We may justify this allocation of the $abn - 1$ total degrees of freedom to the sums of squares as follows: The main effects A and B have a and b levels, respectively; therefore, they have $a - 1$ and $b - 1$ degrees of freedom as shown. The interaction degrees of freedom are simply the number of degrees of freedom for cells (which is $ab - 1$) minus the number of degrees of freedom for the two main effects A and B; that is, $ab - 1 - (a - 1) - (b - 1) = (a - 1)(b - 1)$. Within each of the ab cells, there are $n - 1$ degrees of freedom between the n replicates; thus, there are $ab(n - 1)$ degrees of freedom for error. Note that the number of degrees of freedom on the right-hand side of Eq. (5.5) adds to the total number of degrees of freedom.

Each sum of squares divided by its degrees of freedom is a **mean square**. The expected values of the mean squares are

$$E(MS_A) = E\left(\frac{SS_A}{a-1}\right) = \sigma^2 + \frac{bn\sum_{i=1}^{a}\tau_i^2}{a-1}$$

$$E(MS_B) = E\left(\frac{SS_B}{b-1}\right) = \sigma^2 + \frac{an\sum_{j=1}^{b}\beta_j^2}{b-1}$$

$$E(MS_{AB}) = E\left(\frac{SS_{AB}}{(a-1)(b-1)}\right) = \sigma^2 + \frac{n\sum_{i=1}^{a}\sum_{j=1}^{b}(\tau\beta)_{ij}^2}{(a-1)(b-1)}$$

and

$$E(MS_E) = E\left(\frac{SS_E}{ab(n-1)}\right) = \sigma^2$$

Notice that if the null hypotheses of no row treatment effects, no column treatment effects, and no interaction are true, then MS_A, MS_B, MS_{AB}, and MS_E all estimate σ^2. However, if there are differences between row treatment effects, say, then MS_A will be larger than MS_E. Similarly, if there are column treatment effects or interaction present, then the corresponding mean squares will be larger than MS_E. Therefore, to test the significance of both main effects and their interaction, simply divide the corresponding mean square by the error mean square. Large values of this ratio imply that the data do not support the null hypothesis.

If we assume that the model (Eq. (5.1)) is adequate and that the error terms ε_{ijk} are normally and independently distributed with constant variance σ^2, then each of the ratios of mean squares MS_A/MS_E, MS_B/MS_E, and MS_{AB}/MS_E is distributed as F with $a-1$, $b-1$, and $(a-1)(b-1)$ numerator degrees of freedom, respectively, and $ab(n-1)$ denominator degrees of freedom and the critical region would be the upper tail of the F-distribution. The test procedure is usually summarized in an **analysis of variance table**, as shown in Table 5.3.

Computationally, we almost always use a software package to perform the ANOVA calculations. However, it is occasionally useful to have manual calculation formulas available. The total sum of squares is computed as

$$SS_T = \sum_{i=1}^{a}\sum_{j=1}^{b}\sum_{k=1}^{n} y_{ijk}^2 - \frac{y_{...}^2}{abn} \tag{5.6}$$

The sums of squares for the main effects are

$$SS_A = \frac{1}{bn}\sum_{i=1}^{a} y_{i..}^2 - \frac{y_{...}^2}{abn} \tag{5.7}$$

Table 5.3 The Analysis of Variance Table for the Two-factor Factorial, Fixed-effects Model

Source of variation	Sum of squares	Degrees of freedom	Mean square	F_0
A treatments	SS_A	$a-1$	$MS_A = \frac{SS_A}{a-1}$	$F_0 = \frac{MS_A}{MS_E}$
B treatments	SS_B	$b-1$	$MS_B = \frac{SS_B}{b-1}$	$F_0 = \frac{MS_B}{MS_E}$
Interaction	SS_{AB}	$(a-1)(b-1)$	$MS_{AB} = \frac{SS_{AB}}{(a-1)(b-1)}$	$F_0 = \frac{MS_{AB}}{MS_E}$
Error	SS_E	$ab(n-1)$	$MS_E = \frac{SS_E}{ab(n-1)}$	
Total	SS_T	$abn-1$		

and

$$SS_B = \frac{1}{an}\sum_{j=1}^{b} y_{.j.}^2 - \frac{y_{...}^2}{abn} \tag{5.8}$$

It is convenient to obtain the SS_{AB} in two stages. First we compute the sum of squares between the ab cell totals, which is called the sum of squares due to "subtotals":

$$SS_{\text{Subtotals}} = \frac{1}{n}\sum_{i=1}^{a}\sum_{j=1}^{b} y_{ij.}^2 - \frac{y_{...}^2}{abn}$$

This sum of squares also contains SS_A and SS_B. Therefore, the second step is to compute SS_{AB} as

$$SS_{AB} = SS_{\text{Subtotals}} - SS_A - SS_B \tag{5.9}$$

We may compute SS_E by subtraction as

$$SS_E = SS_T - SS_{AB} - SS_A - SS_B \tag{5.10}$$

or

$$SS_E = SS_T - SS_{\text{Subtotals}}$$

The JMP output from the Fit Model analysis procedure is shown in Figure 5.6. Notice from the effect tests (which are the ANOVA F-statistics) that the main effects of both factors Anchor Type and Foam Density are statistically significant. The interaction is not significant. JMP also produces by default a plot of residuals versus the predicted response. This plot is useful in detecting potential problems with model inadequacy, specifically inequality of variance. The plot exhibits random scatter, so no problems with unequal variance is suspected.

The prediction profiler indicates that Anchor Type C may produce higher pull-out force than the other two types. We may use a multiple comparison procedure to test for differences between the levels of this factor. Since there is no interaction, the test procedure can be applied directly to the means of the anchor types. The engineers used the Fisher LSD technique, which involves testing all pairs of means using a two-sample t-test. JMP will perform the analysis and the results are in the boxed display below. The results indicate that all three anchor types result in different pull-out forces and that Anchor Type C is the best design because it results in the highest pull-out force.

Multiple Comparisons for X1

Least Squares Means Estimates

X1	Estimate	Std Error	DF	Lower 95%	Upper 95%
A	221.50000	2.9261750	6	214.33991	228.66009
B	210.50000	2.9261750	6	203.33991	217.66009
C	232.75000	2.9261750	6	225.58991	239.91009

Student's t All Pairwise Comparisons

Quantile = 2.44691, DF = 6.0

All Pairwise Differences

X1	-X1	Difference	Std Error	t Ratio	Prob>|t|	Lower 95%	Upper 95%
A	B	11.0000	4.138236	2.66	0.0376*	0.8741	21.1259
A	C	-11.2500	4.138236	-2.72	0.0347*	-21.3759	-1.1241
B	C	-22.2500	4.138236	-5.38	0.0017*	-32.3759	-12.1241

Response Y

Actual by Predicted Plot

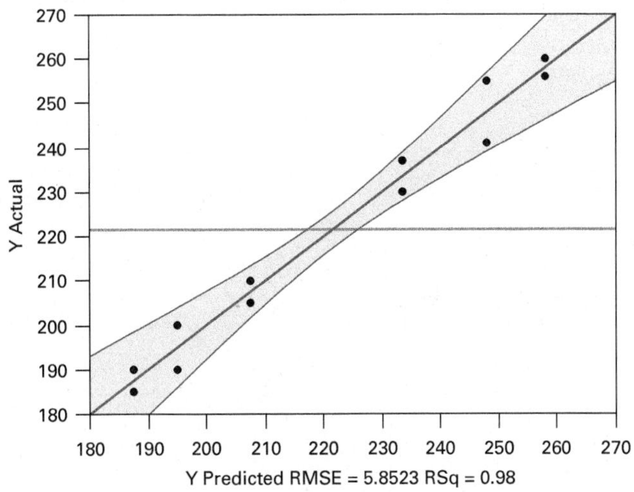

Effect Summary

Source	LogWorth	PValue
X2	5.214	0.00001
X1	2.294	0.00508
X1*X2	0.151	0.70710

Residual by Predicted Plot

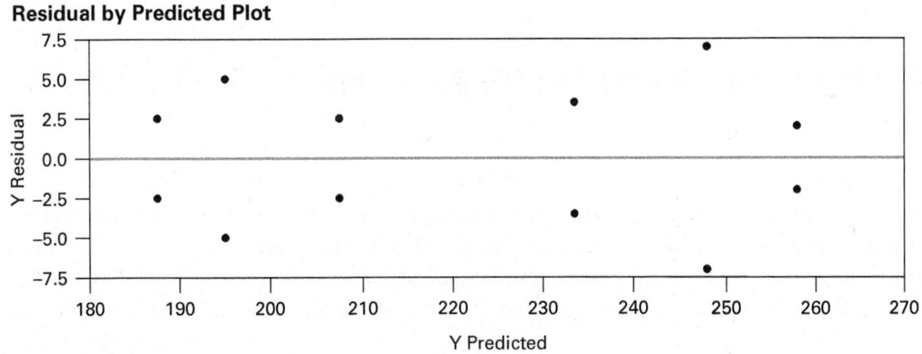

Summary of Fit

RSquare	0.9763
RSquare Adj	0.95655
Root Mean Square Error	5.85235
Mean of Response	221.5833
Observations (or Sum Wgts)	12

FIGURE 5.6 JMP Output for the Bone Anchor Experiment

Effect Tests

Source	Nparm	DF	Sum of Squares	F Ratio	Prob > F
X1	2	2	990.1667	14.4550	0.0051*
X2	1	1	7450.0833	217.5207	<.0001*
X1*X2	2	2	25.1667	0.3674	0.7071

Prediction Profiler

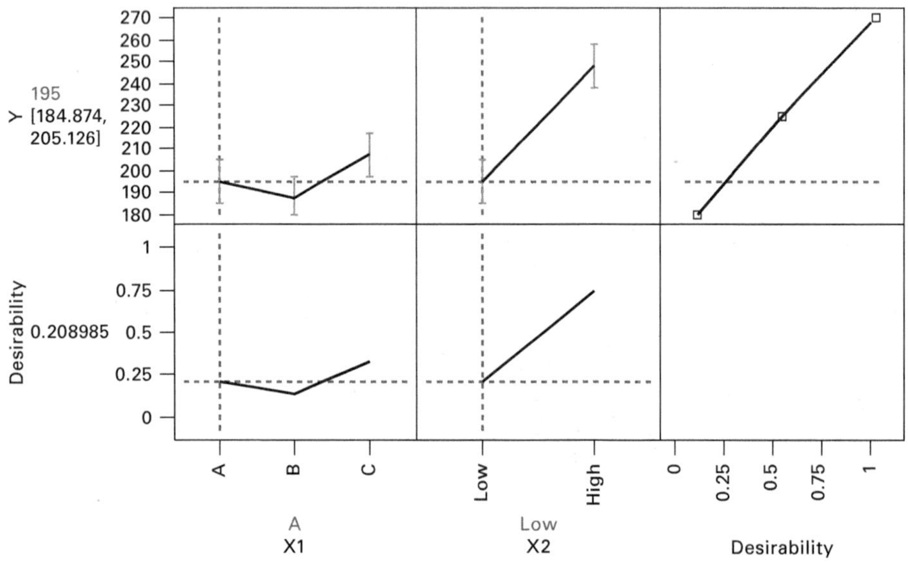

FIGURE 5.6 (*Continued*)

5.4 One Categorical Factor and One Continuous Factor

There are many situations where one or more of the factors in an experiment are continuous. As an example of such an experiment involving two factors, consider a situation where an engineering team is designing a battery for use in a device that will be subjected to some extreme variations in temperature. The only design parameter that can be selected at this point is the plate material for the battery, and there are three possible choices. When the device is manufactured and is shipped to the field, the designers have no control over the temperature extremes that the device will encounter, and from experience it is known that temperature will probably affect the effective battery life. However, temperature can be controlled in the product development laboratory for the purposes of a test.

The engineering team suspects that there may be an interaction involving material type and temperature; that is, some materials may react differently to temperature and exhibit different lifetimes over the range of temperatures likely to be encountered in the field. They are also reluctant to assume that the effect of temperature is linear over its range and that a second-order (quadratic) effect is more likely. This means that temperature should have at least three levels.

JMP Custom Designer was used to construct a D-optimal design for this experiment. The design is shown in the boxed display (see next page). Because the assumed model included both a linear and a quadratic temperature effect, the D-optimal design has three levels for temperature.

The low, medium, and high temperature levels are represented by −1, 0, and +1. Notice that the engineering team also assumed that there could be both linear and quadratic effects of temperature that could interact with the material types. The minimum number of runs required to fit this model is $N = 9$. The recommended nine-run design is a full factorial.

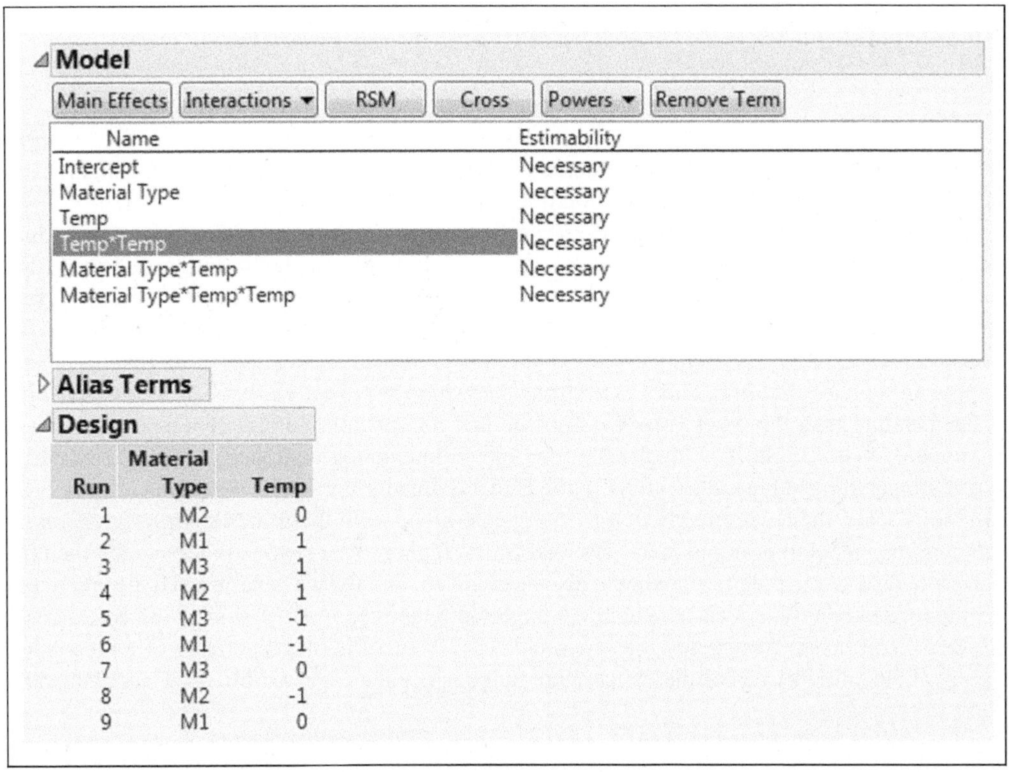

The engineers decided to use 15°F, 70°F, and 125°F for the three temperature levels and to replicate the design four times for a total of $N = 36$ runs. These runs were made in random order, the battery life in hours determined for each run, and the completed experiment is shown in Table 5.4.

Although this experiment has one categorical factor, the analysis of variance procedure traditionally assumes that all factors are categorical. Table 5.5 presents the basic ANOVA that would

Table 5.4 The Battery Life Experiment

	Temperature (°F)					
Material type	**15**		**70**		**125**	
1	130	155	34	40	20	70
	74	180	80	75	82	58
2	150	188	136	122	25	70
	159	126	106	115	58	45
3	138	110	174	120	96	104
	168	160	150	139	82	60

Table 5.5 Analysis of Variance for the Battery Life Experiment

Source of variation	Sum of squares	Degrees of freedom	Mean square	F_0	P-value
Material types	10,683.72	2	5,341.86	7.91	0.0020
Temperature	39,118.72	2	19,559.36	28.97	<0.0001
Interaction	9,613.78	4	2,403.44	3.56	0.0186
Error	18,230.75	27	675.21		
Total	77,646.97	35			

result from that assumption. Both main effects of material type and temperature are significant, and there is a strong two-factor interaction involving these two factors.

It is useful to examine the model between life, material type, and temperature taking into account the continuous nature of the temperature variable. The display in Figure 5.7 shows the output from the JMP Fit Model analysis. The first graph in the output shows the three regression models that describe the relationship between life and temperature for each material type. These curves are very different, showing clearly the two-factor interaction of material type and temperature on battery life. There is a strong indication that Material type 3 is less sensitive to temperature than the other two. We also see that the quadratic effect of temperature is not significant by itself, but it is involved in one of the interaction components (see the Material type*Temperature*Temperature effect in the Effect Summary portion of the output), so we have elected to leave the quadratic effect on temperature along with the Material type*Temperature effect in the model to promote hierarchy; that is, if a higher-order term is in the model, then the corresponding lower-order terms should also be included. Finally, we have targeted the prediction profiler at Material type 3 and the nominal expected temperature of 70°F, showing an expected life of 145.54 hours. Notice how flat the predicted life curve is over the range of temperature, except at the far upper end of the temperature range where the expected life drops more quickly.

5.5 Two Continuous Factors

Some two-factor experiments involve only continuous factors. As an example, consider a chemical process where it is important to maximize the yield of product, and there are two controllable continuous variables—temperature and pressure. The process engineers want to model the relationship between yield and these two factors with a second-order model:

$$y = \beta_0 + \beta_1 x_1 + \beta_{11} x_1^2 + \beta_2 x_2 + \beta_{22} x_2^2 + \beta_{12} x_1 x_2 + \varepsilon$$

where x_1 = temperature and x_2 = pressure. Second-order models are widely used in process modeling and optimization, and there is a field of designed experiments focused on these types of applications called **response surface methodology** or **RSM**. We will discuss RSM optimization techniques and experimental designs in more detail in Chapter 11. In this application, the engineers decided to use two replicates of a factorial experiment with three levels of temperature and pressure to fit the second-order model. The experiment is shown in Table 5.6.

Figure 5.8 shows the JMP Fit Model output for fitting the second-order model to the chemical process yield data in Table 5.6. All model terms are significant although the *P*-value for the interaction term is slightly above 0.05. Notice that JMP codes the variables temperature and pressure when fitting the second-order model. This is a common practice when fitting models with polynomial terms to reduce ill-conditioning. We can see this by observing that the high and low levels of the two factors are shown in the Effects Summary and the Parameter Estimates table in

Response Life

Regression Plot

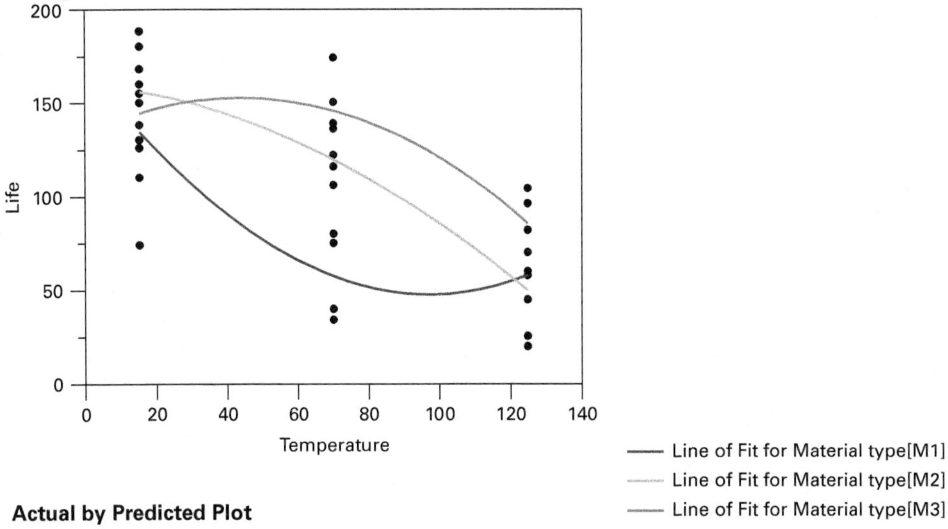

Actual by Predicted Plot

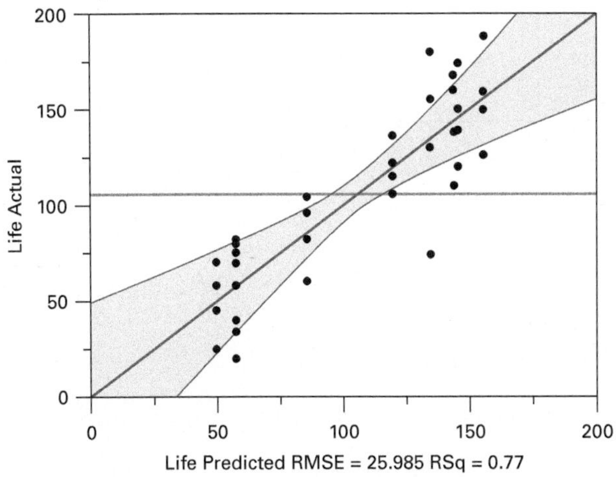

Effect Summary

Source	LogWorth	PValue
Temperature	7.453	0.00000
Material type	3.788	0.00016
Material type*Temperature*Temperature	1.974	0.01061
Material type*Temperature	0.701	0.19911 ^
Temperature*Temperature	0.131	0.73975 ^

FIGURE 5.7 JMP Fit Model Analysis for the Battery Life Experiment

Effect Tests

Source	Nparm	DF	Sum of Squares	F Ratio	Prob > F
Material type	2	2	16552.667	12.2574	0.0002*
Temperature	1	1	39042.667	57.8227	<.0001*
Material type*Temperature	2	2	2315.083	1.7143	0.1991
Temperature*Temperature	1	1	76.056	0.1126	0.7398
Material type*Temperature*Temperature	2	2	7298.694	5.4047	0.0106*

Prediction Profiler

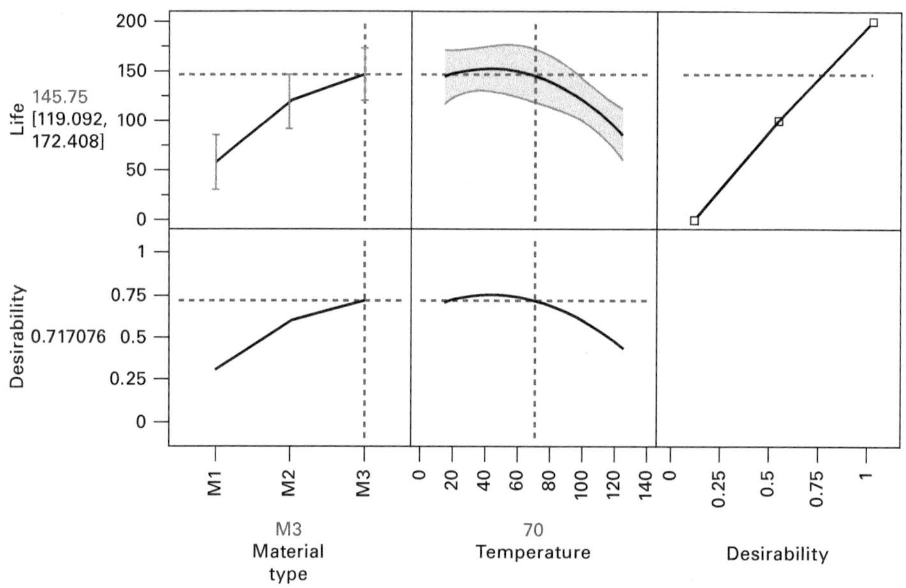

FIGURE 5.7 (*Continued*)

Table 5.6 The Factorial Design for the Chemical Process Optimization Experiment

Temperature/pressure	200	215	230
150	90.4, 90.2	90.7, 90.6	90.2, 90.4
160	90.1, 90.3	90.5, 90.6	89.9, 90.1
170	90.5, 90.7	90.8. 90.9	90.4, 90.1

Figure 5.8. Because of the coding, the standard errors of the linear terms are the same and the standard errors of the pure quadratic terms are the same. The prediction profiler in Figure 5.8 has been set to the levels of temperature and pressure that result in maximum yield. The contour profiler shows the shape of the yield response surface in the temperature–pressure space. This type of surface is called a saddle-point surface, and it is one of several types of surfaces that arise in the application of RSM. This will be discussed more completely in Chapter 11.

Response Yield

Actual by Predicted Plot

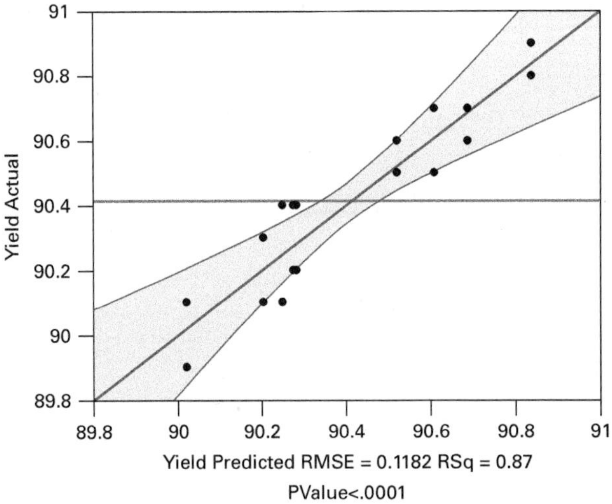

Effect Summary

Source	LogWorth	PValue
Pressure*Pressure	4.788	0.00002
Temperature*Temperature	2.824	0.00150
Pressure(200,230)	1.703	0.01979 ^
Temperature(150,170)	1.316	0.04830 ^
Temperature*Pressure	1.235	0.05817

Residual by Predicted Plot

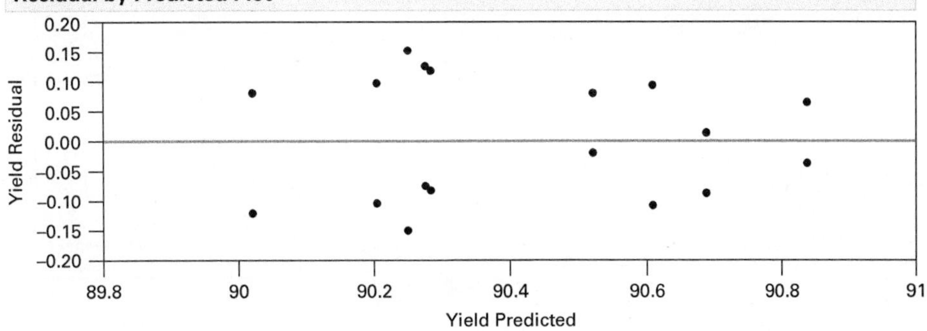

Parameter Estimates

Term	Estimate	Std Error	t Ratio	Prob > |t|
Intercept	90.522222	0.062294	1453.1	<.0001*
Temperature(150,170)	0.075	0.03412	2.20	0.0483*
Pressure(200,230)	−0.091667	0.03412	−2.69	0.0198*
Temperature*Pressure	−0.0875	0.041788	−2.09	0.0582
Pressure*Pressure	−0.408333	0.059097	−6.91	<.0001*
Temperature*Temperature	0.2416667	0.059097	4.09	0.0015*

FIGURE 5.8 JMP Fit Model Output for the Chemical Process Experiment in Table 5.6

Effect Tests

Source	Nparm	DF	Sum of Squares	F Ratio	Prob > F
Temperature(150,170)	1	1	0.06750000	4.8318	0.0483*
Pressure(200,230)	1	1	0.10083333	7.2179	0.0198*
Temperature*Pressure	1	1	0.06125000	4.3844	0.0582
Pressure*Pressure	1	1	0.66694444	47.7415	<.0001*
Temperature*Temperature	1	1	0.23361111	16.7225	0.0015*

Prediction Profiler

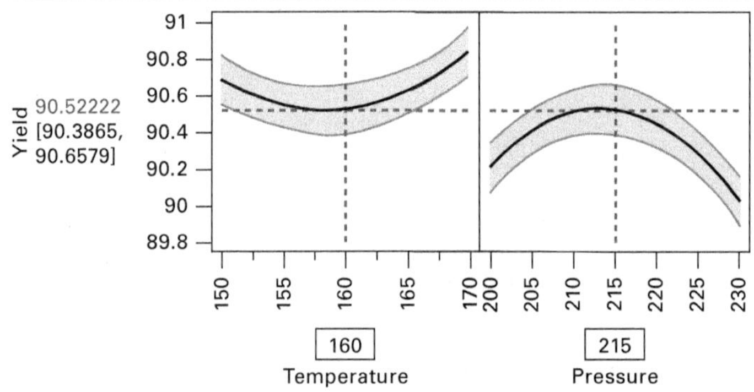

Contour Profiler

Horiz	Vert	Factor		Current X
◉	○	Temperature		160
○	◉	Pressure		215

	Response		Contour	Current Y	Lo Limit	Hi Limit
——	Yield		90.4	90.522222	.	.

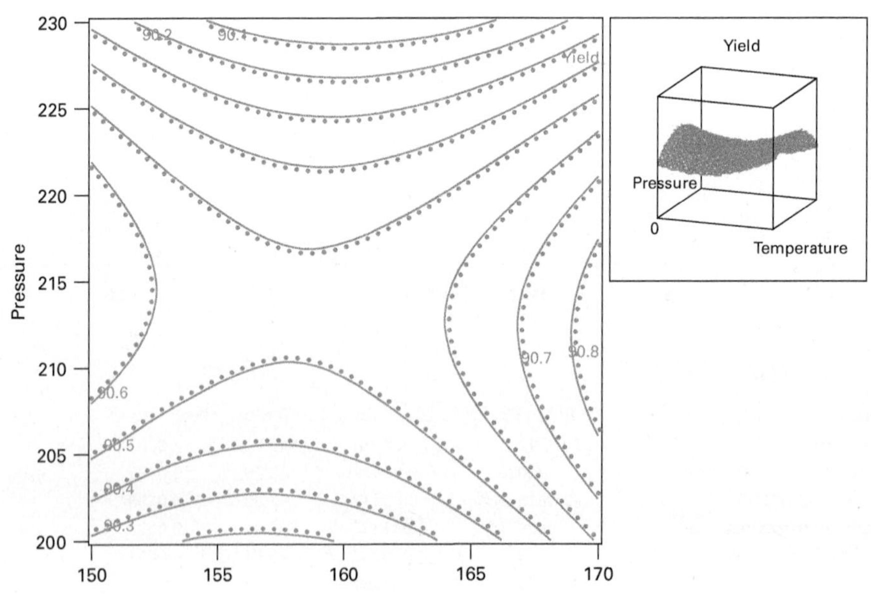

FIGURE 5.8 (*Continued*)

5.6 Design and Analysis When Some Factor Level Combinations Are Infeasible

An engineering team wants to investigate four machines (1, 2, 3, and 4) for measuring a part. There are also four operators (Bob, Carol, Ted, and Alice), and the goal of the study is to compare the deviation from the nominal dimension of a standard part with a known value.

The wrinkle in this scenario is that Alice has not been trained on Machine 4 and is not approved to use it. The model for each run is

$$y_{ij} = \mu + \tau_i + \theta_j + \varepsilon_{ij}$$

where y_{ij} is the deviation for the *ith* machine and the *jth* operator, τ_i is the fixed effect of the ith machine, θ_j is the fixed effect of the *jth* operator, and ε_{ij} is the random error for the ith machine and the jth operator.

Since τ_i and θ_j are deviations from the overall mean, μ , their sums add to zero as described in Section 5.3, which provides the analysis of variance calculations for the case with two categorical factors.

There is a more compact matrix formulation for this scenario using the regression model

$$y = \mathbf{X}\beta + \varepsilon \tag{5.11}$$

The matrix, **X**, requires a column of ones for the intercept, three columns for the machine effects, and three columns for the operator effect. Note that, though there are four machines, we only need three columns for the fixed effects because the last effect is −1 times the sum of the other three.

We need a way to encode the levels for the three columns of each categorical factor. One such way is called **effects coding**. The three columns for a four-level factor are called **dummy columns** or **dummy variables**.

Table 5.7 shows the three columns for each level of a four-level factor using effects coding.

Thus, if a run employs Machine 1, the dummy columns are [1 0 0], for Machine 2, [0 1 0], for Machine 3, [0 0 1], and for Machine 4, [−1 −1 −1]. Note the result that, given this coding, the fixed effect of Machine 4 is −1 times the sum of the fixed effects of the other three machines.

Apply the same scheme for the four operators. Thus, if a run employs Machine 1 and Operator 1, the row of the **X** matrix will be

Intercept	Machine dummy 1	Machine dummy 2	Machine dummy 3	Operator dummy 1	Operator dummy 2	Operator dummy 3
1	1	0	0	1	0	0

Table 5.7 The Dummy Columns for a Four-level Categorical Factor Using Effects Coding

Machines/dummy columns	Dummy 1	Dummy 2	Dummy 3
Machine 1	1	0	0
Machine 2	0	1	0
Machine 3	0	0	1
Machine 4	−1	−1	−1

The infeasible factor combination is Machine 4 and Alice (Operator 4). Thus, the row below is disallowed:

Intercept	Machine dummy 1	Machine dummy 2	Machine dummy 3	Operator dummy 1	Operator dummy 2	Operator dummy 3
1	−1	−1	−1	−1	−1	−1

Table 5.8 shows the 15-run D-optimal design sorted by Operator and Machine with response data. Note that each operator, except for Alice, uses every machine. Alice uses Machines 1, 2, and 3 but not 4.

Note that there are $4 \times 4 = 16$ possible combinations. The design in Table 5.8 has all every combination except for Machine 4 and Alice.

Table 5.9 shows the **X** matrix corresponding to Table 5.8. Note that the first column of Table 5.9 is not a part of the **X** matrix. It is shown to illustrate the matching of the run numbers in Table 5.8. Though Table 5.8 is in sorted order, the run order of the experiment should be randomized.

The vector, $\boldsymbol{y}$, in Eq. (5.11), is shown in the last column of Table 5.8.

Given the **X** matrix and $\boldsymbol{y}$ vector, the matrix equations of (4.3) through (4.6) apply.

That is, Eq. (5.12) shows the estimates of the model coefficients, $\boldsymbol{b}$

$$\boldsymbol{b} = (\mathbf{X}'\mathbf{X})^{-1}\mathrm{X}'y \tag{5.12}$$

Applying Eq. (5.12) to our data, the estimates for the vector, $\boldsymbol{b}$, are (to three decimal places)

Intercept	Machine dummy 1	Machine dummy 2	Machine dummy 3	Operator dummy 1	Operator dummy 2	Operator dummy 3
−0.030	1.006	−0.919	1.231	1.080	−0.769	0.531

Table 5.8 D-optimal 15-run Design with Response Data

Run	Machine	Operator	Deviation from nominal
1	1	Alice	−0.5
2	2	Alice	−1.7
3	3	Alice	0.9
4	1	Bob	2.3
5	2	Bob	0.4
6	3	Bob	2.1
7	4	Bob	−0.6
8	1	Carol	0.7
9	2	Carol	−1.9
10	3	Carol	0.3
11	4	Carol	−2.3
12	1	Ted	1.4
13	2	Ted	−0.6
14	3	Ted	1.5
15	4	Ted	−0.3

Table 5.9 X Matrix for 15-run Design in Table 5.8

Run	Intercept	Machine dummy 1	Machine dummy 2	Machine dummy 3	Operator dummy 1	Operator dummy 2	Operator dummy 3
1	1	1	0	0	−1	−1	−1
2	1	0	1	0	−1	−1	−1
3	1	0	0	1	−1	−1	−1
4	1	1	0	0	1	0	0
5	1	0	1	0	1	0	0
6	1	0	0	1	1	0	0
7	1	−1	−1	−1	1	0	0
8	1	1	0	0	0	1	0
9	1	0	1	0	0	1	0
10	1	0	0	1	0	1	0
11	1	−1	−1	−1	0	1	0
12	1	1	0	0	0	0	1
13	1	0	1	0	0	0	1
14	1	0	0	1	0	0	1
15	1	−1	−1	−1	0	0	1

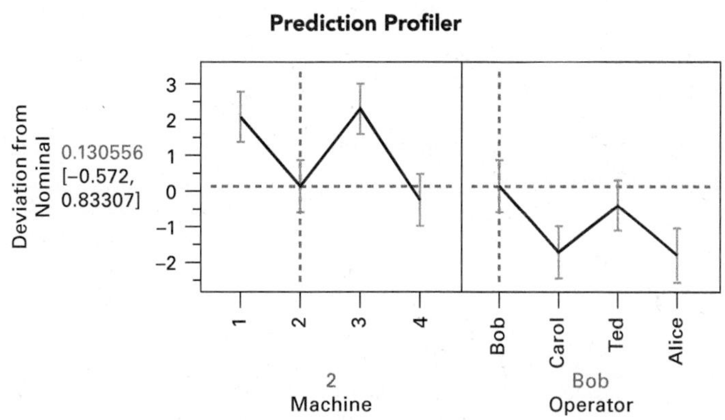

FIGURE 5.9 JMP Prediction Profile Plot Showing the Predicted Deviation from Nominal for Each Machine and Each Operator

Note that the estimate of the effect of Machine 4 is $-1(1.006 - 0.919 + 1.231) = -1.347$. The estimate of Alice's effect is $-1(1.080 - 0.769 + 0.531) = -0.842$.

Equation (5.13) shows the estimates of the predicted values of $\boldsymbol{y}$

$$\hat{y} = \mathbf{X}\boldsymbol{b} \tag{5.13}$$

Equation (5.14) shows the residuals

$$\hat{\varepsilon} = \boldsymbol{e} = \boldsymbol{y} - \hat{y} \tag{5.14}$$

Finally, Eq. (5.15) shows the estimated variance of the residuals

$$\hat{\sigma}^2 = \boldsymbol{e}'\boldsymbol{e}/(n - p) \tag{5.15}$$

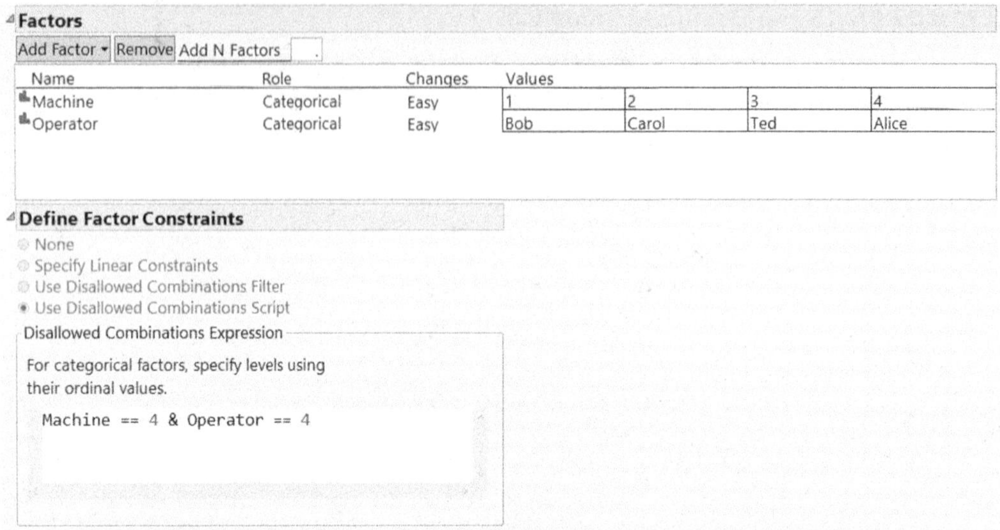

FIGURE 5.10 Factors and Expression for Disallowed Combinations in JMP's Custom Design Tool

In Eq. (5.15), the number of runs, $n = 15$, and the number of parameters, $p = 7$.

Applying Eqs. (5.13) through (5.15) to our data, the **error mean square**, which estimates the error variance, is 0.209.

Figure 5.9 shows the prediction profile plot from the JMP Fit Model analysis of this data.

Figure 5.10 shows the setup for this design in JMP's Custom Design tool. Creating the 15-run design yields the data in Table 5.8.

Blocking

6

6.1 The Randomized Complete Block Design

In any experiment, variability arising from a **nuisance** factor can affect the results. Generally, we define a nuisance factor as a design factor that probably has an effect on the response, but we are not interested in that effect. Sometimes, a nuisance factor is unknown and uncontrolled; that is, we do not know that the factor exists, and it may be changing levels while we are conducting the experiment. **Randomization** is the design technique used to guard against such a "lurking" nuisance factor. In other cases, the nuisance factor is known but uncontrollable. If we can at least observe the value that the nuisance factor takes on at each run of the experiment, we can compensate for it in the statistical analysis by using the **analysis of covariance**. If the source of variability is **known and controllable**, a design technique called **blocking** can be used to systematically eliminate its effect on the statistical comparisons among treatments. Blocking was introduced by Fisher and his coworkers at the Rothamsted Agricultural Research Station and is an extremely important design technique used extensively in both industrial and agricultural experimentation. This chapter provides an introduction to the blocking principle.

Let us start with an example. A medical device manufacturer produces vascular grafts (artificial veins). These grafts are manufactured by extruding billets of polytetrafluoroethylene (PTFE) resin combined with a lubricant into tubes. Frequently, some of the tubes in a production run contain small, hard protrusions on the external surface of the graft. These defects are known as "flicks." The defect is the cause for rejection of the unit.

The product developer responsible for the vascular grafts suspects that the extrusion pressure affects the occurrence of flicks. She intends to conduct an experiment to investigate this hypothesis. However, the resin used in manufacturing the grafts is produced by an external supplier and is delivered to the medical device company in batches. The engineer also suspects that there may be significant batch-to-batch variation, due to possible changes in uncontrolled factors in the supplier's operation from batch to batch. Therefore, the product developer decides to investigate the effect of four different levels of extrusion pressure on flicks, but because several different batches of the resin material must be used, she wants to conduct the experiment in such a way to eliminate or at least minimize the batch to batch variation in the resin. One design that accomplishes this consists of testing all four extrusion pressures in each batch of the material. This type of design is called a **randomized complete block design (RCBD)** and is shown in Table 6.1, where she has decided to use six batches of resin. In this design, the pressures are the treatments and the batches of the resin material are the blocks. In the RCBD, the treatments are tested in random order within each block. The response variable is yield, or the percentage of tubes in the production run that did not contain any flicks.

The RCBD is a widely used experimental design. The RCBD is appropriate for many situations. Units of test equipment or machinery are often different in their operating characteristics and would be a typical blocking factor. Batches of raw material (as in the above example), people, and time are also common nuisance sources of variability in an experiment that can be systematically controlled through blocking.

Table 6.1 Randomized Complete Block Design for the Vascular Graft Experiment

	Batch of resin (blocks)						
Extrusion pressure (psi)	**1**	**2**	**3**	**4**	**5**	**6**	**Treatment totals**
8500	90.3	89.2	98.2	93.9	87.4	97.9	556.9
8700	92.5	89.5	90.6	94.7	87.0	95.8	550.1
8900	85.5	90.8	89.6	86.2	88.0	93.4	533.5
9100	82.5	89.5	85.6	87.2	78.9	90.7	514.6
Block totals	350.8	359.0	364.0	363.2	341.3	377.8	2155.1

Blocking may also be useful in situations that do not necessarily involve nuisance factors. For example, suppose that a chemical engineer is interested in the effect of catalyst feed rate on the viscosity of a polymer. She knows that there are several factors, such as raw material source, temperature, operator, and raw material purity that are very difficult to control in the full-scale process. Therefore, she decides to test the catalyst feed rate factor in blocks, where each block consists of some combination of these uncontrollable factors. In effect, she is using the blocks to test the robustness of her process variable (feed rate) to conditions she cannot easily control. For more discussion of this, see Coleman and Montgomery (1993).

6.2 Statistical Analysis of the RCBD

Suppose we have, in general, *a* treatments that are to be compared and *b* blocks. The RCBD is shown in Figure 6.1. There is one observation per treatment in each block, and the order in which the treatments are run within each block is determined randomly.

The **statistical model** for the RCBD can be written in several ways. The traditional model is an **effects model**:

$$y_{ij} = \mu + \tau_i + \beta_j + \varepsilon_{ij} \begin{cases} i = 1,2,\ldots,a \\ j = 1,2,\ldots,b \end{cases} \tag{6.1}$$

where μ is an overall mean, τ_i is the effect of the *i*th treatment, β_j is the effect of the *j*th block, and ε_{ij} is the usual NID $(0, \sigma^2)$ random error term. We treat both treatments as fixed factors. The important case of random blocks is considered in Chapter 9. Just as in the single-factor experimental design model in Chapter 3, the effects model for the RCBD is an overspecified model. Consequently, we usually think of the treatment and block effects as deviations from the overall mean so that

$$\sum_{i=1}^{a} \tau_i = 0 \quad \text{and} \quad \sum_{j=1}^{b} \beta_j = 0$$

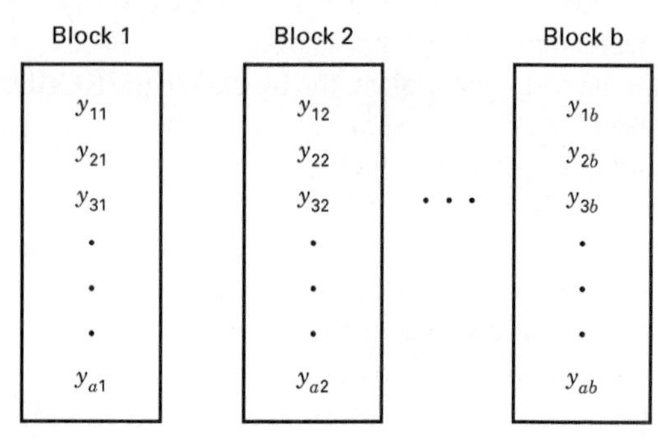

FIGURE 6.1 The Randomized Complete Block Design

In an experiment involving the RCBD, we are interested in testing the equality of the treatment means. Thus, the hypotheses of interest are

$$H_0 : \mu_1 = \mu_2 = \cdots = \mu_0$$
$$H_1 : \text{at least one } \mu_i \neq \mu_j$$

Because the ith treatment mean $\mu_i = (1/b)\sum_{j=1}^{b}(\mu + \tau_i + \beta_j) = \mu + \tau_i$, an equivalent way to write the above hypotheses is in terms of the treatment effects, say

$$H_0 : \tau_1 = \tau_2 = \cdots = \tau_a = 0$$
$$H_1 : \tau_i \neq 0 \text{ at least one } i$$

The analysis of variance can be easily extended to the RCBD. Let $y_{i.}$ be the total of all observations taken under treatment i, $y_{.j}$ be the total of all observations in block j, $y_{..}$ be the grand total of all observations, and $N = ab$ be the total number of observations. Expressed mathematically,

$$y_{i.} = \sum_{j=1}^{b} y_{ij} \qquad i = 1,2, \ldots , a \tag{6.2}$$

$$y_{.j} = \sum_{i=1}^{a} y_{ij} \qquad j = 1,2, \ldots , b \tag{6.3}$$

and

$$y_{..} = \sum_{i=1}^{a}\sum_{j=1}^{b} y_{ij} = \sum_{i=1}^{a} y_{i.} = \sum_{j=1}^{b} y_{.j} \tag{6.4}$$

Similarly, $\bar{y}_{i.}$ is the average of the observations taken under treatment i, $\bar{y}_{.j}$ is the average of the observations in block j, and $\bar{y}_{..}$ is the grand average of all observations. That is,

$$\bar{y}_{i.} = y_{i.}/b \quad \bar{y}_{.j} = y_{.j}/a \quad \bar{y}_{..} = y_{..}/N \tag{6.5}$$

We may express the total corrected sum of squares as

$$\sum_{i=1}^{a}\sum_{j=1}^{b}(y_{ij} - \bar{y}_{..})^2 = \sum_{i=1}^{a}\sum_{j=1}^{b}\left[(\bar{y}_{i.} - \bar{y}_{..}) + (\bar{y}_{.j} - \bar{y}_{..}) + (y_{ij} - \bar{y}_{i.} - \bar{y}_{.j} + \bar{y}_{..}\right]^2 \tag{6.6}$$

By expanding the right-hand side of Eq. (6.6), we obtain

$$\begin{aligned}
\sum_{i=1}^{a}\sum_{j=1}^{b}(y_{ij} - \bar{y}_{..})^2 &= b\sum_{i=1}^{a}(\bar{y}_{i.} - \bar{y}_{..})^2 + a\sum_{j=1}^{b}(\bar{y}_{.j} - \bar{y}_{..})^2 \\
&\quad + \sum_{i=1}^{a}\sum_{j=1}^{b}(y_{ij} - \bar{y}_{i.} - \bar{y}_{.j} + \bar{y}_{..})^2 + 2\sum_{i=1}^{a}\sum_{j=1}^{b}(\bar{y}_{i.} - \bar{y}_{..})(\bar{y}_{.j} - \bar{y}_{..}) \\
&\quad + 2\sum_{i=1}^{a}\sum_{j=1}^{b}(\bar{y}_{.j} - \bar{y}_{..})(\bar{y}_{ij} - \bar{y}_{i.} - \bar{y}_{.j} + \bar{y}_{..}) \\
&\quad + 2\sum_{i=1}^{a}\sum_{j=1}^{b}(\bar{y}_{i.} - \bar{y}_{..})(\bar{y}_{ij} - \bar{y}_{i.} - \bar{y}_{.j} + \bar{y}_{..}).
\end{aligned}$$

Simple but tedious algebra proves that the three cross products are zero. Therefore,

$$\sum_{i=1}^{a}\sum_{j=1}^{b}(y_{ij}-\bar{y}_{..})^2 = b\sum_{i=1}^{a}(\bar{y}_{i.}-\bar{y}_{..})^2 + a\sum_{j=1}^{b}(\bar{y}_{.j}-\bar{y}_{..})^2 + \sum_{i=1}^{a}\sum_{j=1}^{b}(y_{ij}-\bar{y}_{.j}-\bar{y}_{i.}+\bar{y}_{..})^2 \tag{6.7}$$

represents a partition of the total sum of squares. This is the fundamental ANOVA equation for the RCBD. Expressing the sums of squares in Eq. (6.7) symbolically, we have

$$SS_T = SS_{\text{Treatments}} + SS_{\text{Blocks}} + SS_E \tag{6.8}$$

Because there are N observations, SS_T has $N - 1$ degrees of freedom. There are a treatments and b blocks, so $SS_{\text{Treatments}}$ and SS_{Blocks} have $a - 1$ and $b - 1$ degrees of freedom, respectively. The error sum of squares is just a sum of squares between cells minus the sum of squares for treatments and blocks. There are ab cells with $ab - 1$ degrees of freedom between them, so SS_E has $ab - 1 - (a - 1) - (b - 1) = (a - 1)(b - 1)$ degrees of freedom. Furthermore, the degrees of freedom on the right-hand side of Eq. (6.8) add to the total on the left; therefore, making the usual normality assumptions on the errors, one can show that $SS_{\text{Treatments}}/\sigma^2$, $SS_{\text{Blocks}}/\sigma^2$, and SS_E/σ^2 are independently distributed chi-square random variables. Each sum of squares divided by its degrees of freedom is a mean square. The expected value of the mean squares, if treatments and blocks are fixed, can be shown to be

$$E(MS_{\text{Treatments}}) = \sigma^2 + \frac{b\sum_{i=1}^{a}\tau_i^2}{a-1}$$

$$E(MS_{\text{Blocks}}) = \sigma^2 + \frac{a\sum_{j=1}^{b}\beta_j^2}{b-1}$$

$$E(MS_E) = \sigma^2$$

Therefore, to test the equality of treatment means, we would use the test statistic

$$F_0 = \frac{MS_{\text{Treatments}}}{MS_E}$$

which is distributed as $F_{\alpha-1,(a-1)(b-1)}$ if the null hypothesis is true. The critical region is the upper tail of the F-distribution, and we would reject H_0 if $F_0 > F_{\alpha,a-1,(a-1)(b-1)}$. A P-value approach can also be used.

The procedure is usually summarized in an ANOVA table, such as the one shown in Table 6.2. The computing would usually be done with a statistical software package. However, computing

Table 6.2 Analysis of Variance Table for a Randomized Complete Block Design

Source of variation	Sum of squares	Degrees of freedom	Mean square	F_0
Treatments	$SS_{\text{Treatments}}$	$a-1$	$\frac{SS_{\text{Treatments}}}{a-1}$	$\frac{MS_{\text{Treatments}}}{MS_E}$
Blocks	SS_{Blocks}	$b-1$	$\frac{SS_{\text{Blocks}}}{b-1}$	
Error	SS_E	$(a-1)(b-1)$	$\frac{SS_E}{(a-1)(b-1)}$	
Total	SS_T	$N-1$		

formulas for the sums of squares may be obtained for the elements in Eq. (6.7) by working directly with the identity

$$y_{ij} - \bar{y}_{..} = (\bar{y}_{i.} - \bar{y}_{..}) + (\bar{y}_{.j} - \bar{y}_{..}) + (y_{ij} - \bar{y}_{i.} - \bar{y}_{.j} + \bar{y}_{..})$$

These quantities can be computed in the columns of an Excel spreadsheet. Then each column can be squared and summed to produce the sum of squares. Alternatively, computing formulas can be expressed in terms of treatment and block totals. These formulas are

$$SS_T = \sum_{i=1}^{a} \sum_{j=1}^{b} y_{ij}^2 - \frac{y_{..}^2}{N} \tag{6.9}$$

$$SS_{\text{Treatments}} = \frac{1}{b} \sum_{i=1}^{a} y_{i.}^2 - \frac{y_{..}^2}{N} \tag{6.10}$$

$$SS_{\text{Blocks}} = \frac{1}{a} \sum_{j=1}^{b} y_{.j}^2 - \frac{y_{..}^2}{N} \tag{6.11}$$

and the error sum of squares is obtained by subtraction as

$$SS_E = SS_T - SS_{\text{Treatments}} - SS_{\text{Blocks}} \tag{6.12}$$

The ANOVA computations are almost always done using a software package. Figure 6.2 below presents the JMP output for the RCBD vascular grafts experiment. The Effects Tests section of the output summarizes the ANOVA. There is a statistically significant difference between the levels of extrusion pressure. The plot of residuals versus the predicted response does not indicate any problems, so we should be comfortable with the results of the experiment. A key question that the experimenter wanted to answer was which extrusion pressures led to the highest yield. The least squares means table in the output indicates that lower pressures led to higher yields. The Fisher LSD method is displayed in the Student's t All Possible Comparisons at the bottom of the computer output. It indicates that the lowest extrusion pressure 8500 psi is statistically different from the other three pressures, so it would be a good recommended operating condition for the process.

Note that the JMP analysis is done using the matrix analysis approach shown in Section 5.6, Chapter 5. The effects coding for the six batches is shown in Table 6.3.

In general, a categorical factor with k levels, requires $k - 1$ dummy columns. The *ith* row has a 1 in the ith *column* and zeros in the other columns except for the kth row, which has the entry -1 for every column.

The **X** matrix for the vascular graft experiment has one column for the intercept, three dummy columns for the four settings of pressure, and five dummy columns for the six batches, for a total of nine parameters to estimate. Remember that the estimate for the last pressure setting is -1 times the sum of the fixed effect estimates for the other three pressures. The estimate for the last batch setting is -1 times the sum of the fixed effect estimates for the other five batches.

For example, the estimate of the effect for a pressure of 9100 is $-1(3.021 + 1.887 - 0.879) = -4.029$.

The least squares mean estimate for a pressure of 9100 is the intercept estimate, 89.796, plus the effect estimate, -4.029 ($89.796 - 4.029 = 85.767$).

6.3 Blocking and Optimal Designs

The RCBD is widely used. It is easy to construct and run each treatment once in each block. If the size of the blocks is the same as the number of treatments, then the RCBD is an obvious choice. Choosing the sample size is accomplished by choosing the number of blocks. The RCBD is a

Actual by Predicted Plot

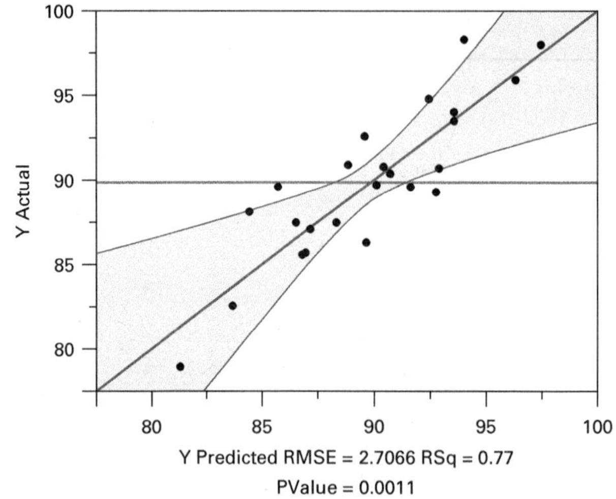

Effect Summary

Source	LogWorth	PValue
Pressure	2.718	0.00192
Resin Batch	2.257	0.00553

Residual by Predicted Plot

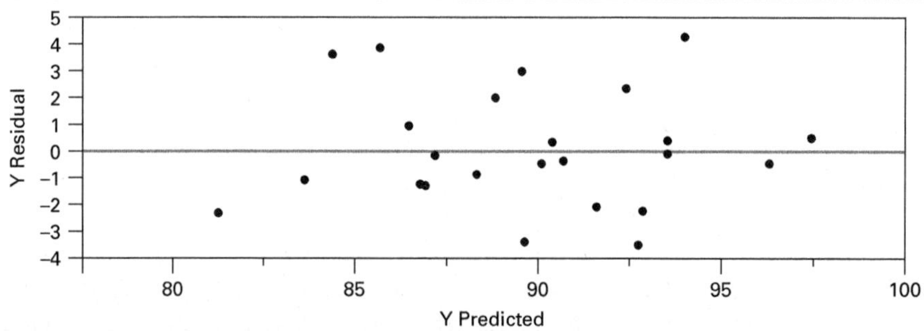

Parameter Estimates

Term	Estimate	Std Error	t Ratio	Prob>|t|
Intercept	89.795833	0.552485	162.53	<.0001*
Pressure[8500]	3.0208333	0.956932	3.16	0.0065*
Pressure[8700]	1.8875	0.956932	1.97	0.0673
Pressure[8900]	−0.879167	0.956932	−0.92	0.3728
Resin Batch[1]	−2.095833	1.235394	−1.70	0.1104
Resin Batch[2]	−0.045833	1.235394	−0.04	0.9709
Resin Batch[3]	1.2041667	1.235394	0.97	0.3452
Resin Batch[4]	0.7541667	1.235394	0.61	0.5507
Resin Batch[5]	−4.470833	1.235394	−3.62	0.0025*

Effect Tests

Source	Nparm	DF	Sum of Squares	F Ratio	Prob > F
Pressure	3	3	178.17125	8.1071	0.0019*
Resin Batch	5	5	192.25208	5.2487	0.0055*

FIGURE 6.2 JMP Output for the Vascular Graft Experiment

Multiple Comparisons for Pressure

Least Squares Means Estimates

Pressure	Estimate	Std Error	DF	Lower 95%	Upper 95%
8500	92.816667	1.1049698	15	90.461479	95.171854
8700	91.683333	1.1049698	15	89.328146	94.038521
8900	88.916667	1.1049698	15	86.561479	91.271854
9100	85.766667	1.1049698	15	83.411479	88.121854

Student's t All Pairwise Comparisons

Quantile = 2.13145, DF = 15.0

All Pairwise Differences

Pressure	-Pressure	Difference	Std Error	t Ratio	Prob> |t|	Lower 95%	Upper 95%
8500	8700	1.133333	1.562663	0.73	0.4795	−2.19740	4.46407
8500	8900	3.900000	1.562663	2.50	0.0247*	0.56926	7.23074
8500	9100	7.050000	1.562663	4.51	0.0004*	3.71926	10.38074
8700	8900	2.766667	1.562663	1.77	0.0970	−0.56407	6.09740
8700	9100	5.916667	1.562663	3.79	0.0018*	2.58593	9.24740
8900	9100	3.150000	1.562663	2.02	0.0621	−0.18074	6.48074

FIGURE 6.2 (*Continued*)

Table 6.3 **The Dummy Columns for a Six-level Categorical Factor Using Effects Coding**

Batches/dummy columns	Dummy 1	Dummy 2	Dummy 3	Dummy 4	Dummy 5
Batch 1	1	0	0	0	0
Batch 2	0	1	0	0	0
Batch 3	0	0	1	0	0
Batch 4	0	0	0	1	0
Batch 5	0	0	0	0	1
Batch 6	−1	−1	−1	−1	−1

D-optimal design. To show this, we will construct the design used in the vascular graft problem using the Custom Designer in JMP. We selected the pressure factor to have four levels and a block size of 4. Then selecting a 24-run design produces the output shown in Figure 6.3. This is the design that we used in the vascular graft example although we have not randomized the extrusion pressure runs within each block. Notice that the D-efficiency of the design is 100%. It is also an A-optimal design and a G-optimal design. All RCBDs will have these properties.

Optimally blocked designs can be extremely useful in situations where the RCBD is not possible due to resource constraints. For example, suppose that in the vascular graft experiment the experimenter still wants to test four extrusion pressures, but there are only four batches of resin material available. Furthermore, each batch is only large enough to test three of the four pressures. Supplying this information into the JMP Custom Designer yields the design shown in Figure 6.4. This design is slightly over 95% D-efficient. This is another example of the fact that reported D-efficiencies are a loose lower bound on the true efficiency because **balanced incomplete block designs (BIBDs)** are D-optimal. Looking at the design display, you will notice that the design has an interesting structure. Every pair of pressures occurs together in the same block exactly twice. This is an example of a BIBD. In general, a BIBD is a blocked design in which any pair of treatments occurs together the same number of times as any other pair. BIBDs have been widely used in agricultural experimentation, because there are often a fairly large number of treatments and block sizes are typically small.

Design

Run	Pressure	Resin Batch
1	8700	1
2	8900	1
3	8500	1
4	9100	1
5	8900	2
6	9100	2
7	8700	2
8	8500	2
9	8700	3
10	8500	3
11	8900	3
12	9100	3
13	9100	4
14	8900	4
15	8700	4
16	8500	4
17	8900	5
18	9100	5
19	8700	5
20	8500	5
21	9100	6
22	8700	6
23	8900	6
24	8500	6

Design Diagnostics

D Optimal Design	
D Efficiency	100
G Efficiency	100
A Efficiency	100
Average Variance of Prediction	0.375
Design Creation Time (seconds)	0.033333

FIGURE 6.3 The D-optimal RCBD Constructed with JMP

As a final example, consider an experiment with five treatments and two different blocking factors. Suppose that each block is of size of 5. Figure 6.5 shows the design from JMP for this situation. It is another incomplete block design, but this one also has an interesting structure. Notice that if we consider two blocking factors as the rows and columns, each treatment (the letters A, B, C, D, and E) appear exactly once in each row and column. This type of design is called a **Latin square design**. Specifically, it is an example of a 5×5 Latin square. The Latin square

Design

Run	Pressure	Resin Batch
1	8700	1
2	8500	1
3	8900	1
4	8700	2
5	9100	2
6	8500	2
7	8900	3
8	8500	3
9	9100	3
10	9100	4
11	8900	4
12	8700	4

Design Evaluation

Design Diagnostics	
D Optimal Design	
D Efficiency	95.07744
G Efficiency	83.666
A Efficiency	90.32258
Average Variance of Prediction	0.645833
Design Creation Time (seconds)	16.01667

FIGURE 6.4 A D-optimal Blocked Design with Four Treatments and Three Treatments per Block

	Block	a	b	c	d	e
1	1	D	E	C	B	A
2	2	B	D	E	A	C
3	3	C	A	B	E	D
4	4	A	B	D	C	E
5	5	E	C	A	D	B

FIGURE 6.5 Latin Square Design with Five Treatments and Two Five-Level Blocking Factors

design is used to eliminate two nuisance sources of variability; that is, it systematically allows blocking in two directions. Thus, the rows and columns actually represent two restrictions on randomization. Latin square designs have been used extensively in agricultural experimentation.

Note that RCBDs, BIBDs, and Latin square designs all have an inflexible structure that requires specific numbers of runs in total and in each block. An optimal design approach allows us to overcome these restrictions. As an example, suppose the problem situation allows the blocks to have more runs than the number of treatments. This would allow for replication in each block and a pure error estimate of the variance.

7 The 2^k Factorial Design

7.1 Introduction

Full-factorial designs are popular for experiments involving a few factors where it is necessary to study the joint effect of the factors on a response. Chapter 5 presented general methods for the construction and analysis of factorial designs. However, there are several important special cases of the general factorial design because they are widely used in industrial research and development work and because they form the basis of other designs of considerable practical value.

The most important of these special cases is when there are k factors and each factor has only two levels. These levels may be **quantitative**, such as two values of temperature or pressure; or they may be **qualitative**, such as two machines or two operators, the "high" and "low" levels of a factor, or perhaps the presence and absence of a factor. A complete replicate of such a design requires $2 \times 2 \times \cdot \cdot \times 2 = 2^k$ observations and is called a **2^k factorial design**. This chapter focuses on these designs. Throughout this chapter, we assume that (1) the factors are fixed, (2) the designs are completely randomized, and (3) the usual normality assumptions required for the analysis are satisfied.

The 2^k design is particularly useful in the early stages of experimental work when experimenters often need to study the effects of many factors and they suspect that not all factors are going to be important. In these situations, the experimenter is usually interested in discovering the set of **active** factors from the original group. The 2^k design provides the smallest number of runs with which k factors can be studied in a complete factorial design. Consequently, these designs are widely used in **factor screening experiments when runs are inexpensive**. It is also easy to develop effective blocking schemes for these designs and to construct them in fractional versions.

7.2 The 2^2 Factorial Design

The first design in the 2^k family has two factors, say A and B, each run at two levels. This design is called a 2^2 factorial design. The levels of the factors may be arbitrarily called "low" and "high." As an example, suppose that a router is used to cut locating notches on a printed circuit board. The vibration level at the surface of the board as it is cut is considered to be a major source of dimensional variation in the notches. Two factors are believed to influence vibration: bit size (A) and cutting speed (B). Two bit sizes (1/16 and 1/8 in.) and two speeds (40 and 90 rpm) are selected, and three boards are cut at each set of conditions as shown in Table 7.1. The response variable is vibration measured as the resultant vector of three accelerometers (x, y, and z) on each of the circuit boards used in the experiment.

A convention for labeling the runs in a 2^k design is shown in Table 7.1. The four treatment combinations in this design are shown in the first two columns of Table 7.1. We denote the effect of a factor by a capital Latin letter. Thus, "A" refers to the effect of factor A, "B" refers to the effect of factor B, and "AB" refers to the AB interaction. In the 2^2 design, the low and high levels of A

Table 7.1 **The Router Experiment**

A (bit size)	B (speed)	Run labels	Replicate 1	Replicate 2	Replicate 3
–	–	(1) (A low, B low)	18.2	18.9	12.9
+	–	*a* (A high, B low)	27.2	24.0	22.4
–	+	*b* (A low, B high)	15.9	14.5	15.1
+	+	*ab* (A high, B high)	41.0	43.9	36.3

and B are denoted by "–" and "+," respectively. Thus, – for factor A axis represents the low level of bit size and + represents the high level;—for factor B axis represents the low level of cutting speed and + denotes the high level. This is sometimes called the "orthogonal coding" for a 2^k design. The four treatment combinations in the design are also represented by lowercase letters, as shown in Table 7.1. The high level of any factor in the treatment combination is denoted by the corresponding lowercase letter and that the low level of a factor in the treatment combination is denoted by the absence of the corresponding letter. Thus, a represents the treatment combination of A at the high level and B at the low level, b represents A at the low level and B at the high level, and ab represents both factors at the high level. By convention, (1) is used to denote both factors at the low level. This notation is used throughout the 2^k series. Often, these lowercase letters represent the total of all observations that are taken at the corresponding treatment combinations.

The analysis of a replicated 2^k design is similar to the analysis of general factorials from Chapter 6. Figure 7.1 is the output from the Fit Model Platform in JMP for the 2^2 factorial for the router experiment in Table 7.1. The analysis of variance portion of the display shows that the overall model is significant, implying that at least one of the two main effects and their interaction are active. The following sections showing the individual effect tests and the parameter estimates indicate that both of the main effects and the bit size–speed interaction are significant. The residuals versus predicted value plot does not indicate any obvious problems with model adequacy and the model R^2 is 95%. The two-factor interaction plot shows that when the small bit is used cutting speed has very little effect on vibration, but when the larger bit is used, increasing cutting speed leads to large increases in vibration. This is confirmed by the Prediction Profiler, which is set at the low level of bit size. This display shows that when using the smaller bit any cutting speed can be used without increasing the vibration level.

7.2.1 How Much Replication Is Necessary?

A standard question that arises in almost every experiment is how much replication is necessary. We have discussed this in previous chapters, but there are some aspects of this topic that are particularly useful in 2^k designs, when they are used for factor screening (i.e. when $k < 4$ or runs are inexpensive). That is, studying a group of k factors to determine which ones are active. Recall from our previous discussions that the choice of an appropriate sample size in a designed experiment depends on how large the effect of interest is, the power of the statistical test, and the choice of type I error rate. While the size of an important effect is obviously problem dependent, in many practical situations experimenters are interested in detecting effects that are at least as large as twice the error standard deviation (2σ). Smaller effects are usually of less interest because changing the factor associated with such a small effect often results in a change in response that is small relative to the background noise in the system. Adequate power is also problem dependent, but in many practical situations achieving power of at least 0.80 or 80% should be the goal.

We will illustrate how an appropriate choice of sample size can be determined using the 2^2 router experiment. Suppose that we are interested in detecting effects of size 2σ. If the basic 2^2 design is replicated twice for a total of eight runs, there will be four degrees of freedom for estimating a model-independent estimate of error (pure error). If the experimenter uses a significance level or type I error rate of $\alpha = 0.05$, this design results in a power of 0.572 or 57.2%. This is too low, and the experimenter should consider more replications. There is another alternative that could be useful in screening experiments, using a higher type I error rate. In screening experiments, type I errors (thinking a factor is active when it really is not) usually do not have the same impact than type II errors (failing to identify an active factor). If a factor is mistakenly thought to be active, that error will be discovered in further work and so the consequences of this

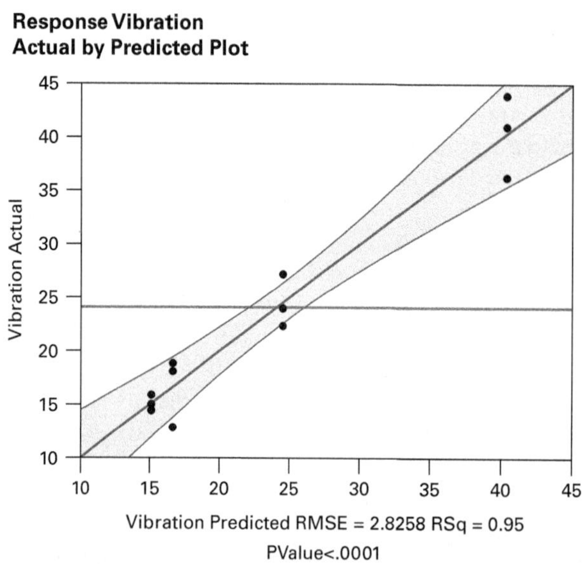

Effect Summary

Source	LogWorth	PValue
Bit Size(0.0625,0.125)	5.118	0.00001
Bit Size*Speed	3.149	0.00071
Speed(40,90)	2.643	0.00228 ^

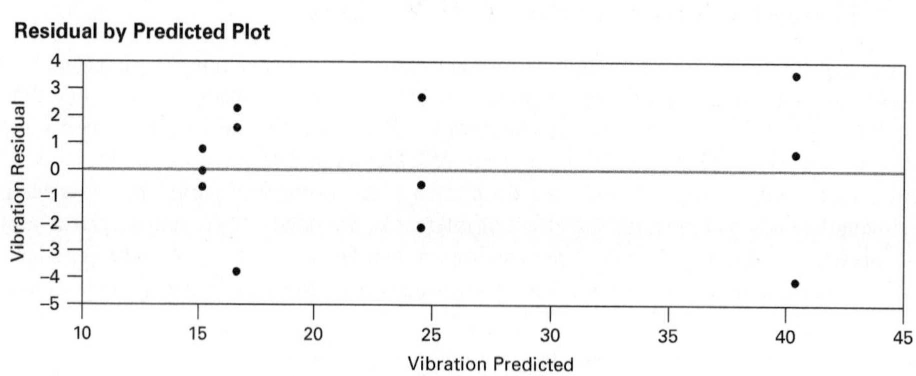

FIGURE 7.1 JMP Fit Model Output for the Router Experiment in Table 7.1

Studentized Residuals

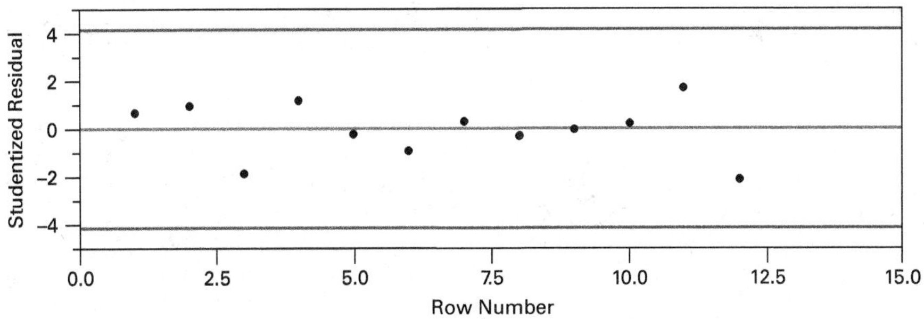

Externally Studentized Residuals with 95% Simultaneous Limits (Bonferroni)

Analysis of Variance

Source	DF	Sum of Squares	Mean Square	F Ratio
Model	3	1202.7092	400.903	50.2070
Error	8	63.8800	7.985	**Prob > F**
C. Total	11	1266.5892		<.0001*

Parameter Estimates

Term	Estimate	Std Error	t Ratio	Prob>|t|
Intercept	24.191667	0.815731	29.66	<.0001*
Bit Size(0.0625,0.125)	8.275	0.815731	10.14	0.0001*
Speed(40,90)	3.5916667	0.815731	4.40	0.0023*
Bit Size*Speed	4.3416667	0.815731	5.32	0.0007*

Effect Tests

Source	Nparm	DF	Sum of Squares	F Ratio	Prob > F
Bit Size(0.0625,0.125)	1	1	821.70750	102.9064	<.0001*
Speed(40,90)	1	1	154.80083	19.3865	0.0023*
Bit Size*Speed	1	1	226.20083	28.3282	0.0007*

Prediction Expression

$$24.191666667$$

$$+ 8.275 \bullet \left(\frac{(\text{Bit Size} - 0.09375)}{0.03125} \right)$$

$$+ 3.5916666667 \bullet \left(\frac{(\text{Speed} - 65)}{25} \right)$$

$$+ \left(\frac{(\text{Bit Size} - 0.09375)}{0.03125} \right) \bullet \left(\left(\frac{(\text{Speed} - 65)}{25} \right) \bullet 4.3416666667 \right)$$

FIGURE 7.1 (*Continued*)

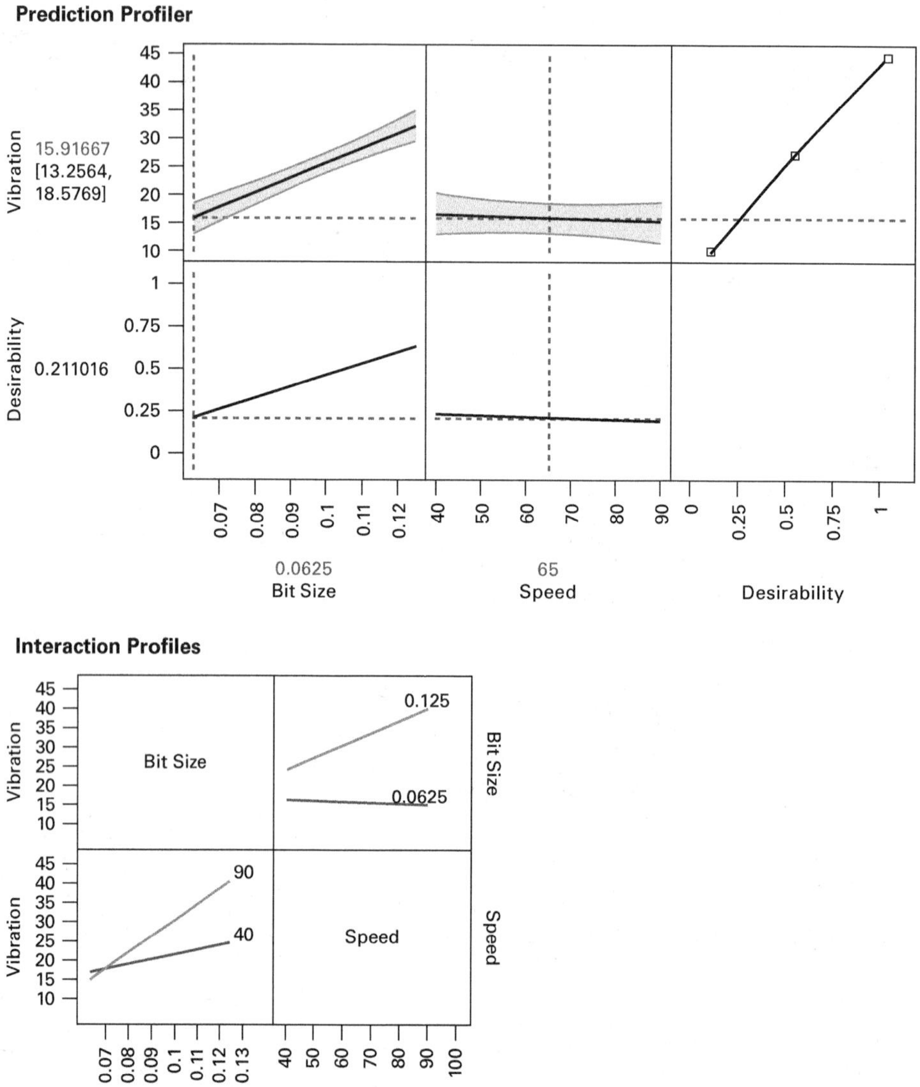

FIGURE 7.1 (*Continued*)

type I error is usually small. However, failing to identify an active factor is usually problematic because that factor is set aside and typically never considered again. So, in screening experiments, experimenters are often willing to consider higher type I error rates, say 0.10 or 0.20. Suppose that we use $\alpha = 0.10$ in our router experiment. This would result in power of 75%. Using $\alpha = 0.20$ increases the power to 89%, a very reasonable value. The other alternative is to increase the sample size by using additional replicates. If we use three replicates, there will be eight degrees of freedom for pure error and if we want to detect effects of size 2σ with $\alpha = 0.05$, this design will result in power of 85.7%. This is a good value for power, so the experimenters decided to use three replicates of the 2^2 design.

Software packages can be used to produce the aforementioned power calculations. The boxed display shows the power calculations from JMP. The model has both main effects and the two-factor interaction and the effects of size 2σ is chosen by setting the square root of mean square error (anticipated RMSE) to 1 and setting the size of each anticipated coefficient in the prediction (regression) model coefficient to 1.

Evaluate Design

Model

Intercept

X1

X2

X1*X2

Power Analysis

Significance Level	0.05
Anticipated RMSE	1

Term	Anticipated Coefficient	Power
Intercept	1	0.857
X1	1	0.857
X2	1	0.857
X1*X2	1	0.857

7.3 The 2^3 Factorial Design

Suppose that three factors, *A*, *B*, and *C*, each at two levels, are of interest. The design is called a 2^3 factorial design, and the eight treatment combinations can now be displayed geometrically as a cube, as shown in Figure 7.2a. Using the "+ and −" orthogonal coding to represent the low and high levels of the factors, we show the eight runs in the 2^3 design in Figure 7.2b. This is sometimes called the **design matrix**. Extending the label notation discussed in Section 7.2, we write the treatment combinations in standard order as (1), *a, b, ab, c, ac, bc,* and *abc*. Remember that these symbols also often represent the total of all *n* observations taken at those specific treatment combinations.

As an example of an experiment involving a 2^3 design, suppose that a beverage bottler is investigating the performance of a bottle-filling machine. These fillers are usually designed to fill to a specific target height, and bottles that deviate significantly from the target are either under- or overfilled. The filler has three controllable factors: carbonation level of the product, filling

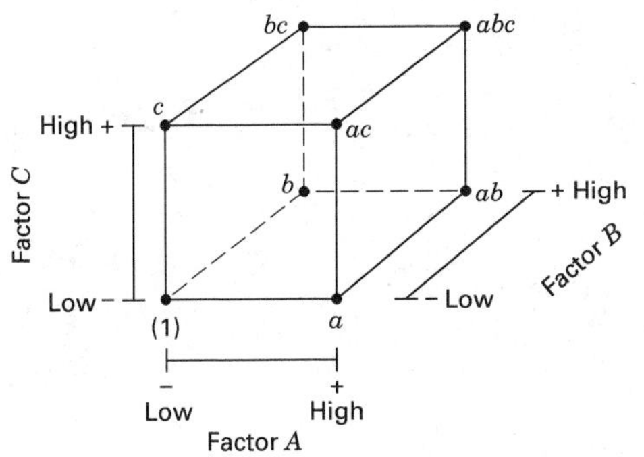

(a)

Run	Factor *A*	Factor *B*	Factor *C*
1	−	−	−
2	+	−	−
3	−	+	−
4	+	+	−
5	−	−	+
6	+	−	+
7	−	+	+
8	+	+	+

(b)

FIGURE 7.2 The 2^3 Factorial Design: (a) Geometric View; (b) Design Matrix

Table 7.2 The Bottle Filling Experiment

Carbonation	Pressure	Speed	Treatment label	Replicate 1	Replicate 2
−	−	−	(1)	−3	−1
+	−	−	a	0	1
−	+	−	b	−1	0
+	+	−	ab	2	3
−	−	+	c	−1	0
+	−	+	ac	2	1
−	+	+	bc	1	1
+	+	+	abc	6	5

pressure, and bottling speed. These three factors were investigated using the 2^3 design shown in Table 7.2. The low and high levels of the three factors are as follows:

Carbonation level 10%, 12%; pressure 25 psi, 30 psi; speed 200 bottles/minute, 250 bottles/minute

The response variable is fill deviation as measured on a coded scale, with 0 indicating a fill height that is exactly on the desired target. Positive values of the response indicate overfilled bottles and negative values indicate underfilled bottles.

We will analyze this experiment by fitting the full three-factor factorial model. This model contains all three main effects, the three two-factor interactions, and the three-factor interaction. Figure 7.3 shows the output from the JMP Fit Model Platform.

Response Fill Deviation
Actual by Predicted Plot

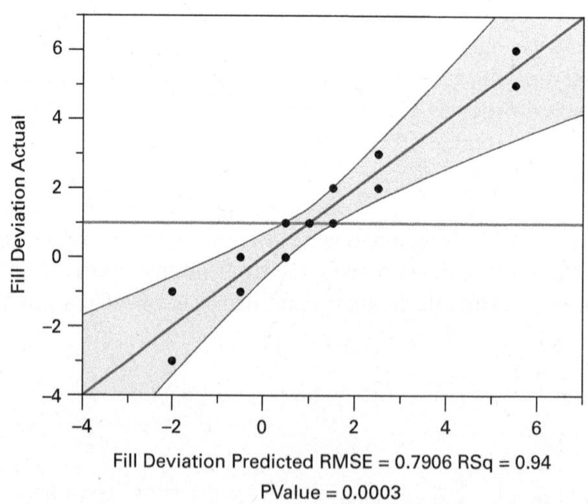

Effect Summary

Source	LogWorth	PValue
Carbonation(10,12)	4.196	0.00006
Pressure(25,30)	3.339	0.00046
Speed(200,250)	2.657	0.00221
Carbonation*Pressure	1.025	0.09435
Pressure*Speed	0.617	0.24150
Carbonation*Pressure*Speed	0.617	0.24150
Carbonation*Speed	0.264	0.54474 ^

FIGURE 7.3
JMP Fit Model Output for the Bottle Filling Experiment

Residual by Predicted Plot

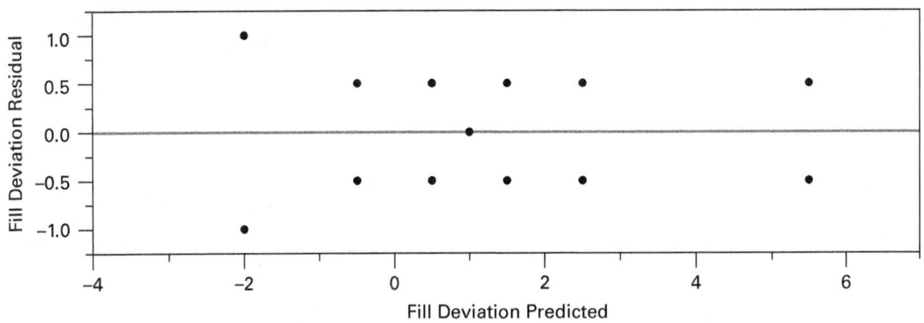

Studentized Residuals

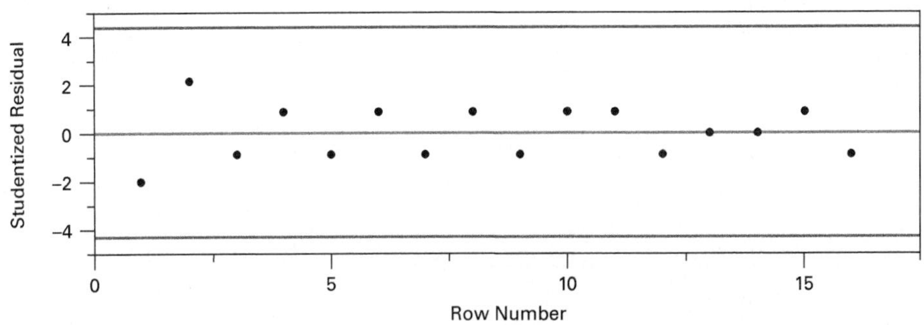

Externally Studentized Residuals with 95% Simultaneous Limits (Bonferroni)

Analysis of Variance

Source	DF	Sum of Squares	Mean Square	F Ratio
Model	7	73.000000	10.4286	16.6857
Error	8	5.000000	0.6250	Prob > F
C. Total	15	78.000000		0.0003*

Parameter Estimates

Term	Estimate	Std Error	t Ratio	Prob>|t|
Intercept	1	0.197642	5.06	0.0010*
Carbonation(10,12)	1.5	0.197642	7.59	<.0001*
Pressure(25,30)	1.125	0.197642	5.69	0.0005*
Speed(200,250)	0.875	0.197642	4.43	0.0022*
Carbonation*Pressure	0.375	0.197642	1.90	0.0943
Carbonation*Speed	0.125	0.197642	0.63	0.5447
Pressure*Speed	0.25	0.197642	1.26	0.2415
Corbonation*Pressure*Speed	0.25	0.197642	1.26	0.2415

Effect Tests

Source	Nparm	DF	Sum of Squares	F Ratio	Prob > F
Carbonation(10,12)	1	1	36.000000	57.6000	<.0001*
Pressure(25,30)	1	1	20.250000	32.4000	0.0005*
Speed(200,250)	1	1	12.250000	19.6000	0.0022*
Carbonation*Pressure	1	1	2.250000	3.6000	0.0943
Carbonation*Speed	1	1	0.250000	0.4000	0.5447
Pressure*Speed	1	1	1.000000	1.6000	0.2415
Corbonation*Pressure*Speed	1	1	1.000000	1.6000	0.2415

FIGURE 7.3 (*Continued*)

The analysis of variance indicates that the overall model is significant, and the effect tests show that all three main effects are highly significant and that one of the two-factor interactions, carbonation–pressure, is significant at approximately the 10% level. The size of this interaction effect is less than half of the smallest main effect.

Figure 7.4 is the result of fitting a reduced model containing only the three main effects and the carbonation–pressure interaction. One again, we see that all three main effects are highly significant and that the carbonation–pressure interaction is only borderline significant. Since the size of the interaction effect is less than half the size of the smallest main effect, it is unlikely that it will play an important role in determining the settings for the bottle filling machine.

Response Fill Deviation
Actual by Predicted Plot

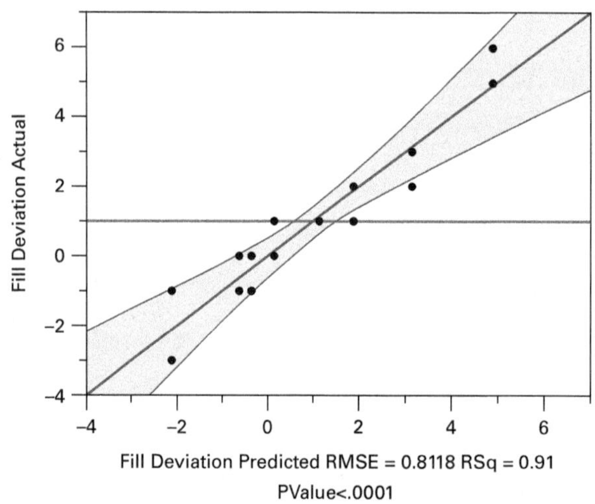

Effect Summary

Source	LogWorth	PValue
Carbonation(10,12)	4.861	0.00001
Pressure(25,30)	3.758	0.00017
Speed(200,250)	2.909	0.00123
Carbonation*Pressure	1.038	0.09170

Residual by Predicted Plot

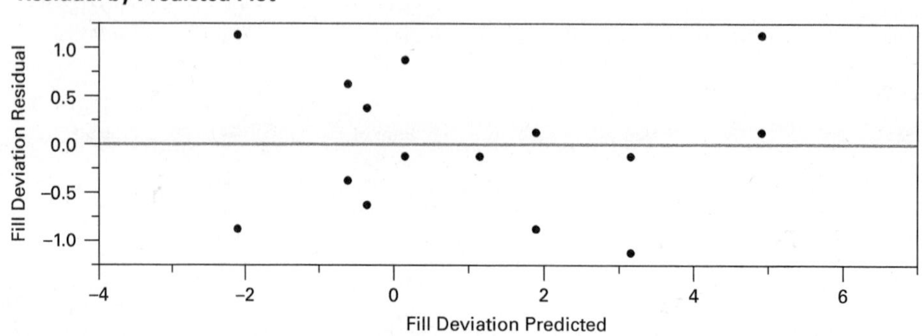

FIGURE 7.4 The Reduced Model for the Bottle Filling Experiment

Studentized Residuals

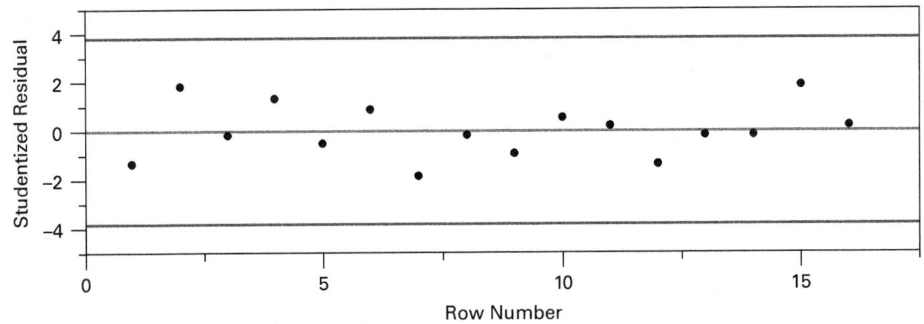

Externally Studentized Residuals with 95% Simultaneous Limits (Bonferroni)

Analysis of Variance

Source	DF	Sum of Squares	Mean Square	F Ratio
Model	4	70.750000	17.6875	26.8362
Error	11	7.250000	0.6591	Prob > F
C. Total	15	78.000000		<.0001*

Parameter Estimates

Term	Estimate	Std Error	t Ratio	Prob>|t|
Intercept	1	0.202961	4.93	0.0005*
Carbonation(10,12)	1.5	0.202961	7.39	<.0001*
Pressure(25,30)	1.125	0.202961	5.54	0.0002*
Speed(200,250)	0.875	0.202961	4.31	0.0012*
Carbonation*Pressure	0.375	0.202961	1.85	0.0917

Effect Tests

Source	Nparm	DF	Sum of Squares	F Ratio	Prob > F
Carbonation(10,12)	1	1	36.000000	54.6207	<.0001*
Pressure(25,30)	1	1	20.250000	30.7241	0.0002*
Speed(200,250)	1	1	12.250000	18.5862	0.0012*
Carbonation*Pressure	1	1	2.250000	3.4138	0.0917

Prediction Profiler

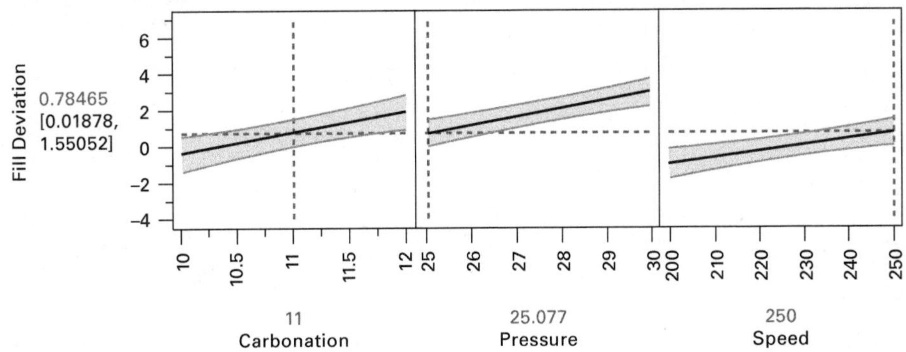

FIGURE 7.4 (*Continued*)

It is desirable to set up the machine so that the average fill deviation is close to zero. The Prediction Profiler in Figure 7.4 displays one possible configuration. In this solution, speed is set at the high level (this is a throughput consideration so that as much product as possible can be filled). Filling pressure is also at the low level (this is known to reduce foaming, a major source of under- and overfilled bottles) and carbonation is at the midpoint of the range since this is the specification for the characteristic.

7.3.1 Replication of the 2^3 Design

The experimenter in the bottle filling experiment used two replicates of the 2^3 design. This will provide eight degrees of freedom for pure error. Suppose that effects of size 2σ are of interest, the experimenter wants to consider all main effects and interactions (the full-factorial model) and use $\alpha = 0.05$. The JMP power calculations are shown in the following display:

Evaluate Design

Model

Intercept

X1

X2

X3

X1*X2

X1*X3

X2*X3

X1*X2*X3

Power Analysis

Significance Level	0.05
Anticipated RMSE	1

Term	**Anticipated Coefficient**	**Power**
Intercept	1	0.937
X1	1	0.937
X2	1	0.937
X3	1	0.937
X1*X2	1	0.937
X1*X3	1	0.937
X2*X3	1	0.937
X1*X2*X3	1	0.937

The power of this design is 93.7%. Even if the experimenter decides to use $\alpha = 0.01$ the power is still 72%. Two replicates of the 2^3 design is a good choice for this experiment.

7.4 A Single Replicate of the 2^k Design

For even a moderate number of factors, the total number of treatment combinations in a 2^k factorial design is large. For example, a 2^5 design has 32 treatment combinations, a 2^6 design has 64 treatment combinations, a 2^{10} has 1024 treatment combinations, and so on. Because resources are usually limited, the number of replicates that the experimenter can employ may be restricted. Frequently, available resources only allow a single replicate of the design to be run, unless the experimenter is willing to omit some of the original factors. This is usually an undesirable solution since it runs the risk of missing an important factor.

An obvious risk when conducting an experiment that has only one run at each test combination is that we may be fitting a model to noise. That is, if the response variable y is highly variable, misleading conclusions may result from the experiment. Of course, if there is less variability in the response, the likelihood of an erroneous conclusion is smaller. One way to ensure that reliable effect estimates are obtained is to increase the distance between the low (−) and high (+) levels of the factor.

The single-replicate strategy is often used in screening experiments when relatively many factors are under consideration. Because we can never be entirely certain in such cases that the experimental error is small, a good practice in these types of experiments is to spread out the factor levels aggressively.

A single replicate of a 2^k design is sometimes called an unreplicated factorial. With only one replicate, there is no internal estimate of error (or "pure error"). One approach to the analysis of an unreplicated factorial is to assume that certain high-order interactions are negligible and combine their mean squares to estimate the error. This is an appeal to the **sparsity of effects principle**; that is, most systems are dominated by some of the main effects and low-order interactions, and most high-order interactions are negligible.

While the effect sparsity principle has been observed by experimenters for many decades, only recently has it been studied more objectively. A paper by Li, Sudarsanam, and Frey (2006) studied 113 response variables obtained from 43 published experiments from a wide range of science and engineering disciplines. All of the experiments were full factorials with between three and seven factors, so no assumptions had to be made about interactions. Most of the experiments had either three or four factors. The authors found that about 40% of the main effects in the experiments they studied were significant, while only about 11% of the two-factor interactions were significant. Three-factor interactions were very rare, occurring only about 5% of the time. The authors also investigated the absolute values of factor effects for main effects, two-factor interactions, and three-factor interactions. The median size of main effects was found to be about four times larger than that of two-factor interactions. The median size of two-factor interactions was more than two times larger than that of three-factor interactions. However, there were many two- and three-factor interactions that were larger than the median main effect. Another paper by Bergquist, Vanhatalo, and Nordenvaad (2011) also studied the effect of the sparsity question using 22 different experiments with 35 responses. They considered both full-factorial and fractional factorial designs with factors at two levels. Their results largely agree with those of Li et al. (2006), with the exception that three-factor interactions were less frequent, occurring only about 2% of the time. This difference may be partially explained by the inclusion of experiments with indications of curvature and the need for transformations in the Li et al. (2006) study. Bergquist et al. (2011) excluded such experiments. Overall, both of these studies confirm the validity of the sparsity of effects principle.

When analyzing data from unreplicated factorial designs, occasionally real high-order interactions occur. The use of an error mean square obtained by pooling high-order interactions is inappropriate in these cases as it will make the error term too large and can lead to failure to identify active effects. A method of analysis proposed by Daniel (1959) provides a simple way to overcome this problem. Daniel suggests examining a normal probability plot of the estimates of the effects. The effects that are negligible are normally distributed, with mean zero and variance σ^2 and will tend to fall along a straight line on this plot, whereas significant effects will have nonzero means and will not lie along the straight line. Alternatively, a half-normal plot can be used. The apparently nonzero effects will be used to form the preliminary model and the remaining negligible effects are then combined as an estimate of error.

As an example of an unreplicated factorial, consider an experiment involving a nickel–titanium alloy that is used to make components for jet turbine aircraft engines. Cracking is a potentially serious defect in the final part because it can lead to nonrecoverable failure. The parts producer performed an experiment to determine the effects of four factors on the crack length. The four factors are pouring temperature (A), titanium content (B), heat treatment method (C), and amount of grain refiner used (D). A single replicate of a 2^4 factorial design is run, and the length of crack (in mm $\times 10^{-2}$) induced in a sample coupon subjected to a standard test is measured. The design and the response data are shown in Table 7.3.

Figure 7.5 is the output from the JMP screening analysis for this experiment. The bottom portion of the display is a half-normal plot of the factor effects. This plot indicates that all four main effects are active and that two of the two-factor interactions, Titanium Content–Temperature and Heat Treatment Method–Temperature, are also active.

The normal (or half-normal) probability plotting method has been criticized because it is subjective, and although it is usually effective in identifying large active effects it may not detect smaller ones as effectively. A very popular analytic approach has been developed by Lenth (1989) that has good power to detect active factors. Lenth's method uses the smallest half of the effect estimates to form a robust "pseudo standard error" that can be used to form a t-like ratio for testing the significance of individual effects. The JMP software package implements Lenth's method as part of the screening platform analysis procedure for two-level designs. In their implementation, P-values for each factor and interaction are computed from

Table 7.3 The Alloy Experiment

A	B	C	D	Treatment combination	Crack length
−	−	−	−	(1)	8.4
+	−	−	−	a	13.3
−	+	−	−	b	10.3
+	+	−	−	ab	19.1
−	−	+	−	c	8.6
+	−	+	−	ac	5.8
−	+	+	−	bc	10.9
+	+	+	−	abc	11.6
−	−	−	+	D	10.3
+	−	−	+	ad	15.3
−	+	−	+	bd	12.2
+	+	−	+	abd	20.3
−	−	+	+	cd	10.5
+	−	+	+	acd	7.6
−	+	+	+	bcd	12.7
+	+	+	+	abcd	13.8

Screening for Crack Length
Contrasts

Term	Contrast	Lenth t-Ratio	Individual p-Value	Simultaneous p-Value
Titanium content	1.94375	29.62	<.0001*	<.0001*
Heat treatment method	−1.73125	−26.38	<.0001*	<.0001*
Temperature	1.43125	21.81	<.0001*	<.0001*
Amt of grain refiner	0.91875	14.00	0.0002*	0.0005*
Titanium content*Heat treatment method	0.11875	1.81	0.0853	0.6011
Titanium content*Temperature	0.90625	13.81	0.0002*	0.0005*
Heat treatment method*Temperature	−1.91875	−29.24	<.0001*	<.0001*
Titanium content*Amt of grain refiner	−0.03125	−0.48	0.6627	1.0000
Heat treatment method*Amt of grain refiner	0.04375	0.67	0.5002	1.0000
Temperature*Amt of grain refiner	−0.01875	−0.29	0.7870	1.0000
Titanium content*Heat treatment method*Temperature	0.03125	0.48	0.6627	1.0000
Titanium content*Heat treatment method*Amt of grain refiner	0.06875	1.05	0.2729	0.9904
Titanium content*Temperature*Amt of grain refiner	−0.01875	−0.29	0.7870	1.0000
Heat treatment method*Temperature*Amt of grain refiner	0.05625	0.86	0.3644	0.9999
Titanium content*Heat treatment method*Temperature*Amt of grain refiner	0.08125	1.24	0.2022	0.9405

Half Normal Plot

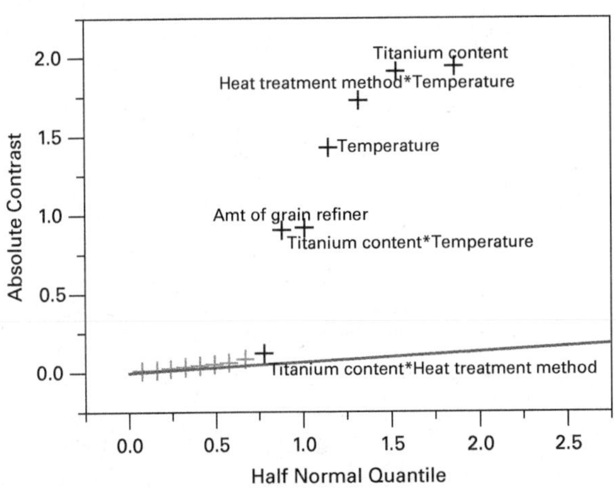

Lenth PSE = 0.06563
P-Values derived from a simulation of 10000 Lenth t ratios.

FIGURE 7.5 JMP Screening Output for the Alloy Experiment

a "real-time" simulation. This simulation assumes that none of the factors in the experiment is significant and calculates the observed value of the Lenth statistic 10,000 times for this null model. Then *P*-values are obtained by determining where the observed Lenth statistics fall relative to the tails of these simulation-based reference distributions. These *P*-values can be used as guidance in selecting factors for the model. The top portion of Figure 7.5 shows the JMP output from the screening analysis platform for the alloy experiment. When the factors are entered into the model, the Lenth procedure would recommend including the same factors in the model that we identified previously from the half-normal plot. Figure 7.6 presents the analysis of the final model. Temperature is a very important variable, both as a main effect and because it is involved in two of the active two-factor interactions. The Prediction Profiler at the bottom of Figure 7.6 indicates a set of process operating conditions that result in small crack lengths. The recommendations are to set temperature at the high level, titanium content at the low level, and use the second of the two heat treating methods. The mount of grain refiner at the above set of conditions on the first three factors seems to have a very small effect, but from a crack length viewpoint and perhaps an operating cost perspective, the lower level is best.

Response Crack Length
Actual by Predicted Plot

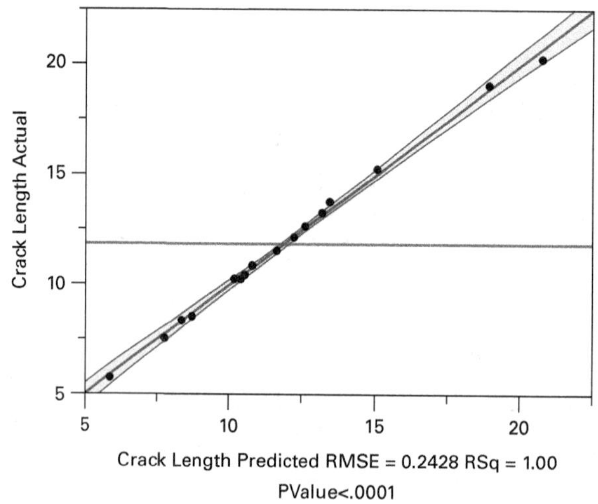

Effect Summary

Source	LogWorth	PValue
Titanium content	9.858	0.00000
Heat treatment method*Temperature	9.807	0.00000
Heat treatment method	9.409	0.00000 ^
Temperature	8.674	0.00000 ^
Amt of grain refiner	6.981	0.00000
Titanium content*Temperature	6.930	0.00000

Residual by Predicted Plot

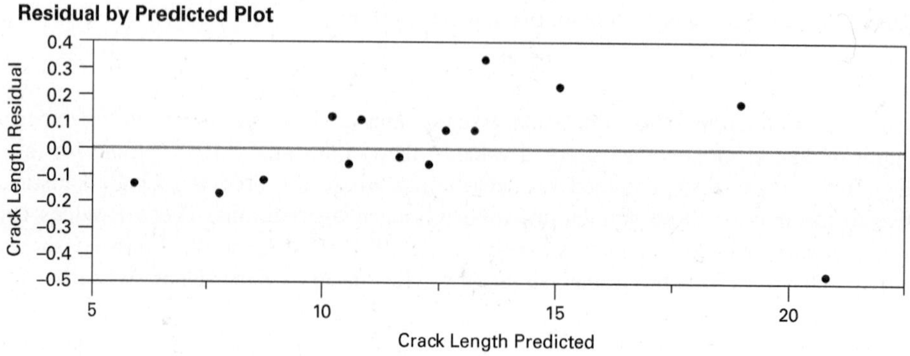

FIGURE 7.6 The Final Model for the Alloy Experiment

Studentized Residuals

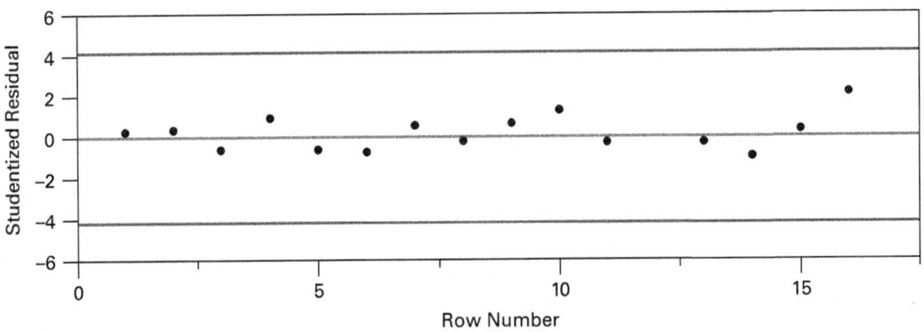

Externally Studentized Residuals with 95% Simultaneous Limits (Bonferroni)

Parameter Estimates

Term	Estimate	Std Error	t Ratio	Prob>|t|
Intercept	11.91875	0.060703	196.34	<.0001*
Titanium content	1.94375	0.060703	32.02	<.0001*
Heat treatment method	−1.73125	0.060703	−28.52	<.0001*
Temperature	1.43125	0.060703	23.58	<.0001*
Amt of grain refiner	0.91875	0.060703	15.14	<.0001*
Titanium content*Temperature	0.90625	0.060703	14.93	<.0001*
Heat treatment method*Temperature	−1.91875	0.060703	−31.61	<.0001*

Prediction Profiler

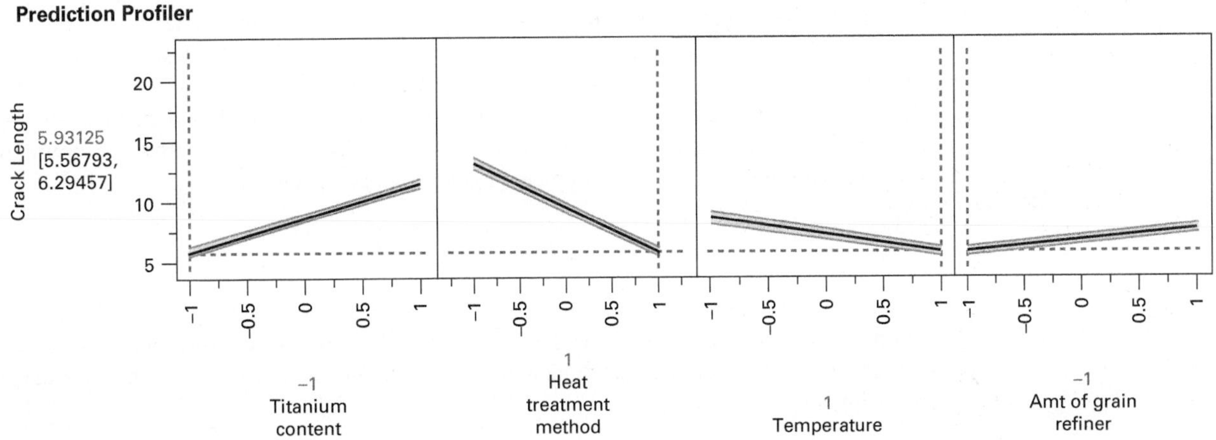

FIGURE 7.6 (*Continued*)

7.5 2^k Designs are Optimal Designs

Two-level factorial designs have many interesting and useful properties. For example, they are orthogonal designs, which makes interpretation of the results of the experiment simpler. They also have several **optimality** properties, which is the subject of this section. We will discuss and illustrate these properties using the 2^2 factorial design, but the results hold for all members of the 2^k full-factorial family.

The full-factorial model for the 2^2 design is

$$y = \beta_0 + \beta_1 x_1 + \beta_2 x_2 + \beta_{12} x_1 x_2 + \varepsilon$$

where x_1 and x_2 are the main effects of the two factors on the ± 1 scale and $x_1 x_2$ is the two-factor interaction. In matrix form, this model is

$$y = \mathbf{X}\beta + \varepsilon$$

where the model matrix is

$$\mathbf{X} = \begin{bmatrix} 1 & -1 & -1 & 1 \\ 1 & 1 & -1 & -1 \\ 1 & -1 & 1 & -1 \\ 1 & 1 & 1 & 1 \end{bmatrix}$$

assuming a single replicate (only four runs) of the design. The least squares estimates of the model parameters are

$$\hat{\beta} = (\mathbf{X}'\mathbf{X})^{-1}\mathbf{X}'\mathbf{y}$$

and

$$\mathbf{X}'\mathbf{X} = \begin{bmatrix} 4 & 0 & 0 & 0 \\ 0 & 4 & 0 & 0 \\ 0 & 0 & 4 & 0 \\ 0 & 0 & 0 & 4 \end{bmatrix}, \quad (\mathbf{X}'\mathbf{X})^{-1} = \begin{bmatrix} 1/4 & 0 & 0 & 0 \\ 0 & 1/4 & 0 & 0 \\ 0 & 0 & 1/4 & 0 \\ 0 & 0 & 0 & 1/4 \end{bmatrix}$$

The $\mathbf{X}'\mathbf{X}$ matrix (and its inverse) is diagonal because the design is orthogonal. If the design is replicated n times, the diagonal elements of $(\mathbf{X}'\mathbf{X})$ are $n2^k$. Now the determinant of $\mathbf{X}'\mathbf{X}$ and the variance of any model regression coefficient are $4^4 = 256$ and $\frac{\sigma^2}{4}$ respectively.

This is the maximum possible value of the determinant for a four-run design on the design space bounded by ± 1 and the smallest possible variance of any regression coefficient. Consequently, the 2^2 design (and all 2^k designs) is D-optimal. Furthermore, the sum of the main diagonal elements of $(\mathbf{X}'\mathbf{X})^{-1}$ is 1, which is the minimum possible value, so the 2^2 design is A-optimal.

Now consider the variance of the predicted response in the 2^2 design

$$V(\hat{y}) = V(\hat{\beta}_0 + \hat{\beta}_1 x_1 + \hat{\beta}_2 x_2 \hat{\beta}_{12} x_1 x_2)$$

The variance of the predicted response is a function of the point in the design space where the prediction is made (x_1 and x_2) and the variance of the model regression coefficients. The estimates of the regression coefficients are independent because the 2^2 design is orthogonal and they all have variance $\sigma^2/4$, so

$$\begin{aligned} V(\hat{y}) &= V(\hat{\beta}_0 + \hat{\beta}_1 x_1 + \hat{\beta}_2 x_2 \hat{\beta}_{12} x_1 x_2) \\ &= \frac{\sigma^2}{4}(1 + x_1^2 + x_2^2 + x_1^2 x_2^2) \end{aligned}$$

The maximum prediction variance occurs when $x_1 = x_2 = \pm 1$ and is equal to σ^2. To determine how good this is, we need to know the best possible value of prediction variance that we can attain. It turns out that the smallest possible value of the maximum prediction variance over the design space is $p\sigma^2/N$, where p is the number of model parameters and N is the number of runs in the design. The 2^2 design has $N = 4$ runs and the model has $p = 4$ parameters, so the model that we fit to the data from this experiment minimizes the maximum prediction variance over the design region. A design that has this property is called a G-optimal design. In general, 2^k designs are G-optimal designs for fitting the first-order model or the first-order model with interaction.

We can also evaluate the prediction variance at any point of interest in the design space. For example, when we are at the center of the design where $x_1 = x_2 = 0$, the prediction variance is

$\frac{\sigma^2}{4}$, and when we are at the point where $x_1 = 1$ and $x_2 = 0$ the prediction variance is $\frac{\sigma^2}{2}$. Instead of evaluating the prediction variance at a lot of points in the design space, it is better to consider the **average prediction variance** over the design space. One way to calculate this average prediction variance is

$$I = \frac{1}{A}\int_{-1}^{1}\int_{-1}^{1} V(\hat{\beta}_0 + \hat{\beta}_1 x_1 + \hat{\beta}_2 x_2 + \hat{\beta}_{12} x_1 x_2) dx_1 dx_2$$

where A is the area (or volume for more than two factors) of the design space. To compute the average, we are integrating the variance function over the design space and dividing by the area (volume) of the region.

Sometimes I is called the integrated variance or average variance criterion. Now for a 2^2 design, the area of the design region is $A = 4$ and $I = \frac{4\sigma^2}{9}$. It turns out that this is the smallest possible value of the average prediction variance that can be obtained from a four-run design used to fit a first-order model with interaction on this design space. A design with this property is called an I-optimal design. In general, 2^k designs are I-optimal designs for fitting the first-order model or the first-order model with interaction.

The JMP software will construct D, A, and I-optimal designs. This can be very useful in constructing designs when a standard design is not readily available.

7.6 More About Replication of 2^k Designs

This chapter has focused on the full 2^k factorial design. When we have illustrated a replicated version of these designs, we have assumed that the experimenter has employed a complete replicate. The 2^2, 2^3, and 2^4 designs with 4, 8, and 16 runs, respectively, that we have illustrated are popular because they employ all possible combinations of the levels of the factors. Such full-factorial designs allow for fitting models having an intercept, all the main effects, and two-way up to k-way interactions, where k is the number of factors. Frequently, practitioners limit consideration to either the main effects model or the model containing main effects and two-factor interactions (2FIs). This practice allows for using higher-order interaction effects as lack-of-fit degrees of freedom for estimating the error variance. There are two problems with this approach for small full-factorial designs. First, there are too few higher-order effects to provide a precise estimate of the error variance. Second, if one of the higher-order effects used to estimate the error variance is not negligible, then the resulting variance estimate will be too large.

Replicating runs of a full-factorial design avoids these problems by providing pure error degrees of freedom for estimating the error variance. Such estimates of the error variance are model independent. That is, they will not be biased as a result of missing a higher-order effect. Two considerations are important in choosing the number of runs to replicate. First, estimates of the error variance based on a small number of degrees of freedom are imprecise. A second consideration is the resulting power of the design to detect effects of a practically significant magnitude. While what constitutes practical significance is obviously problem dependent, we have remarked previously that in many industrial settings effect sizes of two standard deviations are large enough that experimenters should be interested in discovering them. Note that an effect size of 2σ is equivalent to a parameter size of one standard deviation in the underlying regression model for the design, assuming the usual -1 to $+1$ coding convention for the design factors.

Once an experimenter settles on the magnitude of effects of interest, the next issue is to decide what power is adequate. Adequate power could also certainly depend on the situation. We believe that a target power of 0.75–0.80 is a reasonable goal achievable without extensive experimentation.

Jones and Montgomery (2017) studied replication strategies for the 2^2, 2^3, and 2^4 designs with 4, 8, and 16 runs and provided useful recommendations. Jones and Montgomery (2017)

Table 7.4 **The 2^3 Design Augmented with Four Additional Runs**

Run	X1	X2	X3
*1	−1	−1	−1
*2	−1	−1	−1
*3	−1	−1	1
*4	−1	−1	1
5	−1	1	−1
*6	−1	1	1
*7	−1	1	1
8	1	−1	−1
*9	1	−1	1
*10	1	−1	1
11	1	1	−1
12	1	1	1

The rows marked with asterisks indicate replicated pairs of runs.

also investigated partial replication of the 2^4, considering augmentations of four, six, and eight runs. There are $C(16, 4) = 1820$ ways to replicate 4 of the 16 runs in the full-factorial design. The D-efficiencies of the resulting designs range from 96.2 to 96.7. A-efficiencies range from 93.19 to 93.81. We summarize their results here. They recommended that the 2^2 design be replicated three times to provide eight pure degrees of freedom for pure error. As observed earlier in this chapter, the power for detecting main effects and an interaction effect of two standard deviations is 0.857. Sixteen of the designs have both the maximum D- and A-efficiencies as well as have the same relative standard errors for all the main effects and 2FIs (0.231). We show one of the 16 designs in Table 7.5.

The more interesting cases are the 2^3 and the 2^4. Consider first the 2^3 design. Jones and Montgomery investigated augmentations of both two and four runs. Their recommendation was augmenting the basic design with four runs. There are $C(8, 4) = 70$ ways to choose four of the eight runs to replicate. All 70 have the same D-efficiency of 95.08 and A-efficiency of 90.32, and they can be found by using a D-optimal design construction tool. Jones and Montgomery (2017) recommended the unbalanced design shown in Table 7.4. The rows marked with asterisks indicate replicated pairs of runs. This design does not have orthogonal main effects. If only the main effects were active, their relative standard errors would be 0.293 compared to 0.289 for a balanced design. However, for estimating the interactions model, the unbalanced design provides an added benefit. The main effects all have the same relative standard error of 0.306, which is the same as for the balanced design. However, the 2FIs also have the same relative standard errors but they are smaller (0.302 versus 0.306). The standard error of the intercept increases from 0.289 to 0.302. The power for detecting any main effect of size two standard deviations is 0.743. The power for detecting any 2FI of the same size is 0.754. These are reasonable values for power.

Now consider the four-run augmentation of the 2^4 design shown in Table 7.5. Note that this design is not balanced; the first three columns (X1–X3) have 11 runs at the low level (−) and only 9 runs at the high level (+). The last column has 9 runs at the low level and 11 runs at the high level. Consequently, the design is not orthogonal. However, the absolute correlation for all pairs of main effect columns is only 0.0101. For a model consisting of only main effects, the variance inflation factors are 1.0003, so the nonorthogonality of the main effects has no practical consequence. Every main effect has absolute correlations with all the 2FIs of 0.1005. The correlation between 12 of the 15 pairs of 2FIs is zero. The other three pairs have absolute correlations of 0.2.

Table 7.5 **The 2^4 Design with Four Replicated Runs**

Run	X1	X2	X3	X4
*1	−1	−1	−1	−1
*2	−1	−1	−1	−1
3	−1	−1	−1	1
4	−1	−1	1	−1
*5	−1	−1	1	1
*6	−1	−1	1	1
7	−1	1	−1	−1
*8	−1	1	−1	1
*9	−1	1	−1	1
10	−1	1	1	−1
11	−1	1	1	1
12	1	−1	−1	−1
*13	1	−1	−1	1
*14	1	−1	−1	1
15	1	−1	1	−1
16	1	−1	1	1
17	1	1	−1	−1
18	1	1	−1	1
19	1	1	1	−1
20	1	1	1	1

Replicated pairs of rows are marked with asterisks.

The additional four runs result in power for detecting effects of size two standard deviations of 0.97. For detecting effect sizes of one standard deviation, the power is approximately 0.49. The four-run augmentation is recommended. In general, the authors concluded that although their work focuses on full-factorial designs, the basic methodology would apply for a partial replication of any design. As long as the base design has more than 12 runs, a partial replication of 4 runs should be sufficient to produce adequate power for detecting factor effects of at least two standard deviations in magnitude.

7.6.1 Adding Center Runs to a 2^k Design

Another strategy for adding replicate runs to a 2^k design instead of replicating all of the runs is to add between three and five runs at the center of the design region (all factors at the 0 level in coded units). This strategy usually assumes that all of the design factors are continuous so that a "middle" level exists, but it can be employed if some variables are categorical. In that case, the experimenter adds the center runs in the subspaces of the continuous factors at each level of the categorical variable. Experimenters often use this approach to obtain an estimate of pure error in single-replicate designs with four or more factors. This also allows the experimenter to test for curvature or nonlinearity in the response function over the design region. Comparing the average of the center runs to the average of all of the factorial runs with a two-sample t-test is a simple way to test for the presence of curvature. If the test of equal means indicates that there is curvature, this implies that the underlying model with only main effects and interactions is probably inadequate and a model with higher-order terms, such as quadratic, should be considered. Fitting such a model would require that the design be augmented with a sufficient number of runs to support a quadratic model.

7.7 Blocking in 2^k Designs

In many situations, it is impossible to perform all of the runs in a 2^k factorial experiment under homogeneous conditions. For example, a single batch of raw material might not be large enough to make all of the required runs. In other cases, it might be desirable to deliberately vary the experimental conditions to ensure that the treatments are equally effective (i.e. robust) across many situations that are likely to be encountered in practice. For example, a chemical engineer may run a pilot plant experiment with several batches of raw material because he knows that different raw material batches of different quality grades are likely to be used in the actual full-scale process. Blocking is a design technique that can be used in these situations.

We consider two cases. The first is the case of a replicated full-factorial 2^k design. If there are n replicates of the design, then each set of nonhomogeneous conditions defines a block, and each replicate is run in one of the blocks. The runs in each block (or replicate) would be made in random order. The analysis of the design is similar to that of any blocked experiment.

The second case is that of an unreplicated design. Let us assume that we require two blocks. We cannot assign each replicate to a block since there is only one replicate, so our only option is to divide the 2^k runs up between the two blocks in some manner that has as little negative impact on the conclusions that can be drawn as possible. The preferred approach is to use the minus and plus signs associated with the highest-order interaction effect to form the two blocks, with the minus runs associated with that factor in one block and the plus runs associated with that factor on the other block.

For example, consider the 2^3 design shown in Table 7.6. This table shows the basic design in columns A, B, and C, and it also shows the plus and minus signs that form the contrasts for estimating the interactions. From the table of plus and minus signs, notice that we have assigned the treatment combinations that are minus on ABC to block 1 and those that are plus on ABC to block 2. The resulting design is shown in Figure 7.7. Because the equation for computing the ABC interaction effect is now exactly the same as the equation for computing the block effect, the interaction is indistinguishable from the blocks. We say that the blocks and ABC are **confounded**.

This general idea is used to block any unreplicated 2^k design. By confounding the block with the highest-order interaction, a design with two blocks can be created. Confounding schemes that result in more than two blocks are also available. For discussion of this and some examples, refer to Montgomery (2020). Blocking schemes with restrictions on either the number of blocks or

Table 7.6 The 2^3 Design Showing All Contrasts for Estimating the Factor Effects

Treatment combination	Factorial effect								
	I	*A*	*B*	*AB*	*C*	*AC*	*BC*	*ABC*	Block
(1)	+	−	−	+	−	+	+	−	1
a	+	+	−	−	−	−	+	+	2
b	+	−	+	−	−	+	−	+	2
ab	+	+	+	+	−	−	−	−	1
c	+	−	−	+	+	−	−	+	2
ac	+	+	−	−	+	+	−	−	1
bc	+	−	+	−	+	−	+	−	1
abc	+	+	+	+	+	+	+	+	2

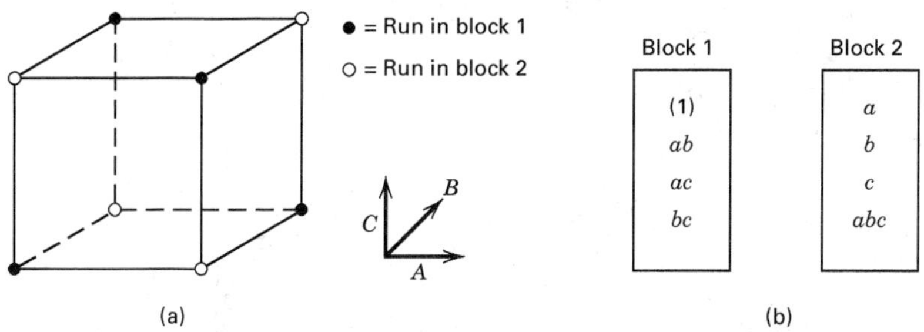

FIGURE 7.7 The 2^3 Design in Two Blocks with ABC Confounded: (a) Geometric View; (b) Assignment of the Eight Runs to Two Blocks

Table 7.7 The 2^4 Design in Eight Blocks of Size 3

X1	X2	X3	X4	Treatment combinations	Blocks
1	−1	−1	−1	a	1
−1	1	−1	1	bd	1
−1	−1	1	−1	c	1
−1	1	−1	−1	b	2
1	−1	1	1	acd	2
−1	1	1	−1	bc	2
−1	−1	1	1	cd	3
−1	−1	1	1	cd	3
1	1	−1	−1	ab	3
−1	−1	−1	1	d	4
−1	1	1	1	bcd	4
1	−1	−1	−1	a	4
−1	−1	1	−1	c	5
1	1	1	−1	abc	5
1	−1	−1	1	ad	5
1	−1	1	−1	ac	6
1	−1	1	1	acd	6
−1	1	−1	1	bd	6
1	1	−1	1	abd	7
−1	−1	−1	−1	(1)	7
1	1	1	1	abcd	7
−1	−1	−1	1	d	8
−1	1	1	−1	bc	8
1	1	1	1	abcd	8

their size can be found using an optimal design tool. To illustrate this situation, suppose that we must run a 2^4 design with 16 runs, but blocking is required, and the block size is restricted to three runs. If the block size was four runs, a standard design confounding two of the three-factor interactions is available. But since the block size is restricted to three runs, we will need to use an optimal design. The 24-run design in Table 7.7 was found by using the optimal design tool in JMP and the D-optimality criteria. The design has eight blocks, each of size 3, and is 96.15% D-efficient.

8 Screening Experiments

8.1 Introduction

This chapter introduces arguably the most common type of industrial experiment—a screening experiment. Investigators employ screening experiments in the early stages of process characterization when the relative importance of the factors is unknown. Generally, a brainstorming session involving the engineering team will generate a list of 5–12 or more potential factors of interest. The assumption that underpins screening is that only a few of these potential factors have sizeable effects on the response(s). The goal, then, of a screening experiment is to identify such few vitally important factors.

This begs the question, "what makes a factor important?" Clearly, having a large main effect suffices. Also, being involved in a large two-factor interaction would make both factors important. Finally, having a large nonlinear effect over the range of experimentation could make a difference affecting future experimentation.

For many traditional screening designs, all the factors have two levels only. This helps to keep the required number of runs manageable. In our experience, industrial screening experiments typically employ between 12 and 32 runs. With only two levels per factor, the models for screening experiments are limited to main effects and two-factor interactions.

The assumption that only a fraction (often less than a half) of the potential factors of interest are important is consistent with, or perhaps a restatement of, the **Pareto Principle**. According to Wikipedia, "The **Pareto principle** (also known as the **80/20** rule, the law of the vital few, or the **principle** of factor sparsity) states that, for many events, roughly 80% of the effects come from 20% of the causes." The screening assumption of factor sparsity is likely the most notable. However, there are a few other assumptions that have proven useful. First is the assumption of **effect hierarchy**. The hierarchy of effects assumption states that main effects tend to be more common and larger than second-order effects such as two-factor interactions and quadratic effects, which, in turn, tend to be larger and more common than three-factor interactions. Second is the assumption of **effect heredity**. The effects heredity assumption is that a two-factor interaction is most probable if both of its associated factors have significant main effects. Such models exhibit **strong heredity**. Much less probable is a two-factor interaction where only one of the associated main effects is significant. These models exhibit **weak heredity**. Least probable are two-factor interactions that present themselves when neither associated main effect is significant. Every very experienced experimenter has at least one story validating that this happens. However, it is important to remember that this happenstance is an exception that actually proves the rule. People tend to remember (and overemphasize) exceptional events.

Li et al. (2006) made a powerful argument supporting these three assumptions. They analyzed 113 sets of responses from full-factorial designs published in engineering journals having three to seven factors. They found that 170 of 410 or 41% of the main effects were active and that main effects averaged twice as large as two-factor interactions. This result is supportive of the effect sparsity and hierarchy assumptions. There were 569 potential two-factor interactions, of which 63 were significant. So, the percentage of active two-factor interactions was 11% overall. However, if both main effects were active, the conditional probability of a two-factor interaction involving these active main effects increased from 11% to 33%. This result supports the effect heredity assumption. The conditional probability for an active two-factor interaction having only one associated active main effect was 4.5%. This means that systems exhibiting weak heredity do exist but are not common. The conditional probability for an active two-factor interaction where neither of the associated main effects was active was a paltry 0.5%. So, though this kind of effect does occur, it is a comparatively rare event.

8.2 Regular Fractional Factorial Designs for Factor Screening

We observed in the previous chapter that as the number of factors in a 2^k factorial design increases, the number of runs required for a complete replicate of the design rapidly outgrows the resources of most experimenters. For example, a complete replicate of the 2^6 design requires 64 runs. In this design, only 6 of the 63 degrees of freedom correspond to main effects and only 15 degrees of freedom to two-factor interactions. Only 21 degrees of freedom are associated with effects that are likely to be of major interest. The remaining 42 degrees of freedom are associated with three-factor and higher-order interactions.

If the experimenter can reasonably assume that certain high-order interactions are negligible (that is, sparsity of effects), information on the main effects and low-order interactions may be obtained by running only a fraction of the complete factorial experiment. In this section, we present the class of 2^{k-p} **fractional factorial designs**. These are fractional factorials for k factors, which are $1/2^p$ fractions of the original full factorial. These are usually called one-half fractions ($p = 1$), one-quarter fractions ($p = 2$), one-eighth fractions ($p = 3$), and so on, and, historically, fractional factorial designs are among the most widely used for product and process design, process improvement, and industrial/business experimentation.

An important application of fractional factorials is in screening experiments—experiments in which many factors are considered and the objective is to identify the factors (if any) that have large effects. Screening experiments are usually performed in the early stages of a project when many of the factors initially considered likely have little or no effect on the response. The factors identified as important are then investigated more thoroughly in subsequent experiments.

As an example of a fractional factorial design, Table 8.1 shows the 2^{5-1}, a one-half fraction of the 2^5. This design was constructed by first writing down the full factorial that has the desired number of runs (in this case, the 16-run full 2^4 factorial) and then equating the levels of the fifth factor to the product of the signs in the first four columns (that is, the four-factor interaction of those main effects). The equation E = ABCD is called the **generator** of the fraction, and this approach to design construction is called the column generator approach. This design is D-optimal for either the main effects model or the model containing the main effects and any number of two-factor interactions, so it can also be constructed algorithmically.

Multiplying both sides of the column generator by E produces

$$\mathrm{E}^2 = \mathrm{ABCDE}$$

or

$$\mathrm{I} = \mathrm{ABCDE}$$

Since the square of column E contains only plus signs and by convention a column of all plus signs is called the **identity** column I. The relationship I = ABCDE is called the **defining relation** of the fraction.

Table 8.1 **The One-Half Fraction of the $2^5(2^{5-1})$ Design**

	Basic design					
Run	A	B	C	D	E = ABCD	Treatment combination
1	−	−	−	−	+	e
2	+	−	−	−	−	a
3	−	+	−	−	−	b
4	+	+	−	−	+	abe
5	−	−	+	−	−	c
6	+	−	+	−	+	ace
7	−	+	+	−	+	bce
8	+	+	+	−	−	abc
9	−	−	−	+	−	d
10	+	−	−	+	+	ade
11	−	+	−	+	+	bde
12	+	+	−	+	−	abd
13	−	−	+	+	+	cde
14	+	−	+	+	−	acd
15	−	+	+	+	−	bcd
16	+	+	+	+	+	abcde

This 16-run design has 15 degrees of freedom that can be used to estimate factor effects. However, with 5 factors, there are 5 main effects, 10 two-factor interactions, 10 three-factor interactions, 5 four-factor interactions, and 1 five-factor interaction. Clearly, not all of these effects can be uniquely estimated. Suppose that we estimate the five main effects. We can show that when we estimate the main effects we also estimate one of the four-factor interactions along with each main effect:

$$A = A + BCDE$$
$$B = B + ACDE$$
$$C = C + ABDE$$
$$D = D + ABCE$$
$$E = E + ABCD$$

Thus, each main effect estimate is **aliased** by a four-factor interaction. Notice that each alias is the four-factor interaction that does not contain that main effect. The alias relationships can be found from the defining relation for the design. Simply multiply each main effect by both sides of the defining relation by the main effect and simplify, as in the following:

$$A(I) = A(ABCDE)$$
$$A = A^2BCDE$$
$$A = BCDE$$

Now suppose that we estimate the two-factor interactions. The aliases can be found by the same technique. For example, the alias of the AB interaction is AB (I) = AB (ABCDE), or AB = A^2B^2CDE = CDE. It is easy to see that the alias of any two-factor interaction is the three-factor interaction that does not involve the two letters that define the original two-factor interaction.

The 2^{5-1} fractional factorial design is an example of **a resolution V** design, that is, a design in which the main effects are clear of **aliasing** or **confounding** with two-factor interactions and two-factor interactions are aliased or confounded with interactions of order three or higher. If cost is not an object, a resolution V design is ideal for factor screening, as in many of these studies interest is focused on main effects and two-factor interactions and these are estimated clear of any other effects assuming that interactions of order three and higher are not active. The design in Table 8.1 is a **principal fraction**. An **alternate** fraction could be obtained by switching the signs in column E. This alternate fraction would have a column generator E = −ABCD, and a defining relation I = −ABCDE. Using the alternate fraction simply changes the sign in the alias relationship; for example, in the alternate fraction the alias of A is A = −BCDE. The two fractions of the 2^5 with I = ±ABCDE are **regular** fractions, because the constants in the alias relationships are either ±1. This means that aliased terms are **completely confounded**. If two effects are confounded, there is no data driven way to determine which effect is causal.

There is no six-factor regular fractional factorial design that is resolution V. However, the six-factor 2^{6-1} design, which is a one-half fraction with 32 runs, is resolution VI. This design is constructed by starting with the 2^5 full factorial as the basic design and using the column generator F = ±ABCDE. The defining relation is I = ±ABCDEF, and each main effect is aliased with a single five-factor interaction. The two-factor interactions are aliased with one of the four-factor interactions. Similarly, there is no seven-factor resolution V design, but the 64-run seven-factor 2^{7-1} one-half fraction is resolution VII. This design starts with the full 2^6 as the basic design and using the column generator G = ±ABCDEF, resulting in the defining relation I = ±ABCDEFG.

The column generators that we have shown are selected to ensure that the designs created have the highest possible resolution for the specified number of runs. These designs are widely available in software. Most software packages default to the principal fraction of highest possible resolution when a regular fractional factorial design is selected.

As another example of a fractional factorial design, consider the eight-run design for five two-level factors in Table 8.2. This design was constructed using the column generator method starting with the full 2^3 factorial design and using the generators D = +ABC and E = +BC, or I = +ABCD and I = +BCE. The complete defining relation consists of I = +ABCD and I = +BCE and their product, or

$$I = +ABCD = +BCE = +ADE$$

The complete defining relation has three "words," so each effect will have three aliases. For example, the alias of A is A = +BCD = +ABCE = +DE, so the main effect of A is aliased with a two-factor interaction, DE. It is easy to see that every main effect is aliased with at least one two-factor interaction.

Table 8.2 The 2^{5-2} Resolution III Regular Fractional Factorial Design

A	B	C	D = +ABC	E = +BC
1	1	−1	−1	−1
1	−1	1	−1	−1
1	1	−1	1	−1
−1	−1	1	1	−1
−1	−1	−1	−1	1
−1	1	1	−1	1
1	−1	−1	1	1
1	1	1	1	1

This is an example of a **resolution III design**. In general, a resolution III design aliases at least one two-factor interaction with at least one main effect. The aliases of the design in Table 8.2 are

$$
\begin{aligned}
A &= +DE \\
B &= +CE \\
C &= +BE \\
D &= +AE \\
E &= +AD = +BC \\
AB &= +CD \\
AC &= +BD
\end{aligned}
$$

where we ignore interactions of order three and higher. All the column generators used in this example were positive, so this is a **principal fraction**. There are three other alternate fractions that are obtained by simply changing the signs on the column generators. This would produce corresponding sign changes on some of the aliases.

The resolution of a regular fractional factorial design can be found directly from the number of letters in the shortest word in the complete defining relation. The generators selected in this example guarantee that the design has the highest possible resolution. Note also that the three-word lengths in the defining relation for this this design have the **pattern** {4, 3 3}. This design minimizes the number of words in the defining relation that are of minimum length. Such a design is called a **minimum aberration design**. Minimizing aberration in a design of resolution R ensures that the design has the minimum number of main effects aliased with interactions of order $R-1$, the minimum number of two-factor interactions aliased with interactions of order $R-2$, and so on. It is a desirable property. When computer software is used to select a regular fractional factorial design, the generators are chosen to maximize resolution and minimize aberration.

Regular resolution III designs are available for 5, 6, and 7 factors in 8 runs and for 9–15 factors in 16 runs. The column generators for these designs are shown in Table 8.3. These generators ensure that the resulting deigns maximize resolution and minimize aberration. In some cases, the generators are not unique; that is, a different set of generators will produce a design having the same resolution and aberration.

As another example of a fractional factorial design, consider the one-half fraction of the 2^4 (a 2^{4-1} design) in Table 8.4. This design was constructed using the column generator D = ABC, so the defining contrast is I = ABCD. Therefore, the design is resolution IV. In resolution IV designs, main effects are aliased with three-factor and higher-order interactions, and two-factor interactions are aliased with each other. For example, in the design in Table 8.4, the alias of the main effect A is A = BCD and the alias of the AB interaction is AB = CD.

The 2^{4-1} is the only resolution IV fractional factorial with eight runs. Resolution IV fractions with 16 runs are available for 6, 7, and 8 factors. The 2^{9-3} design, the 2^{10-4} and the 2^{11-5} design with 64 runs and the 2^{7-3} design, the 2^{8-3} design, the 2^{9-4} design, the 2^{10-5} design, and the 2^{11-6} design with 32 runs are resolution IV designs. The generators for these designs are shown in Table 8.5. These generators assure maximum resolution and minimum aberration, although other choices of generators will produce equivalent designs.

8.2.1 A General Method for Finding the Alias Relationships in Fractional Factorial Designs

We have shown how to find the alias relationships in a 2^{k-p} fractional factorial design by using the complete defining relation. This method works well in regular fractional factorial designs,

Table 8.3 **Regular Resolution III Designs in 8 and 16 Runs**

Number of factors	Fraction	Number of runs	Column generators
5	2^{5-2}	8	D = ±ABC, E = ±BC
6	2^{6-3}	8	D = ±AB, E = ±AC, F = ±BC
7	2^{7-4}	8	D = ±AB, E = ±AC, F = ±BC, G = ±ABCD
8	2^{8-4}	16	E = ±BCD, F = ±ACD G = ±ABC, H = ±ABD
9	2^{9-5}	16	E = ±ABC, F = ±BCD G = ±ACD, H = ±ABD, J = ±ABCD
10	2^{10-6}	16	E = ±ABC, F = ±BCD G = ±ACD, H = ±ABD J = ±ABCD, K = ±AB
11	2^{11-7}	16	F = ±BCD, G = ±ACD H = ±ABD, J = ±ABCD K = ±AB, L = ±AC
12	2^{12-8}	16	E = ±ABC, F = ±ABD G = ±ACD, H = ±BCD J = ±ABCD, K = ±AB L = ±AC, M = ±AD
13	2^{13-9}	16	E = ±ABC, F = ±ABD G = ±ACD, H = ±BCD J = ±ABCD, K = ±AB L = ±AC, M = ±AD N = ±BC
14	2^{14-10}	16	E = ±ABC, F = ±ABD G = ACD, H = ±BCD J = ±ABCD, K = ±AB L = ±AC, M = ±AD N = ±BC, O = ±BD
15	2^{15-11}	16	E = ±ABC, F = ±ABD G = ±ACD, H = ±BCD J = ±ABCD, K = ±AB L = ±AC, M = ±AD N = ±BC, O = ±BD P = ±CD

Table 8.4 **The 2^{4-1} Fractional Factorial Design**

	Basic design			
Run	**A**	**B**	**C**	**D = ABC**
1	−	−	−	−
2	+	−	−	+
3	−	+	−	+
4	+	+	−	−
5	−	−	+	+
6	+	−	+	−
7	−	+	+	−
8	+	+	+	+

Table 8.5 Some Regular Resolution IV Fractional Factorial Designs

Number of factors	Fraction	Number of runs	Column generators
4	2^{4-1}	8	D = ±ABC
6	2^{6-2}	16	E = ±ABC, F = ±BCD
7	2^{7-2}	32	F = ±ABCD, G = ±ABDE
7	2^{7-3}	16	E = ±ABC, F = ±BCD G = ±ACD
8	2^{8-3}	32	F = ±ABC, G = ±ABD H = ±BCDE
8	2^{8-4}	16	E = ±BCD, F = ±ACD G = ±ABC, H = ±ABD
9	2^{9-3}	64	G = ±ABCD. H = ±ACEF J = ±CDEF
9	2^{9-4}	32	F = ±BCDE, G = ±ACDE H = ±ABDE, J = ±ABCE
10	2^{10-4}	64	G = ±BCDF, H = ±ACDF J = ±ABDE, K = ±ABCE
10	2^{10-5}	32	F = ±ABCD, G = ±ABCE H = ±ABDE, J = ±ACDE K = ±BCDE
11	2^{11-5}	64	G = ±CDE, H = ±ABCD J = ±ABF, K = ±BDEF L = ±ADEF
11	2^{11-6}	32	F = ±ABC, G = ±BCD H = ±CDE, J = ±ACD K = ±ADE, L = ±BDE

but it does not work in other cases, such as some of the nonregular fractions that we will discuss subsequently. Furthermore, there are some fractional factorials that do not have defining relations.

Fortunately, there is a general method available that works satisfactorily in many situations. The method uses the polynomial or regression model representation of the model, say

$$y = \mathbf{X}_1\boldsymbol{\beta}_1 + \varepsilon$$

where $\boldsymbol{y}$ is an $n \times 1$ vector of the responses, $\mathbf{X}_1$ is an $n \times p$ matrix containing the design matrix expanded to the form of the model that the experimenter is fitting, $\boldsymbol{\beta}_1$ is a $p \times 1$ vector of the model parameters, and ε is an $n \times 1$ vector of errors. The least squares estimate of $\boldsymbol{\beta}_1$ is

$$\widehat{\beta}_1 = (\mathbf{X}_1'\mathbf{X}_1)^{-1}\mathbf{X}_1'\boldsymbol{y}$$

Suppose that the **true** model is

$$y = \mathbf{X}_1\boldsymbol{\beta}_1 + \mathbf{X}_2\boldsymbol{\beta}_2 + \boldsymbol{\varepsilon}$$

where $\mathbf{X}_2$ is an $n \times p_2$ matrix containing additional variables that are not in the model and $\boldsymbol{\beta}_2$ is a $p_2 \times 1$ vector of the parameters associated with these variables. It can be shown that

$$\begin{aligned} E(\widehat{\beta}_1) &= \boldsymbol{\beta}_1 + (\mathbf{X}_1'\mathbf{X}_1)^{-1}\mathbf{X}_1'\mathbf{X}_2\boldsymbol{\beta}_2 \\ &= \boldsymbol{\beta}_1 + \mathbf{A}\boldsymbol{\beta}_2 \end{aligned}$$

The matrix **A** is the **alias matrix** introduced in Chapter 4. The elements of this matrix operating on $\boldsymbol{\beta}_2$ identify the alias relationships for the parameters in the vector $\boldsymbol{\beta}_1$.

We illustrate the application of this procedure with a simple example. Suppose that we have conducted a 2^{3-1} design with defining relation I = ABC. The model that the experimenter plans to fit is the main-effects-only model

$$y = \beta_0 + \beta_1 x_1 + \beta_2 x_2 + \beta_3 x_3 + \varepsilon$$

In the notation defined above

$$\boldsymbol{\beta}_1 = \begin{bmatrix} \beta_0 \\ \beta_1 \\ \beta_2 \\ \beta_3 \end{bmatrix} \quad \text{and} \quad \mathbf{X}_1 = \begin{bmatrix} 1 & -1 & -1 & 1 \\ 1 & 1 & -1 & -1 \\ 1 & -1 & 1 & -1 \\ 1 & 1 & 1 & 1 \end{bmatrix}$$

Suppose that the true model contains all the two-factor interactions, so that

$$y = \beta_0 + \beta_1 x_1 + \beta_2 x_2 + \beta_3 x_3 + \beta_{12} x_1 x_2 + \beta_{13} x_1 x_3 + \beta_{23} x_2 x_3 + \varepsilon$$

and

$$\boldsymbol{\beta}_2 = \begin{bmatrix} \beta_{12} \\ \beta_{13} \\ \beta_{23} \end{bmatrix} \quad \text{and} \quad \mathbf{X}_2 = \begin{bmatrix} 1 & -1 & -1 \\ -1 & -1 & 1 \\ -1 & 1 & -1 \\ 1 & 1 & 1 \end{bmatrix}$$

Now

$$\mathbf{X}_1'\mathbf{X}_1 = 4\,\mathbf{I}_4 \quad \text{and} \quad \mathbf{X}_1'\mathbf{X}_2 = \begin{bmatrix} 0 & 0 & 0 \\ 0 & 0 & 4 \\ 0 & 4 & 0 \\ 4 & 0 & 0 \end{bmatrix}$$

Therefore,

$$(\mathbf{X}_1'\mathbf{X}_1)^{-1} = \frac{1}{4}\mathbf{I}_4$$

and

$$E(\hat{\boldsymbol{\beta}}_1) = \boldsymbol{\beta}_1 + \mathbf{A}\boldsymbol{\beta}_2$$

$$E\begin{bmatrix} \hat{\beta}_0 \\ \hat{\beta}_1 \\ \hat{\beta}_2 \\ \hat{\beta}_3 \end{bmatrix} = \begin{bmatrix} \beta_0 \\ \beta_1 \\ \beta_2 \\ \beta_3 \end{bmatrix} + \frac{1}{4}\mathbf{I}_4 \begin{bmatrix} 0 & 0 & 0 \\ 0 & 0 & 4 \\ 0 & 4 & 0 \\ 4 & 0 & 0 \end{bmatrix} \begin{bmatrix} \beta_{12} \\ \beta_{13} \\ \beta_{23} \end{bmatrix}$$

$$= \begin{bmatrix} \beta_0 \\ \beta_1 \\ \beta_2 \\ \beta_3 \end{bmatrix} + \begin{bmatrix} 0 & 0 & 0 \\ 0 & 0 & 1 \\ 0 & 1 & 0 \\ 1 & 0 & 0 \end{bmatrix} \begin{bmatrix} \beta_{12} \\ \beta_{13} \\ \beta_{23} \end{bmatrix}$$

$$= \begin{bmatrix} \beta_0 \\ \beta_1 \\ \beta_2 \\ \beta_3 \end{bmatrix} + \begin{bmatrix} 0 \\ \beta_{23} \\ \beta_{13} \\ \beta_{12} \end{bmatrix}$$

$$= \begin{bmatrix} \beta_0 \\ \beta_1 + \beta_{23} \\ \beta_2 + \beta_{13} \\ \beta_3 + \beta_{12} \end{bmatrix}$$

The interpretation of this, of course, is that each of the main effects is aliased with one of the two-factor interactions, which we know to be the case for this design. Notice that every row of the alias matrix represents one of the factors in β_1 and every column represents one of the factors in β_2. While this is a very simple example, the method is very general and can be applied to much more complex designs.

Software can display the alias matrix for fractional factorial designs. The display below shows the alias matrix from JMP for the 2^{3-1} fractional factorial design with I = ABC. Notice that as we saw above that every main effect is aliased with the two-factor interaction with which it is not involved. Additionally, we see that the intercept estimate is biased by the three-factor interaction ABC.

Alias matrix

Effect	A*B	A*C	B*C	A*B*C
Intercept	0	0	0	1
A	0	0	1	0
B	0	1	0	0
C	1	0	0	0

8.2.2 Dealiasing Effects

When analyzing a regular fractional factorial design, it is not unusual to find that one or more factor effects of potential interest are aliased. In regular designs, aliased effects are completely confounded, and their individual effects cannot be estimated from the data supplied by the design. For example, in the 2^{5-2} fractional factorial design in Table 8.2, suppose that the largest effect estimates are for A, B, C, and the AB interaction. A relatively simple interpretation of these results is that these are the active factors, but AB is aliased with CD, so it is certainly possible that the large AB effect is attributable to either AB or CD or to both AB and CD. It is possible to dealias effects in situations like this by augmenting the original fraction with additional runs.

A simple type of augmentation scheme that can be very effective for some resolution III designs is **foldover**. A **full** foldover of a resolution III design consists of performing an additional group of runs in which the sign of every run in the original fraction is reversed. This produces a combined design with $2(2^{k-p}) = 2^{k-p+1}$ runs. In this combined set of runs, all main effects are completely separated from their two-factor interaction aliases. This procedure results in a combined design that is now resolution IV. Unfortunately, this method will not be useful here, because we have a pair of aliased two-factor interactions, so after a full foldover they will still be aliased. Also, the full foldover technique will only work with resolution III designs. There are variations of the foldover technique that can be useful. The **single-factor foldover** reverses signs in a single column only. This can be useful in resolution IV designs to dealias all of the two-factor interactions associated with that column. A significant disadvantage of foldovers is that **they require a follow-on experiment that is as large as the original one. Partial foldovers** can alleviate that problem, as they use only half of the runs associated with a full fold-over. For a complete discussion of the foldover technique and some examples, see Montgomery (2020).

Alternatively, the runs to add to a fractional factorial design to dealias terms of interest can be determined by using a D-optimal augmentation procedure. This uses the D-criterion to determine the set of runs to add to the original fraction to fit the desired model, which would not include the terms originally aliased. The output from JMP illustrating this procedure, where we have elected to add four runs to the original eight-run fraction and to place the additional four runs in a separate block, is shown in Figure 8.1. The alias matrix indicates that AB and CE are no longer

Augment Design

Design

Run	A	B	C	D	E	Block
1	1	1	−1	−1	−1	1
2	1	−1	1	−1	−1	1
3	−1	1	−1	1	−1	1
4	−1	−1	1	1	−1	1
5	−1	−1	−1	−1	1	1
6	−1	1	1	−1	1	1
7	1	−1	−1	1	1	1
8	1	1	1	1	1	1
9	−1	−1	1	−1	1	2
10	−1	1	1	1	−1	2
11	1	−1	−1	1	−1	2
12	1	1	−1	−1	1	2

Design Evaluation

Power Analysis

Significance Level	0.05
Anticipated RMSE	1

Term	Anticipated Coefficient	Power
Intercept	1	0.597
A	1	0.597
B	1	0.597
C	1	0.597
D	1	0.573
E	1	0.573
A*B	1	0.573
C*E	1	0.597
Block	1	0.597

Alias Matrix

Effect	A*C	A*D	A*E	B*C	B*D	B*E	C*D	D*E	A*B	C*E
Intercept	−0.5	0	0	0	0	0	0	−0.5	0	0
A	0	0	0	0	−0.25	0.5	0	0.75	0	0
B	0	−0.5	0.5	0	0	0	0.5	0	0	0
C	0	0	0	0	0.25	0.5	0	0.25	0	0
D	0	0.2	0.8	0.4	0	0	0.2	0	0	0
E	0	0.8	0.2	0.6	0	0	−0.2	0	0	0
A*B	0	−0.2	0.2	−0.4	0	0	0.8	0	1	0
C*E	0	0.5	−0.5	0	0	0	−0.5	0	0	1
Block	0.5	0	0	0	0	0	0	0.5	0	0

FIGURE 8.1 JMP Output for Design Augmentation

completely confounded, and that all of the main effects are partially confounded with two-factor interactions. The power to detect effect sizes of two standard deviations is approximately 0.6. This is relatively low power, but about all one can expect from an eight-run design.

8.3 Nonregular Orthogonal Designs

Over the last decade or so massive computer searches have discovered many different designs with orthogonal main effects. These designs exist for numbers of runs that are multiples of four.

In Section 8.2, we introduced the regular fractional factorial designs with orthogonal main effects. The last design in Table 8.3 is the 2^{15-11} fractional factorial design. The main effects are uncorrelated in this design. Along with all the other designs in Table 8.3, it is D-optimal for the main effects model. However, every main effect is confounded with seven different two-factor interactions.

From the bottom row of Table 8.6, we see that there are five different 16-run designs with 15 factors that have orthogonal main effects. Each of these designs is also D-optimal for the main effects model. Since there are multiple D-optimal designs, we can look for a secondary criterion to choose the best of the five designs. It turns out that one of these five has much less confounding than the regular fractional factorial design. The first seven factors are each confounded with three two-factor interactions. The remaining factors are correlated but not confounded with several two-factor interactions. We call an orthogonal main effects plan a nonregular orthogonal design when some factorial effects are correlated rather than confounded.

We would not recommend running either of the above designs unless budget constraints were extreme. There are no degrees of freedom for error, so any estimate of variance would have to come from pooling the small main effect estimates. However, if we were forced to choose, we would pick the nonregular design.

Each of the orthogonal designs in Table 8.6 is a projection of one of the five orthogonal designs with 15 factors. That is, each of these designs can be found by choosing specific columns from one of these five designs. Also, every design in Table 8.6 is D-optimal for the main effects model.

Of special interest are the designs with six, seven, or eight factors. Jones and Montgomery (2010) identified alternatives to the resolution IV designs where none of the two-factor interactions is confounded with each other. Correlations between main effects and two-factor interactions and between pairs of two-factor interactions are ± 0.5.

Table 8.7 displays the factor setting for a six-factor 16-run nonregular orthogonal design. It is one of the 27 six-factor orthogonal designs in Table 8.6. The alias matrix for this design is given in Table 8.8.

Table 8.6 The Number of Statistically Distinguishable Orthogonal Designs for 6–15 Factors in 16 Runs

Number of factors	Number of orthogonal designs
6	27
7	55
8	80
9	87
10	78
11	58
12	36
13	18
14	10
15	5

Table 8.7 Six-factor 16-run Design with No Confounding of Two-factor Interactions

Run	A	B	C	D	E	F
1	+	+	+	+	+	+
2	+	+	+	−	−	−
3	+	−	−	+	+	−
4	+	−	−	−	−	+
5	+	+	−	+	−	+
6	+	+	−	−	+	−
7	+	−	+	+	−	−
8	+	−	+	−	+	+
9	−	+	+	+	−	+
10	−	+	+	−	+	−
11	−	−	−	+	−	−
12	−	−	−	−	+	+
13	−	+	−	+	+	−
14	−	+	−	−	−	+
15	−	−	+	+	+	+
16	−	−	+	−	−	−

Note that factor A is not aliased with any two-factor interaction. Factors B to F are each aliased with a pair of two-factor interactions, while factor F is aliased with four two-factor interactions. The resolution IV fractional factorial design has no aliasing of any main effect by a two-factor interaction, so one might wonder why we are interested in the nonregular orthogonal design in Table 8.7. The answer lies in the correlations between the pairs of interaction columns of both designs. In the resolution IV design, eight pairs of two-factor interactions are confounded. If any of these interaction contrasts is active, then the result of the data analysis will be ambiguous. This is because the observed effect could be caused by either a two-factor interaction acting alone or by any linear combination of the aliased two-factor interactions. For the nonregular design, 12 of the 105 pairs of two-factor interactions have correlations of ± 0.5. The remaining pairs of two-factor interactions are uncorrelated. In the analysis, if only a few two-factor interactions are active, it is possible to identify them without ambiguity. In this case, we would prefer the nonregular orthogonal design.

Table 8.8 Aliasing of Main Effects by Two-factor Interactions for the Design in Table 8.7

Alias matrix

Effect	A*B	A*C	A*D	A*E	A*F	B*C	B*D	B*E	B*F	C*D	C*E	C*F	D*E	D*F	E*F
Intercept	0	0	0	0	0	0	0	0	0	0	0	0	0	0	0
A	0	0	0	0	0	0	0	0	0	0	0	0	0	0	0
B	0	0	0	0	0	0	0	0	0	0	0	0	0	0.5	−0.5
C	0	0	0	0	0	0	0	0	0	0	0	0	0	0.5	0.5
D	0	0	0	0	0	0	0	0	0.5	0	0	0.5	0	0	0
E	0	0	0	0	0	0	0	0	−0.5	0	0	0.5	0	0	0
F	0	0	0	0	0	0	0.5	−0.5	0	0.5	0.5	0	0	0	0

In Section 8.2, it was mentioned that there is no resolution V regular fractional factorial design for seven factors. The 32-run regular fractional factorial design is resolution IV. It is impossible to estimate the main effects plus two-factor interactions model with this design. To fit the main effects plus two-factor interactions model with a regular fractional factorial design requires 64 runs. However, the design in Table 8.9 is a nonregular orthogonal design in 32 runs that can fit the main effects plus two-factor interactions model.

The largest absolute correlation between any pair of effects in the model is only 0.25. We would again prefer this nonregular orthogonal design to any regular alternative for seven factors.

Table 8.9 Six Factor 32-run Orthogonal Main Effects Design Able to Estimate All Two-factor Interactions

Run	A	B	C	D	E	F	G
1	−	−	−	−	−	−	−
2	−	−	−	−	−	−	+
3	−	−	−	−	+	+	−
4	−	−	−	+	−	+	+
5	−	−	−	+	+	−	+
6	−	−	+	−	−	+	+
7	−	−	+	−	+	−	−
8	−	−	+	+	−	+	−
9	−	+	−	−	+	+	+
10	−	+	−	+	−	−	−
11	−	+	−	+	+	+	−
12	−	+	+	−	−	+	−
13	−	+	+	−	+	−	+
14	−	+	+	+	−	−	+
15	−	+	+	+	+	−	−
16	−	+	+	+	+	+	+
17	+	−	−	−	+	+	+
18	+	−	−	+	−	−	−
19	+	−	−	+	+	+	−
20	+	−	+	−	−	+	−
21	+	−	+	−	+	−	+
22	+	−	+	+	−	−	+
23	+	−	+	+	+	−	−
24	+	−	+	+	+	+	+
25	+	+	−	−	−	−	+
26	+	+	−	−	−	+	+
27	+	+	−	−	+	−	−
28	+	+	−	+	−	+	−
29	+	+	−	+	+	−	+
30	+	+	+	−	−	−	−
31	+	+	+	−	+	+	−
32	+	+	+	+	−	+	+

Table 8.10 **Factor Settings for the 11-factor 12-run Plackett–Burman design**

Factor settings										
+	+	+	+	+	+	+	+	+	+	+
−	+	−	+	+	+	−	−	−	+	−
−	−	+	−	+	+	+	−	−	−	+
+	−	−	+	−	+	+	+	−	−	−
−	+	−	−	+	−	+	+	+	−	−
−	−	+	−	−	+	−	+	+	+	−
−	−	−	+	−	−	+	−	+	+	+
+	−	−	−	+	−	−	+	−	+	+
+	+	−	−	−	+	−	−	+	−	+
+	+	+	−	−	−	+	−	−	+	−
−	+	+	+	−	−	−	+	−	−	+
+	−	+	+	+	−	−	−	+	−	−

The well-known Plackett–Burman designs are nonregular orthogonal designs for numbers of runs that are multiples of four but not powers of two. The designs with 12, 20, 24, or 28 runs are D-optimal for the main effects model for any number of factors less than the number of runs in the design. The 12-run Plackett–Burman design is shown in Table 8.10. There are several six-factor projections of the 24-run Plackett–Burman design that can fit the main effects plus two-factor interactions model. Jones, Schoen, and Montgomery (2016) pointed out that the most efficient nonregular orthogonal design for fitting the main effects plus two-factor interactions model is not a projection of the 24-run Plackett–Burman design.

Except for the 12-run Plackett–Burman design, there are always more efficient nonregular orthogonal designs for fitting some combination of main effects and two-factor interactions. We would recommend the use of the 12-run Plackett–Burman design for five to eight factors, if most or all the factors are categorical and runs are expensive.

D-optimal designs of five to eight factors result from using any five to eight columns of the design in Table 8.10.

8.4 Nonorthogonal Screening Designs

In Section 8.3, we mentioned that there are often many D-optimal main effects designs for numbers of runs that are multiples of four and numbers of factors that are fewer than the number of runs. Many of these designs are plagued by correlations between main effects and two-factor interactions. When a two-factor interaction that is correlated with a main effect has a substantial effect, then the main effect estimate will be biased, which is undesirable.

In some cases, it may be acceptable to allow the main effects to have small correlations in exchange for removing any aliasing between main effects and two-factor interactions. Jones and Nachtsheim (2011) addressed this issue and proposed a constrained optimality criterion that explicitly trades D-efficiency for a reduction in aliasing. The optimization is

$$min_{d \in D} Tr[\mathbf{A}(d)'\mathbf{A}(d)] \text{ subject to } D_e(d) \geq l_d$$

where $\mathbf{A}$ is the alias matrix of design d for a group of potential terms, $D_e(d)$ is the D-efficiency of d, and l_d is an acceptable lower bound on the efficiency. Designs created using the above optimization are called minimum aliasing designs.

Table 8.11 Alias Optimal Design with Six Factors and 12 Runs

Design

Run	X1	X2	X3	X4	X5	X6
1	−1	−1	1	1	1	1
2	1	1	1	1	−1	1
3	−1	1	−1	1	1	1
4	−1	1	1	1	1	−1
5	1	1	−1	−1	−1	−1
6	1	−1	1	−1	−1	−1
7	−1	−1	−1	−1	1	−1
8	1	−1	−1	−1	−1	1
9	−1	−1	−1	1	−1	−1
10	1	1	1	−1	1	1
11	−1	1	1	−1	−1	1
12	1	−1	−1	1	1	−1

Table 8.12 Alias Matrix of Alias Optimal Design with Six Factors and 12 Runs

Alias matrix

Effect	X1*X2	X1*X3	X1*X4	X1*X5	X1*X6	X2*X3	X2*X4	X2*X5	X2*X6	X3*X4	X3*X5	X3*X6	X4*X5	X4*X6	X5*X6
Intercept	0	0	−0.33	−0.33	0	0.333	0	0	0.333	0	0	0.333	0.333	0	0
X1	0	0	0	0	0	0	0	0	0	0	0	0	0	0	0
X2	0	0	0	0	0	0	0	0	0	0	0	0	0	0	0
X3	0	0	0	0	0	0	0	0	0	0	0	0	0	0	0
X4	0	0	0	0	0	0	0	0	0	0	0	0	0	0	0
X5	0	0	0	0	0	0	0	0	0	0	0	0	0	0	0
X6	0	0	0	0	0	0	0	0	0	0	0	0	0	0	0

As an alternative to the Plackett–Burman design of 6 factors and 12 runs where each main effect has aliases of $\pm\frac{1}{3}$ with 10 two-factor interactions, they found the design given in Table 8.11. They generated the design by specifying a main effects model as the *a priori* model and all the two-factor interactions as aliasing terms. Their lower bound for D-efficiency was 90%. The design in Table 8.11 has a D-efficiency of 91.8% where the maximum possible D-efficiency is 100%.

Its alias matrix appears in Table 8.12.

This design is nonorthogonal as each factor has correlations of $\pm\frac{1}{3}$ with two other factors. The variance inflation factor (VIF) is a statistic that is a measure of nonorthogonality. The VIF for every factor in the minimal aliasing design is 1.2, which means that the variance is 20% larger than it would be if the design were orthogonal. The VIF for every term in a Plackett–Burman design is 1.0.

For comparison, the alias matrix of the Plackett–Burman design is shown in Table 8.13.

Here the trade-off is a 20% increase in the variance of each main effect in exchange for the complete removal of aliasing between main effects and two-factor interactions.

The minimal aliasing design has an undesirable characteristic. Eighteen of the pairs of two-factor interactions have absolute correlations greater than 0.7. This could lead to substantial variance inflation if we need to add some of these effects to the model.

Table 8.13 Alias Matrix of Plackett–Burman Design with Six Factors and 12 Runs

Alias matrix

Effect	X1*X2	X1*X3	X1*X4	X1*X5	X1*X6	X2*X3	X2*X4	X2*X5	X2*X6	X3*X4	X3*X5	X3*X6	X4*X5	X4*X6	X5*X6
Intercept	0	0	0	0	0	0	0	0	0	0	0	0	0	0	0
X1	0	0	0	0	0	0.333	−0.33	0.333	0.333	−0.33	0.333	−0.33	−0.33	0.333	0.333
X2	0	0.333	−0.33	0.333	0.333	0	0	0	0	0.333	−0.33	0.333	0.333	0.333	0.333
X3	0.333	0	−0.33	0.333	−0.33	0	0.333	−0.33	0.333	0	0	0	0.333	0.333	−0.33
X4	−0.33	−0.33	0	−0.33	0.333	0.333	0	0.333	0.333	0	0.333	0.333	0	0	−0.33
X5	0.333	0.333	−0.33	0	0.333	−0.33	0.333	0	0.333	0.333	0	−0.33	0	−0.33	0
X6	0.333	−0.33	0.333	0.333	0	0.333	0.333	0.333	0	0.333	−0.33	0	−0.33	0	0

Table 8.14 Six-factor 12-run EFD

Design

Run	A	B	C	D	E	F
1	1	−1	−1	−1	−1	1
2	−1	−1	−1	−1	−1	−1
3	1	−1	−1	−1	1	−1
4	−1	−1	1	1	1	1
5	1	−1	1	−1	−1	−1
6	−1	1	1	−1	1	1
7	−1	1	1	1	1	−1
8	1	1	1	1	1	1
9	−1	1	1	1	−1	1
10	1	1	−1	−1	−1	−1
11	−1	1	−1	1	1	1
12	1	−1	−1	1	−1	−1

Errore et al. (2017) found a way to avoid aliasing of main effects and two-factor interactions and at the same time minimizing the maximum absolute correlation of pairs of two-factor interactions. They call these design-efficient foldover designs (EFDs). Table 8.14 shows their EFD for 6 factors and 12 runs. Figure 8.2 is the correlation cell plot.

To have orthogonal main effects, the number of runs in a two-level screening design must be a multiple of four. Choosing any seven columns of a 12-run Plackett–Burman design yields a seven-factor orthogonal and D-optimal design for the main effects model. However, the power to detect a main effect that is twice the error standard deviation is 0.737 for every effect. Suppose there is a mandate that every model coefficient in a designed experiment must have a power to detect a 2σ effect that is greater than or equal to 0.8. One way to increase the power is to add runs to an experiment. A 16-run experiment adds 33% to the cost of the runs, which may not be acceptable.

A 14-run D-optimal design is shown in Table 8.15.

This design is not orthogonal, but every main effect is estimated 5.4% more efficiently than the orthogonal Plackett–Burman design with 12 runs. Also, the power to detect each main effect greater in magnitude than 2σ in this design is 0.858. So, though not orthogonal, this D-optimal design supplies the desired power at a minimum cost.

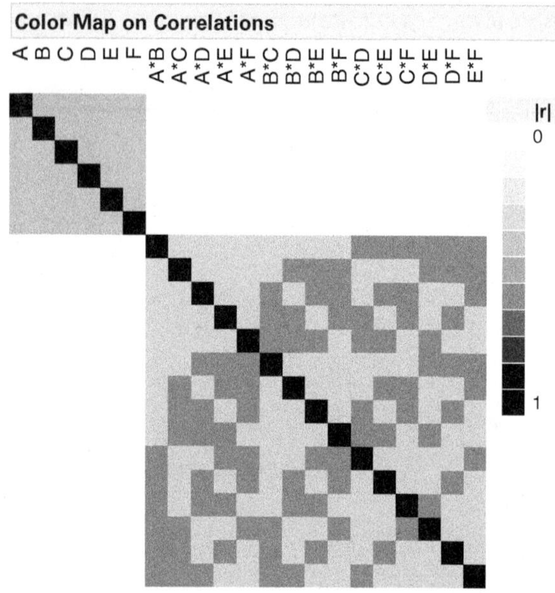

FIGURE 8.2 The Correlation Cell Plot of the EFD

Table 8.15 **Seven-factor 14-run D-optimal Design**

X1	X2	X3	X4	X5	X6	X7
1	1	1	1	1	−1	−1
−1	1	1	−1	1	1	1
1	−1	1	−1	1	1	1
1	−1	1	1	−1	−1	1
1	1	−1	−1	−1	−1	1
−1	−1	1	−1	1	1	−1
−1	1	1	1	−1	1	−1
1	−1	−1	1	1	1	−1
−1	−1	−1	−1	1	−1	−1
1	1	−1	−1	−1	1	−1
−1	1	−1	−1	1	−1	1
−1	1	−1	1	1	−1	1
−1	−1	−1	1	−1	1	1
−1	−1	1	−1	−1	−1	−1

8.5 Definitive Screening Designs

Definitive screening designs (DSDs) are designs for problems where most, if not all, the factors are continuous. Every continuous factor in a DSD has three levels. This makes a DSD able to identify quadratic factor effects, which could encourage engineers to feel less uncomfortable about using larger ranges between factor levels. The discovery of DSDs is relatively recent. Jones and Nachtsheim (2011) introduced the family of DSDs, provided tables of DSDs for 6–30 factors, and discussed their desirable statistical properties.

One such desirable characteristic of a DSD is that it is a foldover design. As mentioned in Section 8.2, main effects and two-factor interactions are uncorrelated in a foldover design.

Table 8.16 The Six-factor DSD with a Minimum Number of Runs

Run	A	B	C	D	E	F
1	0	1	1	1	1	1
2	0	−1	−1	−1	−1	−1
3	1	0	−1	1	1	−1
4	−1	0	1	−1	−1	1
5	1	−1	0	−1	1	1
6	−1	1	0	1	−1	−1
7	1	1	−1	0	−1	1
8	−1	−1	1	0	1	−1
9	1	1	1	−1	0	−1
10	−1	−1	−1	1	0	1
11	1	−1	1	1	−1	0
12	−1	1	−1	−1	1	0
13	0	0	0	0	0	0

Therefore, an active two-factor interaction does not bias (or alias) any main effect. If the *a priori* model is the main effects model and the aliasing terms are all the two-factor interactions, the DSD is a special case of the alias optimal design introduced in Section 8.4.

Table 8.16 shows the structure of the six-continuous-factor DSD, which it shares with DSDs having any number of factors. The tabled design consists of six pairs of rows. The second row in each pair results from multiplying each element in the first row by −1. This is what makes the DSD a self-foldover design. Another structural element of the DSD is that each row has one and only one zero value. Every column in the first 12 rows has two zero values. The last row is a center run.

Figure 8.3 shows a correlation color cell plot of the six-factor DSD of Table 8.16. It shows that the main effects of the DSD are uncorrelated with each other and with all two-factor interactions.

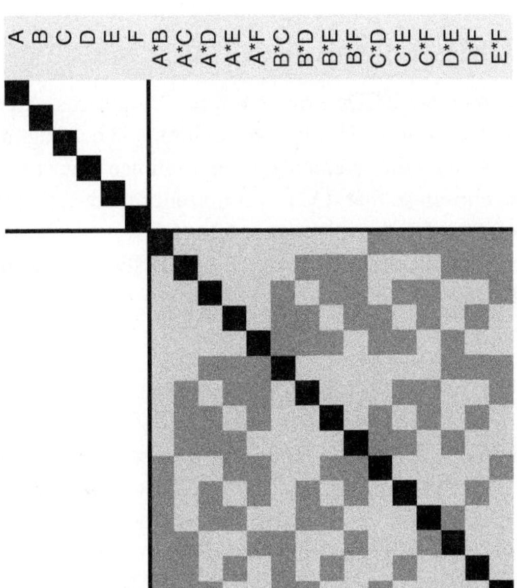

FIGURE 8.3 Correlation Cell Plot of the Six-factor DSD in Table 8.16

The absolute values of correlations between pairs of two-factor interactions are either 1/4 (light gray cells) or ½ (darker gray cells). Although pairs of two-factor interactions are correlated, none is confounded as in the resolution III and resolution IV regular fractional factorial designs introduced in Section 8.2. As a result, if only a few two-factor interactions are active, the DSD has a high probability of identifying them.

8.5.1 Statistical Properties of a DSD

As screening designs, DSDs have many desirable statistical characteristics:

1. Main effects are orthogonal to each other.
2. Main effects are orthogonal to two-factor interactions and quadratic effects.
3. All the quadratic effects of continuous factors are estimable.
4. No two-factor interaction is confounded with any other two-factor interaction or quadratic effect although these effects may be correlated.
5. For DSDs with 13 or more runs, it is possible to fit the full quadratic model in any three-factor subset.
6. DSDs are inexpensive to field requiring only a minimum of $2m + 1$ runs, where m is the number of factors.
7. You can add runs to a DSD by creating a DSD with more factors than necessary and dropping the extra factors. The resulting design has the first five properties above and has more power to identify second-order effects. Jones and Nachtsheim (2017) strongly advocated adding at least two unused factors, which they called "fake" factors. They pointed out that fake factors are orthogonal to all effects lower than third-order effects, so if you assume that third- and higher-order effects are negligible, these extra columns are useful for computing a lack-of-fit estimate of the error variance.

8.5.2 Constructing DSDs Using Conference Matrices

Xiao et al. (2012) showed how to create DSDs using conference matrices. A conference matrix is a square matrix having an even number of rows and columns. The diagonal elements of a conference matrix are all zeros. All the other elements in a conference matrix are either +1 or −1. These off-diagonal elements are chosen so that if **C** is a conference matrix, then $\mathbf{C}'\mathbf{C} = (m-1)\,\mathbf{I}$, where m is the number of factors and $\boldsymbol{I}$ is the identity matrix.

Xiao et al. provided a simple and elegant construction of a DSD using conference matrices. A DSD, **D**, can be constructed using

$$\mathbf{D} = \begin{pmatrix} \mathbf{C} \\ -\mathbf{C} \\ \mathbf{0} \end{pmatrix}$$

where 0 is a row of zeros. That is, a DSD is a conference matrix vertically concatenated with the same matrix where every element is multiplied by −1. That operation makes the DSD a self-foldover design. The final operation in the construction is to add a final row of zeros.

If $m - 1$ is a prime number, an m row and m column conference matrix always exists and is easy to generate. Fortunately, for the construction of DSDs, many small integers are prime numbers (e.g. 5, 7, 11, 13, 17). There are more complicated constructions for other cases (e.g. 10-factor and 16-factor DSDs).

8.5.3 Constructing DSDs with Additional Two-level Categorical Factors

The construction of DSDs when two-level factors are added starts by using the above construction of Xiao et al. where the number of columns in the conference matrix is greater than or equal to the total number of factors. To keep the foldover structure, however, you must add a second row of zeros so that if m is the number of columns in the conference matrix, then the number of runs in the design is $2m + 2$ rather than $2m + 1$. In these last two rows, the categorical factors must be 1 or -1, not zero like all the continuous factors. If a categorical factor takes the value -1 in the next to last row, it must take the value 1 in the last row.

The main problem of adding two-level factors to a conference matrix construction of a DSD is figuring out what to do with the zeros in the columns for the categorical factors. Jones and Nachtsheim (2013) pointed out that to keep the self-foldover structure it is necessary to change the zero value in the column in **C** associated with a two-level categorical factor to either $+1$ or -1. Then, the same element in the same column of **–C** must be -1 or $+1$, respectively.

The two-level categorical columns constructed this way have a small correlation with each other and with all the columns of continuous factors. However, all the main effects remain uncorrelated with any second-order effect.

8.5.4 Constructing Orthogonally Blocked DSDs

The corresponding rows of **C** and **–C** form foldover pairs. For example, the first row of **–C** is a foldover of the first row of **C**. To create an orthogonally blocked DSD, it is sufficient to put both rows of a foldover pair in the same block. To maintain the ability to estimate the quadratic effects of all the factors, Jones and Nachtsheim (2016) advocated adding a center run to each block.

It is easy to create orthogonally blocked DSDs with 2 to m blocks, where m is the number of columns in the conference matrix used to create the DSDs. For example, the six-factor screening design shown in Table 8.16 can have two blocks of seven runs, three blocks of five runs, four blocks of four runs, five blocks where one block has five runs and the remaining blocks have three runs, and, finally, six blocks with three runs. Thus, blocking DSD is much more flexible than blocking other screening designs.

The general construction procedure for creating an orthogonally blocked DSD is given in the following text.

Let the desired number of blocks be b (where b is between 2 and m inclusive). Assign the first b rows of **C** and the first b rows of **–C** to blocks 1, 2, …, b. Now start over by assigning rows $b + 1$ to $2b$ of **C** and **–C** to blocks 1, 2, …, b if $2b < m$. Otherwise, stop when you run out of rows in **C** and **–C**. If $m > 2b$ continue assigning rows to blocks sequentially for both again until you run out of rows in **C** and **–C**. Add b rows of zeros and assign each of these to a separate block. If m is not divisible by b, some of the blocks will have more runs than others although the difference in the number of runs in different blocks never exceeds 2.

Blocks constructed using this approach are orthogonal to the main effects of all the factors. However, the block effects and second-order effects of the factors are sometimes correlated, so block effects may be biased by active two-factor interactions or quadratic effects.

8.5.5 Situations When You Should Use a Screening Design Other Than a DSD

1. When there are constraints on the design region or some factor-level combinations are infeasible;
2. When some of the factors are ingredients in a mixture;

3. When there are categorical factors with more than two levels;
4. When the *a priori model* of interest has three-factor interactions or other higher-order polynomial terms.

8.5.6 Recommendations

A DSD is the design of choice for an initial factor screening study with mostly continuous factors and insufficient budget to run a design capable of fitting all the two-factor interactions. We recommend adding four extra runs to the minimum run DSD by adding unused factors. If the number of factors is odd, we recommend using the conference matrix construction with one extra unused factor, except in the case where the budget for runs precludes having any more than the bare minimum of runs.

8.6 Screening Summary

This chapter introduced several different approaches to design for factor screening. Historically, the regular fractional factorial designs introduced in Section 8.2 have been standard operating procedure for implementing a screening study. However, since the beginning of this century, new approaches have substantially enriched the choice available to practitioners who need to deploy a screening study. Sometimes having too many choices becomes a burden instead of an opportunity, so the purpose of this section is to provide statistical perspective on the available choices and make recommendations about the appropriate use of the different screening approaches introduced in this chapter.

The two most important statistical considerations for evaluating and choosing a design are variance and bias. Reducing variance increases precision, which reduces the length of confidence intervals on coefficient estimates. The effect of reducing or eliminating bias is that estimated coefficients really mean what you intend them to mean.

First, we consider variance. It is desirable that the estimates of the coefficients of the *a priori* model be as precise as possible. Often, the primary intent is to fit the main effects model. For a given run budget, any two-level orthogonal design minimizes the variance of all the main effects. You can also reduce the variance of the estimates of the main effects by increasing the number of runs beyond the bare minimum required to fit the main effects model. For a two-level orthogonal design and main effects model, the variance of any effect is $\frac{\sigma^2}{n}$, where σ is the error variance and n is the number of runs. Doubling the number of runs cuts the variance in half but also increases (possibly doubles) the cost of the experiment, so there is always a dynamic tension between cost and information.

Bias is an equally important concern in screening experiments. Though the primary intent may be to identify the large main effects, for some screening designs active two-factor interactions can bias the estimates of the main effects. An active two-factor interaction can make its correlated or confounded main effect look larger than it really is, causing investigators to waste resources on follow-up research. Worse yet, an active two-factor interaction can make its correlated or confounded main effect look smaller than it really is, causing the engineering team to dismiss, perhaps forever, an important factor.

Resolution III fractional factorial designs are the two-level orthogonal designs with the fewest runs. When each run is expensive, it is tempting to use these designs as an initial screening study. However, these designs are the most prone to bias since each main effect is confounded with at least one and, sometimes, several two-factor interactions. We would recommend these designs only when management is unwilling to budget for a larger study.

Resolution IV fractional factorial designs require twice as many runs as a resolution III design but avoid confounding main effects with two-factor interactions. Unfortunately, these designs confound two-factor interactions with other two-factor interactions. If the estimated effect due to a group of two-factor interactions is large, the results of the experiment will be ambiguous. Further experimentation will be necessary to resolve this ambiguity. For this reason, we do not recommend the use of resolution IV fractional factorial design, because there are other design choices that require the same or fewer resources that do not suffer from confounding of two-factor interactions.

Resolution V (or higher) fractional factorial designs are variance optimal for the main effects plus two-factor interactions model. As screening designs, they provide the greatest protection against bias. However, these benefits come at a cost of substantially more runs. However, if runs are inexpensive in time and money, then we can recommend their use. However, we caution that these designs are not capable of modeling curvature in the effects of factors.

Nonregular orthogonal designs are an alternative to the regular fractional factorial designs just discussed. They avoid complete confounding of effects, which can allow for identifying more active effects than were included in the *a priori* model when there are only a few of these. Nonregular designs may also be able to fit the main effects plus two-factor interactions model in fewer runs than a resolution V requires. One example is the 32-run nonregular orthogonal design in seven factors introduced in Section 8.3. The smallest regular fractional factorial design for this case requires 64 runs. Another example is a 24-run nonregular orthogonal design for six factors. This design is an alternative to the 32-run regular fractional factorial design. Model selection using these designs requires more expertise, which should be taken into consideration before employing them in practice.

Alias optimal and efficient foldover designs may sometimes sacrifice the orthogonality of main effects in exchange for eliminating bias due to active two-factor interactions. When there is substantial concern about the possible presence of active two-factor interactions, these designs deserve consideration. They are especially useful as alternatives to definitive screening designs when most of the factors are categorical.

DSDs would be our recommended screening design when most of the factors are continuous, there are no restrictions on randomization, there are no infeasible factor-level combinations, and the budget for runs is tight.

Finally, D-optimal designs generated using a computer algorithm offer the most flexibility for fielding a factor screening study. They can accommodate categorical factors at more than two levels. They offer the most flexibility in the choice of the number of runs to perform. They can produce a design given any polynomial model. Finally, they can avoid infeasible factor-level combinations. However, all this flexibility can sometimes come at the cost of sacrificing the orthogonality of the main effects. On the other hand, the aim of the D-optimality criterion is to approach orthogonality as much as possible given the constraints of the problem statement.

We have elevated considerations of variance and bias in this discussion. However, the utility of orthogonality deserves some discussion. In most cases, when we say that a design is orthogonal we mean that the main effects are orthogonal or uncorrelated with each other. If we are fitting the main effects model using an orthogonal design, we can remove any term from the model and the coefficients of the other terms will not change. This is a desirable feature. Also, two-level orthogonal designs are guaranteed to be D-optimal for the main effects model. In the case of resolution V (or higher) fractional factorial designs, both the main effects and all the two-factor interactions are orthogonal. Many nonregular orthogonal designs have orthogonal main effects, but these effects may be correlated with two-factor interactions. Some nonregular orthogonal designs can fit the main effects plus two-factor interactions model, but the two-factor interactions may be correlated, which results in variance inflation when they appear together in a regression model. As Jones et al. (2016) showed algorithmically generated D-optimal designs for the main effects plus two-factor interactions model may be nonorthogonal for the main effects while still having a lower average variance of the effects than a nonregular orthogonal design. It should be noted that definitive screening designs have orthogonal main effects. However, they are not D-optimal for the main effects model, because the three zeros in each column are not useful to lower the variance of the main effect estimates.

Historically, textbooks have put a great deal of emphasis on orthogonal designs. Given this emphasis, a student might wrongly conclude that orthogonality is essential for successful experimentation. Before the advent of computers, the heavy emphasis on orthogonal design was warranted, because these designs allow for hand calculation of the effects. Today, computers make fitting regression models for nonorthogonal designs easy. We do not deny that orthogonal designs provide benefits in terms of mathematical elegance and simplicity of model fitting. However, orthogonal design does not provide a panacea for all problems, and orthogonality can be sacrificed in exchange for eliminating bias or lowering cost.

9 Experiments with Random Blocks

9.1 Introduction

Chapter 6 dealt with design where the blocks are **fixed** factors. Fixed blocks are a special case of a categorical factor where there is a restriction on the number of runs that can be performed in a block. For example, if the blocking factor is Day, there is an upper bound on the number of runs that can be performed in a day. If an engineering team wants to run an experiment for five days and day-to-day variation is expected, then, historically, Day has been treated as a fixed blocking factor. The number of degrees of freedom for this factor is the number of blocks, b, minus 1, i.e. $b - 1$. In this case of the five-day experiment, the number of block degrees of freedom is $5 - 1 = 4$. So, estimating the block effect requires four extra runs. The motivation for running experiments with a fixed blocking factor is to separate the variability due to blocks from the run-to-run variability or error variance. In the event of substantial day-to-day variability, blocking the experiment by day results in a great reduction in the variance of the other fixed effect estimates. This is because the analysis removes the day-to-day variability, which, without blocking, would add to the error variance. When we treat blocks as fixed, we assume that the block effects are unknown constants. In principle, this means that there is no basis for predicting the result of a run in a future block. This is because the block effect of a future block is another unknown constant, which, in theory, could be any value.

An alternative to treating blocks as fixed is to consider the experimental blocks as a random sample from a population of blocks. In this case, we assume that the block effects and the random errors are mutually independent and normally distributed random variables with zero mean. The block variance is σ_b^2, while the error variance is σ_ε^2. The block variance is a single unknown parameter to be estimated in the analysis, which requires only one degree of freedom. This provides an advantage over the fixed blocks analysis especially when the experiment requires many blocks. From a design standpoint, experimentation using random blocks does not require as many runs as experimentation using fixed blocks. Another advantage of designing for random blocks is that it is possible to make predictions of responses in new blocks not part of the experiment. To do this, we assume that the new blocks are in the same population of blocks as the blocks in the experiment.

Treating both the blocks and the errors as random indicates that the experiment has two random effects as well as the fixed effects of the other factors. We call models for designs having at least one fixed effect and more than one random effect **mixed models**. Fitting mixed models is more difficult than least-squares regression. This may explain why most textbooks on design of experiments only introduce design for fixed factors. Nowadays, there is commercial software that can fit mixed models with ease. One analytical approach supported by JMP is **restricted maximum likelihood** (REML).

Ideally, the randomization for a design with random blocks has two steps. The order of the blocks is randomized and the order of the runs within each block is randomized. A good name for such a design is a **bi-randomization design**. However, when the random block factor is the days of experimentation, this two-step randomization is impossible. However, when the blocks are batches of experimental material, then the above procedure for randomization is feasible.

This chapter first provides a small example of an optimal random block design and its analysis. Later sections provide examples of optimal random block designs for factor screening. A final section deals with suggestions for practical use of random block designs.

9.2 Motivating Example: Design and Analysis

The first example in Chapter 6 was a randomized complete block design (RCBD) for investigating the process of pressurized extrusion of vascular grafts. The experiment involved changing the pressure from 8500 to 9100 psi with two intermediate values at 8700 psi and 8900 psi. There was some concern that the batch of resin used in the makeup of the graft might add to the variability, so resin batch became a blocking factor. The experimental runs consisted of one run at each of the four pressures for each of six batches for a total of 24 runs. Table 9.1 reproduces the data in Table 6.1.

In Chapter 6, the pressure variable was treated as a categorical treatment factor at four levels and the six batches of resin comprised a six-level blocking factor. Defined in this way, the RCBD is a D-optimal design for the fixed-block case.

The JMP output below shows the effect test for Pressure and Least Squares Means Table for the analysis treating the blocks as random rather than fixed. All the values in the output match the values in the output in Chapter 6. This is a consequence of the fact that the RCBD is a balanced design.

Fixed Effect Tests

Source	Nparm	DF	DFDen	F Ratio	Prob > F
Pressure	3	3	15	8.1071	0.0019

Least Squares Means Table

Level	Least Sq Mean	Std Error
8500	92.816667	1.5867638
8700	91.683333	1.5867638
8900	88.916667	1.5867638
9100	85.766667	1.5867638

The differences in the JMP REML analysis for random blocks and the output in Chapter 6 come in the treatment of the inference for the random blocking factor, resin.

Table 9.1 RCBD for the Vascular Graft Experiment

	Batch of resin (block)					
Extrusion pressure (psi)	**1**	**2**	**3**	**4**	**5**	**6**
8500	90.3	89.2	98.2	93.9	87.4	97.9
8700	92.5	89.5	90.6	94.7	87.0	95.8
8900	85.5	90.8	89.6	86.2	88.0	93.4
9100	82.5	89.5	85.6	87.4	78.9	90.7

The output below shows the estimates of $\sigma_b^2 = 7.78$ and $\sigma_\varepsilon^2 = 7.33$. We see that more than half of the variability in the response is due to the variation in the batches of resin.

REML Variance Component Estimates

Random Effect	Var Ratio	Var Component	Std Error
Resin	1.06	7.78	6.12
Residual		7.33	2.67
Total		15.1	6.40

The estimates of the parameters in the REML model are for the intercept, the three degrees of freedom for the pressure, and the two variance components.

The REML analysis also allows for making predictions of the effects of each block. The JMP output below shows the predictions of all the block effects.

Random Effect Predictions

Term	BLUP	Std Error	DFDen
Resin[Resin Batch 1]	−1.70	1.59	8.621
Resin[Resin Batch 2]	−0.04	1.59	8.621
Resin[Resin Batch 3]	0.97	1.59	8.621
Resin[Resin Batch 4]	0.61	1.59	8.621
Resin[Resin Batch 5]	−3.62	1.59	8.621
Resin[Resin Batch 6]	3.77	1.59	8.621

Note that the values of the Best Linear Unbiased Predictors (BLUP) column are smaller in magnitude than the estimates of the fixed-block parameters shown in Chapter 6.

9.3 Matrix Formulation of the Model for an Experiment with Random Blocks

The matrix form of the model for random block experiments is concise. The general form is in Eq. (9.1).

$$\mathbf{y} = \mathbf{X}\boldsymbol{\beta} + \mathbf{Z}\boldsymbol{\gamma} + \boldsymbol{\varepsilon} \tag{9.1}$$

where $\mathbf{y}$ is the $n \times 1$ vector of responses, $\mathbf{X}$ is the $n \times p$ model matrix for the fixed effects, $\boldsymbol{\beta}$ is a $p \times 1$ vector of unknown parameters to be fit, $\mathbf{Z}$ is an $n \times b$ indicator matrix where b is the number of blocks, $\boldsymbol{\gamma}$ is a $b \times 1$ vector of random block effects, and $\boldsymbol{\varepsilon}$ is an $n \times 1$ vector of random errors. We assume that $\boldsymbol{\gamma}$ is distributed as a normal random variable with zero mean and variance σ_b^2, $\boldsymbol{\varepsilon}$ is also normally distributed with zero mean and variance σ_ε^2, and $\boldsymbol{\gamma}$ and $\boldsymbol{\varepsilon}$ are uncorrelated.

Given these assumptions, the variance of $\mathbf{y}$ is

$$\mathbf{V} = \sigma_\varepsilon^2 \mathbf{I}_n + \sigma_b^2 \mathbf{Z}\mathbf{Z}' \tag{9.2}$$

When we analyze a random block experiment, we need to estimate β, σ_b^2, and σ_ε^2.

The modern way to analyze designs with random blocks uses the **REML** to obtain estimates of σ_b^2 and σ_ε^2. These estimates are then plugged into Eq. (9.2) to get an estimate of the variance matrix of $\mathbf{y}$, $\widehat{\mathbf{V}}$. Using this estimate, we can use **generalized least squares** to estimate $\boldsymbol{\beta}$. The equation for the estimate of $\boldsymbol{\beta}$ is

$$b = \widehat{\boldsymbol{\beta}}_{\text{gls}} = (\mathbf{X}'\widehat{\mathbf{V}}^{-1}\mathbf{X})^{-1}\mathbf{X}'\widehat{\mathbf{V}}^{-1}\mathbf{y} \tag{9.3}$$

The variance of the estimate of $\boldsymbol{\beta}$ is $\mathbf{X}'\widehat{\mathbf{V}}^{-1}\mathbf{X}$, which is useful for testing hypotheses about $\boldsymbol{b}$. For testing hypotheses, the current state of the art is the approach of Kenward and Roger (1997). The specifics of this methodology are beyond the scope of this book; however, it is supported in the statistical software, JMP.

9.4 Design Considerations

The D-optimal design of a random block experiment maximizes $|\mathbf{X}'\mathbf{V}^{-1}\mathbf{X}|$. This determinant depends on prespecifying the number of blocks, n_b, and the number of runs in each block, n_s. The total number of runs, n, is the product $n_b \times n_s$.

It is also necessary to decide on an *a priori* model. For screening experiments using random blocks, this model is usually the main effects model or the model having the main effects and two-factor interactions. The model for response surface experiments using random blocks generally includes all the main effects, two-factor interactions, and quadratic effects.

The optimal split-plot design also depends on the ratio $\eta = \frac{\sigma_b^2}{\sigma_\varepsilon^2}$; however, simulation studies have shown that the statistical efficiency of optimal random block designs is relatively insensitive to this ratio, which can be set to 1 as a rule of thumb.

9.5 A Screening Design with a Random Blocking Factor

Montgomery (2020) introduces a 16-run experiment to study the effect of four factors, temperature, pressure, concentration of formaldehyde, and stirring rate, and all their two-factor interactions on the filtration rate for a reaction in a pressure vessel. We adapt this study to accommodate a random blocking factor having two blocks of eight runs. The data and design are in Table 9.2.

In Table 9.3, the first eight rows in the first column are 1 while the last eight are 0, indicating that the first eight runs are in Block 1. For the second column, the first eight rows are 0 and the last eight rows are 1, indicating that the last eight runs are in Block 2.

In general, for a design with b blocks, the $\mathbf{Z}$ matrix has b columns and n rows, where n is the number of runs. Each row of $\mathbf{Z}$ has exactly one value of 1, and the rest of the columns in that row take the value 0. If the ith column is 1, that means the row is in the ith block.

The matrix in Table 9.4 has two 8×8 blocks of 1s that are diagonally offset from each other. We call such a matrix **block diagonal.**

If σ_b^2 and σ_ε^2 are both equal to 1, then the $\mathbf{V}$ matrix is shown in Table 9.5.

If a matrix is block diagonal, its inverse is also block diagonal with the same structure. Therefore, the matrix $\mathbf{V}^{-1}$ has two diagonally opposite blocks of 8×8 nonzero values. The diagonal values are 8/9 while the values corresponding to the ones in the $\mathbf{V}$ matrix are $-1/9$.

Table 9.2 D-optimal Random Block Experiment Data

Random block	Temperature	Pressure	Concentration	Stirring rate	Filtration rate
1	−1	−1	−1	−1	45
1	−1	1	−1	1	39
1	1	−1	1	1	92
1	−1	1	1	−1	74
1	1	−1	−1	−1	67
1	1	1	1	−1	61
1	1	1	−1	1	102
1	−1	−1	1	1	68
2	1	1	1	1	98
2	−1	−1	1	−1	80
2	−1	−1	−1	1	54
2	1	1	−1	−1	79
2	−1	1	−1	−1	54
2	1	−1	−1	1	105
2	1	−1	1	−1	67
2	−1	1	1	1	81

Table 9.3 Z Matrix for the Random Block Design Having 16 Runs in Two Blocks of Eight

Block 1	Block 2
1	0
1	0
1	0
1	0
1	0
1	0
1	0
1	0
0	1
0	1
0	1
0	1
0	1
0	1
0	1
0	1

Table 9.4 ZZ′ Matrix for the 16-run Random Block Design

1	1	1	1	I	1	1	1	0	0	0	0	0	0	0	0
1	1	1	1	I	1	1	1	0	0	0	0	0	0	0	0
1	1	1	1	i	1	1	1	0	0	0	0	0	0	0	0
1	1	1	1	I	1	1	1	0	0	0	0	0	0	0	0
1	1	1	1	I	1	1	1	0	0	0	0	0	0	0	0
1	1	1	1	I	1	1	1	0	0	0	0	0	0	0	0
1	1	1	1	1	1	1	1	0	0	0	0	0	0	0	0
1	1	1	1	1	1	1	1	0	0	0	0	0	0	0	0
0	0	0	0	0	0	0	0	1	1	1	1	1	1	1	1
0	0	0	0	0	0	0	0	1	1	1	1	1	1	1	1
0	0	0	0	0	0	0	0	1	1	1	1	1	1	1	1
0	0	0	0	0	0	0	0	1	1	1	1	1	1	1	1
0	0	0	0	0	0	0	0	1	1	1	1	1	1	1	1
0	0	0	0	0	0	0	0	1	1	1	1	1	1	1	1
0	0	0	0	0	0	0	0	1	1	1	1	1	1	1	1

Table 9.5 V Matrix for the Random Block Design Assuming the Variance of Both Random Effects Is 1

2	1	1	1	1	1	1	1	0	0	0	0	0	0	0	0
1	2	1	1	1	1	1	1	0	0	0	0	0	0	0	0
1	1	2	1	1	1	1	1	0	0	0	0	0	0	0	0
1	1	1	2	1	1	1	1	0	0	0	0	0	0	0	0
1	1	1	1	2	1	1	1	0	0	0	0	0	0	0	0
1	1	1	1	1	2	1	1	0	0	0	0	0	0	0	0
1	1	1	1	1	1	2	1	0	0	0	0	0	0	0	0
1	1	1	1	1	1	1	2	0	0	0	0	0	0	0	0
0–	0	0	0	0	0	0	0	2	1	1	1	1	1	1	1
0	0	0	0	0	0	0	0	1	2	1	1	1	1	1	1
0	0	0	0	0	0	0	0	1	1	2	1	1	1	1	1
0	0	0	0	0	0	0	0	1	1	1	2	1	1	1	1
0	0	0	0	0	0	0	0	1	1	1	1	2	1	1	1
0	0	0	0	0	0	0	0	1	1	1	1	1	2	1	1
0	0	0	0	0	0	0	0	1	1	1	1	1	1	2	1
0	0	0	0	0	0	0	0	1	1	1	1	1	1	1	2

The JMP output for the analysis of the model containing all the intercepts, the four main effects, and six two-factor interaction effects is shown below.

Parameter Estimates

Term	Estimate	Std Error	DFDen	t Ratio	Prob>|t|
Intercept	72.875	4.375	1	16.66	0.0382*
Temperature	11	0.852386	4	12.90	0.0002*
Pressure	0.625	0.852386	4	0.73	0.5041
Concentration	4.75	0.852386	4	5.57	0.0051*
Stirring Rate	7	0.852386	4	8.21	0.0012*
Temperature*Pressure	0.5	0.852586	4	0.59	0.5890
Temperature*Concentration	−9.125	0.852386	4	−10.71	0.0004*
Temperature*Stirring Rate	8.375	0.852386	4	9.83	0.0006*
Pressure*Concentration	0.25	0.852386	4	0.29	0.7839
Pressure*Stirring Rate	−0.5	0.852386	4	−0.59	0.5890
Concentration*Stirring Rate	0.125	0.852386	4	0.15	0.8905

Random Effect Predictions

Term	BLUP	Std Error	DFDen	t Ratio	Prob>|t|
Random Block[1]	−4.208929	4.371847	1	−0.96	0.5121
Random Block[2]	4.2089286	4.371847	1	0.96	0.5121

REML Variance Component Estimates

Random Effect	Var Ratio	Var Component	Std Error	95% Lower	95% Upper	Wald p-Value	Pct of To
Random Block	3.1680108	36.828125	54.147613	−69.29925	142.9555	0.4964	76.0
Residual		11.625	8.2201163	4.172916	95.991368		23.9
Total		48.453125	54.613567	11.831833	4527.5919		100.0

Note that the parameter estimates for pressure, all the two-factor interactions involving pressure, and the concentration*stirring rate interaction are not significant. It is standard practice to remove these terms and refit the model. The results of these changes are shown on the next page. Note that the fixed effect estimates have not changed since the design is balanced. However, all of the remaining fixed effects are much more statistically significant because the standard errors are smaller and the denominator degrees of freedom are larger.

The variance estimates and the random effect predictions have changed slightly. However, there is no practically important difference.

Parameter Estimates

Term	Estimate	Std Error	DFDen	t Ratio	Prob>|t|
Intercept	72.875	4.375	1	16.66	0.0382*
Temperature	11	0.656167	9	16.76	<.0001*
Concentration	4.75	0.656167	9	7.24	<.0001*
Stirring Rate	7	0.656167	9	10.67	<.0001*
Temperature*Concentration	−9.125	0.656167	9	−13.91	<.0001*
Temperature*Stirring Rate	8.375	0.656167	9	12.76	<.0001*

Random Effect Predictions

Term	BLUP	Std Error	DFDen	t Ratio	Prob>|t|
Random Block[1]	−4.276587	4.373893	1	−0.98	0.5072
Random Block[2]	4.2765873	4.373893	1	0.98	0.5072

REML Variance Component Estimates

Random Effect	Var Ratio	Var Component	Std Error	95% Lower	95% Upper	Wald p-Value	Pct of To
Random Block	5.4319556	37.420139	54.139385	−68.69111	143.53138	0.4895	84.4
Residual		6.8888889	3.274534	3.2592523	22.959651		15.5
Total		44.309028	54.212383	10.056356	8609.4419		100.0

9.6 Recommendations for Use of Designs with Random Blocks

In industrial applications, the most obvious factors to include are control factors—the switches or dials on the processing equipment. Less obvious are environmental factors that can increase the variability of the response. Thinking carefully about which of these may have the most deleterious effect and including it in the design is a good practice. Often a surrogate for this approach is to group runs by their time order under the assumption that runs that are far from each other in time are more likely to be affected by changes in lurking variables.

An advantage of using a random block factor is that the experiment can employ many blocks without losing a degree of freedom for each block. However, the estimate of the variance when the fixed effects take up most of the available degrees of freedom is highly variable. Especially when the number of experimental runs is only two or three more than the number of fixed effect parameters, there is a high probability that REML will produce a negative variance component for the block variance. In experiments with random blocks, it is desirable to have four or more runs than fixed effect parameters including the intercept.

Finally, even though care has been taken in design to use enough blocks to allow for good inference, there is always the chance that the actual experimental runs will not happen as specified. That is, if a person other than you is responsible for doing the runs, he/she may sort the runs to do all the low values of some factor first followed by all the high values (or vice versa). It is important that everyone involved in the experiment understands the importance of keeping the order of the runs as specified. If possible, the best practice is to observe the experiment as it is being run to make sure that all the factor settings are correctly set and that runs within a block actually have the appropriate level of the blocking factor.

10 Split-plot Experiments

10.1 Introduction

Chapter 9 dealt with design for mixed models. They are designs having at least one fixed effect and more than one random effect. Examples of mixed models are blocked designs where the blocks are viewed as a random sample from a population of blocks. When there is one random blocking factor, fitting the model involves estimating two variances (or random effects), the error variance and the block variance. When there is a random blocking factor, optimal designs generally vary the levels of other factors within each block.

It is common in industrial design of experiments for one or more factors to be hard to change from one run to the next. For example, the temperature of an industrial furnace may take several hours to stabilize after a change from one temperature to another. In such a case, it is simpler and less expensive to perform processing runs in a row while keeping the temperature of the furnace at one setting, but varying the settings of other factors that are easy to change.

In experimenting with mixtures of ingredients, it is often desirable to mix large several batches, each having a different recipe. Afterward these batches can be split into smaller subbatches. Each of these subbatches may be subjected to different processing conditions.

Complete randomization is impossible in both scenarios. In the furnace example, the randomization is restricted because the temperature is kept constant for many runs in a row before it is changed. If the order of the treatment combinations were randomized, it would be necessary to change the temperature much more often, which would make the experiment take longer and be more expensive. An alternative is to perform two randomizations. Suppose the engineering team wants to understand the effect of changing the furnace temperature from 300 degrees to 400 degrees. There is a budget for 16 processing runs, but the experimental plan is to perform four sets of four runs in a row at a fixed temperature while changing the temperature between each set of runs. You could think of this as a four-run experiment in the factor, temperature, while the experiment has 16 runs in the other factors. Here, the order of the four temperatures is randomized. Then the order of each group of runs within a fixed temperature is separately randomized. A good name for such a design is a **bi-randomization design**.

In the mixture experiment, complete randomization is impossible because a batch for each recipe is used several times under different processing conditions. To perform this experiment as a completely randomized design, it would be necessary to separately mix the ingredients for each processing run. Suppose that there were four processing factors of interest, each at two levels, and it was feasible to perform the full-factorial design (16 runs) in the processing factors for each recipe. Suppose also that there were six recipes of interest. The bi-randomization design would involve mixing six batches, splitting these batches into 16 subbatches, and subjecting each subbatch to one of the 16 processing conditions. A completely randomized design would require creating $16*6 = 96$ separate mixtures but only employing six different recipes. Clearly, this means more processing work, which will add to the cost of the experiment. Also, there is an increased potential for error in mixing and correctly labeling each mixture.

In agriculture, a factor such as irrigation method may only be applied to a whole field. Other factors could be accommodated by splitting a field into several plots and applying, for example, a different seed variety or fertilizer to each plot. As it happens, agriculture was the first application of this design approach. Hence, the common term for this type of design is **split-plot design**. Factors applied to a whole field are called **whole-plot** factors. Factors that change across the several subdivisions of the field are called **subplot** factors. In industrial applications of split-plot design, the factors that remain constant for a group of runs are whole-plot factors, while factors with levels that change within a group of runs are called subplot factors. The groups of runs are called **whole plots**, while the individual runs within a whole plot are called **subplots**.

A split-plot design is a special case of a random block design. The difference is that in the split-plot design one or more factors remain fixed within each block. There is also a difference in terminology, since the blocks in a split-plot design are called whole plots.

This chapter first provides a small example of an optimal split-plot design and its subsequent analysis. Later sections provide examples of optimal split-plot designs for factor screening. A final section deals with optimal split-plot design for applications of response surface methodology (RSM), which will be covered in more detail in chapter 11.

10.2 Motivating Example: Design and Analysis

Box, Hunter, and Hunter (2005) (BHH) introduced a split-plot design with one hard-to-change factor (temperature) and one easy-to-change factor (coating). The purpose of the experiment was to improve the corrosion resistance of steel bars treated with four different coatings and heated at three different temperatures (360, 370, and 380 degrees). The design involved six whole plots. Each of the three temperatures appears in two whole plots. The temperature for whole plots 1 and 6 was 360 degrees. For whole plots 2 and 5, the temperature was set to 370 degrees. Finally, for whole plots 3 and 4, the temperature was set to 380 degrees. Table 10.1 shows the corrosion resistance response across whole plots (labeled WP 1–WP 6) with the relevant values of temperature and coating.

BHH do not discuss their choice of design, although it is certainly reasonable. Each combination of temperature and coating appears twice (in different whole plots). In their analysis, BHH treat temperature as categorical even though it is a numeric quantity. The model they fit

Table 10.1 Corrosion Resistance Data (Larger Is Better)

Temperature	Coating	WP 1	WP 2	WP 3	WP 4	WP 5	WP 6
360	C1	67	•	•	•	•	33
360	C2	73	•	•	•	•	8
360	C3	83	•	•	•	•	46
360	C4	89	•	•	•	•	54
370	C1	•	65	•	•	140	•
370	C2	•	91	•	•	142	•
370	C3	•	87	•	•	121	•
370	C4	•	86	•	•	150	•
380	C1	•	•	155	108	•	•
380	C2	•	•	127	100	•	•
380	C3	•	•	147	90	•	•
380	C4	•	•	212	153	•	•

Table 10.2 ANOVA Results Treating Temperature as a Categorical Factor

Fixed Effect Tests

Source	DF	DFDen	F Ratio	Prob > F
Temperature	2	3	2.7548	0.2093
Coating	3	9	11.4798	0.0020*
Temperature*Coating	6	9	.3757	0.0241*

consisted of the main effects of temperature and pressure as well as their two-factor interaction. Symbolically,

$$y_i = \beta_0 + \sum_{j=1}^{2} \beta_j T_{ij} + \sum_{k=1}^{3} \beta_{k+2} C_{ik} + \sum_{j=1}^{2} \sum_{k=1}^{3} \beta_{k+(3)(j-1)+5} T_{ij} C_{ik} + \gamma_l + \varepsilon_i \tag{10.1}$$

where i denotes the ith run, j is the jth level of temperature, k is the kth level of coating, and γ_l is lth whole-plot random effect.

Thus, the categorical temperature effect has two degrees of freedom, the coating effect has three degrees of freedom, and the two-factor interaction of temperature and coating has $2 \times 3 = 6$ degrees of freedom. Treating temperature as categorical makes the BHH design D-optimal for their model.

Table 10.2 shows the ANOVA table for the data in Table 10.1.

We would prefer to identify temperature as a continuous factor. There is an equivalent model to the BHH model, when temperature is treated as continuous. We can write this model as

$$y_i = \beta_0 + \beta_1 T_i + \beta_2 T_i^2 + \sum_{j=2}^{4} \beta_{j+1} C_j + \sum_{j=2}^{4} \beta_{j+4} C_j T_i + \sum_{j=2}^{4} \beta_{j+7} C_j T_i^2 + \gamma_l + \varepsilon_i \tag{10.2}$$

Figure 10.1 shows the resistance versus temperature across the four coatings.

The graph would indicate that model 10.2 provides more clarity about the effects of temperature and coating.

Table 10.3 shows the ANOVA table for the model of Eq. (10.2).

Although the two ANOVA tables are different due to the different model parameterizations, the predicted values for the corrosion resistance are the same. Both models provide the same estimates of the whole-plot variance (1172.2) and the error variance (124.5). The whole-plot variance is more than nine times larger than the error variance. If this experiment had been run as a completely randomized design, the estimate of the error variance would have been the sum of the whole-plot and error variance (1296.7). Fielding this experiment as a split-plot design has dramatically improved the precision for estimating the coating effect and the two-factor interaction of coating and temperature.

10.3 Matrix Formulation of the Model for a Split-plot Experiment

In Section 10.2 we provided two regression models for the corrosion resistance split-plot design. The matrix form of the model for split-plot experiments is much more concise and has the same form as the mixed model for experiments with random blocks introduced in Chapter 9. The general form is shown in Eq. (10.3).

$$\mathbf{y} = \mathbf{X}\boldsymbol{\beta} + \mathbf{Z}\boldsymbol{\gamma} + \boldsymbol{\varepsilon} \tag{10.3}$$

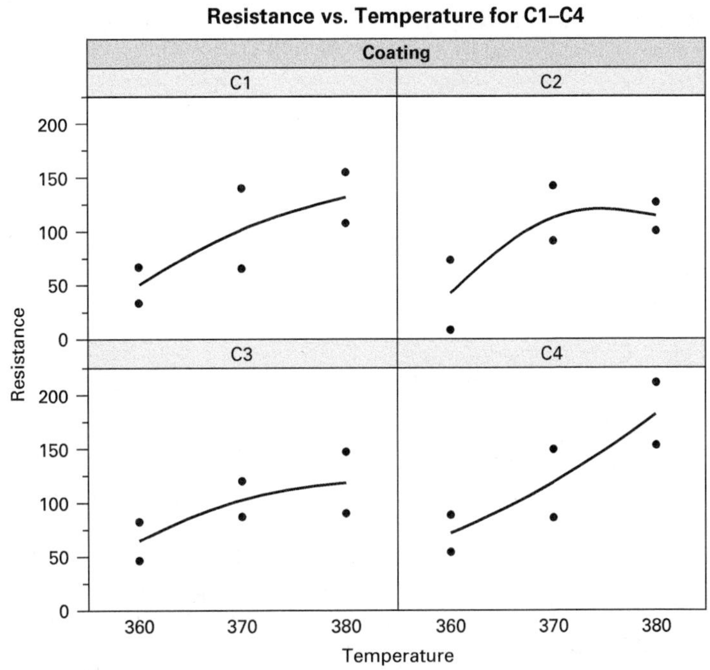

FIGURE 10.1 Resistance Versus Temperature Across the Four Coatings. Note that the quadratic effect of temperature changes as the coating changes.

Table 10.3 **ANOVA Results Treating Temperature as a Continuous Factor**

Fixed Effect Tests

Source	DF	DFDen	F Ratio	Prob > F
Temperature (360,380)	1	3	5.3021	0.1047
Temperature*Temperature	1	3	0.2076	0.6796
Coating	3	9	1.0612	0.4126
Temperature*Coating	3	9	4.5184	0.0340*
Temperature*Temperature*Coating	3	9	4.2330	0.0400*

where $\mathbf{y}$ is the $n \times 1$ vector of responses, $\mathbf{X}$ is the $n \times p$ model matrix for the whole-plot and subplot fixed effects, $\boldsymbol{\beta}$ is a $p \times 1$ vector of unknown parameters to be fit, $\mathbf{Z}$ is an $n \times b$ indicator matrix, where b is the number of whole plots, $\boldsymbol{\gamma}$ is a $b \times 1$ vector of whole-plot random effects, and $\boldsymbol{\varepsilon}$ is an $n \times 1$ vector of random errors. We assume that $\boldsymbol{\gamma}$ is distributed as a normal random variable with zero mean and variance σ_w^2, $\boldsymbol{\varepsilon}$ is also normally distributed with zero mean and variance σ_ε^2, and $\boldsymbol{\gamma}$ and $\boldsymbol{\varepsilon}$ are uncorrelated.

Given these assumptions, the variance of $\mathbf{y}$ is

$$\mathbf{V} = \sigma_\varepsilon^2 \mathbf{I}_n + \sigma_w^2 \mathbf{ZZ}' \tag{10.4}$$

When we analyze a split-plot experiment, we want to estimate $\boldsymbol{b}$, σ_w^2, and σ_ε^2.

The modern way to analyze split-plot designs uses the **restricted maximum likelihood (REML)** to obtain estimates of σ_w^2 and σ_ε^2. These estimates are then plugged into Eq. (10.4) to get an estimate of the variance matrix of $\boldsymbol{y}$, $\widehat{\mathbf{V}}$. Using this estimate, we can use **generalized least squares** to estimate $\boldsymbol{b}$. The equation for the estimate of $\boldsymbol{b}$ is

$$\boldsymbol{b} = \widehat{\boldsymbol{\beta}}_{\text{gls}} = (\mathbf{X}'\widehat{\mathbf{V}}^{-1}\mathbf{X})^{-1}\mathbf{X}'\widehat{\mathbf{V}}^{-1}\boldsymbol{y} \tag{10.5}$$

The variance of the estimate of $\boldsymbol{b}$ is $\mathbf{X}'\widehat{\mathbf{V}}^{-1}\mathbf{X}$, which is useful for testing hypotheses about $\boldsymbol{b}$. For testing hypotheses, the current state of the art is the approach of Kenward and Roger (1997). The specifics of this methodology are beyond the scope of this book; however, it is supported in the statistical software JMP.

10.4 Design Considerations

The D-optimal design of split-plot experiments maximizes $|\mathbf{X}'\mathbf{V}^{-1}\mathbf{X}|$. This determinant depends on prespecifying the number of whole plots, n_w, and the number of runs in each whole plot, n_s. The total number of runs, n, is the product $n_w \times n_s$ assuming that there is the same number of runs in each whole plot.

It is also necessary to decide on an *a priori* model. For screening split-plot experiments, this model is usually the main effects model or the model having the main effects and two-factor interactions. The model for response surface split-plot experiments generally includes all the main effects, two-factor interactions, and quadratic effects.

The optimal split-plot design also depends on the ratio $\eta = \frac{\sigma_w^2}{\sigma_\varepsilon^2}$. However, simulation studies have shown that the statistical efficiency of optimal split-plot designs is relatively insensitive to this ratio, which can be set to 1 as a rule of thumb.

10.5 Split-plot Screening Design

Chapter 8 introduced regular fractional factorial designs for screening and showed how to generate them using column generators. Bingham and Sitter (1999) provided a method for generating split-plot fractional factorial designs with the least amount of confounding of low-order effects. Designs constructed using their methods are D-optimal for any model consisting of terms that are uncorrelated with each other. Using their approach, it is possible, though tedious, to create D-optimal split-plot designs by hand. The strongest feature of fractional factorial split-plot designs (FFSPDs) is that all the effects in any model that is estimable by the design are orthogonal. As a result, the coefficient estimates using least squares are the same as the coefficient estimates produced using generalized least squares. Designs having this property are called **equivalent estimation** designs.

FFSPDs have some weaknesses. One weakness is a lack of flexibility in the number of runs required, which must always be a power of two. The number of whole plots must also be a power of two. These two restrictions may make FFSPDs infeasible for either logistical or budgetary reasons. A more serious statistical problem is that every factorial effect of an FFSPD is either orthogonal or confounded with other factorial effects. Confounded effects cannot appear in a model together as such models are not estimable. The analysis of FFSPDs will be ambiguous if a significant contrast estimates a linear combination of many effects of the same order. For example, if a resolution IV FFSPD has a significant two-factor interaction contrast, it is impossible to determine using a data-driven method which of the confounded two-factor interactions defined by the significant contrast is causing the observed effect.

Optimal split-plot designs generated using a computer algorithm provide flexibility in the choice of the number of whole plots and the total number of runs. Suppose an engineering team had two factors that were hard to change and five factors that were comparatively easy to change from run to run. The team, therefore, needs to run a split-plot experiment. The design budget

Table 10.4 D-optimal Split-plot Design for Seven Factors in Eight Whole Plots of Three Runs

Whole plots	w1	w2	s1	s2	s3	s4	s5	Y
1	−1	−1	−1	1	−1	−1	1	47.9
1	−1	−1	−1	−1	1	1	−1	45.5
1	−1	−1	1	1	−1	−1	1	52.3
2	1	−1	1	1	1	−1	−1	70.7
2	1	−1	−1	−1	−1	−1	1	46.5
2	1	−1	1	−1	1	1	1	46.7
3	−1	1	−1	1	1	1	1	37.4
3	−1	1	1	−1	1	−1	1	46.9
3	−1	1	1	1	−1	1	−1	51.5
4	1	1	1	1	−1	−1	−1	64.9
4	1	1	−1	1	−1	1	1	40.8
4	1	1	1	−1	1	−1	−1	63.1
5	−1	−1	−1	−1	−1	−1	−1	50.9
5	−1	−1	1	−1	−1	1	1	39.5
5	−1	−1	1	1	1	1	−1	53.4
6	−1	1	1	−1	−1	1	−1	47.6
6	−1	1	−1	−1	1	−1	−1	50.2
6	−1	1	−1	1	1	−1	1	43.6
7	1	1	−1	−1	1	1	1	36.2
7	1	1	1	−1	−1	−1	1	49.1
7	1	1	−1	1	−1	1	−1	50.6
8	1	−1	−1	−1	−1	1	−1	48.2
8	1	−1	−1	1	1	−1	−1	60.2
8	1	−1	1	1	1	1	1	49.4

is for 24 runs and the team is willing to field 8 whole plots of 3 runs each. Table 10.4 shows a D-optimal split-plot design for this case where the *a priori* model is the main effects model.

The columns of the design are orthogonal. Assuming σ_w^2 and σ_ε^2 are both equal to 1, the variance matrix of the coefficients is diagonal. See Table 10.5.

The y-values in Table 10.4 were generated by simulating the equation

$$y = 50 + 3w_1 - 2w_2 + 3s_1 + 2s_2 + 1s_3 - 3s_4 - 3s_5 \tag{10.6}$$

with $\sigma_w^2 = 1$ and $\sigma_\varepsilon^2 = 1$. The analysis from JMP is shown in Figure 10.2.

Note that the standard errors for the whole-plot factor coefficient estimates are about 0.378, which are larger than the standard errors of the subplot factor coefficient estimates (0.2098). This is generally true for split-plot designs. The variance of whole-plot factors is larger because in effect the whole-plot sample size is 8 while there are 24 subplot runs.

10.6 Recommendations for Use of Split-plot Designs

In industrial applications, whole-plot factors are hard to change. Often, there is a strong logistical impetus to keep the number of whole plots to an absolute minimum. For example, if there is only one whole-plot factor, the temptation would be to limit the experiment to two whole plots. If this happens, it is impossible to estimate both the whole-plot fixed effect and the whole-plot variance.

Table 10.5 Variance Matrix of the Coefficients

Intercept	w1	w2	s1	s2	s3	s4	s5
6	0	0	0	0	0	0	0
0	6	0	0	0	0	0	0
0	0	6	0	0	0	0	0
0	0	0	22	0	0	0	0
0	0	0	0	22	0	0	0
0	0	0	0	0	22	0	0
0	0	0	0	0	0	22	0
0	0	0	0	0	0	0	22

Parameter Estimates

Term	Estimate	Std Error	DFDen	t Ratio	Prob>|t|
Intercept	49.7125	0.378472	4.675	131.35	<.0001^
w1	2.4875	0.378472	4.675	6.57	0.0016*
w2	−1.220833	0.378472	4.675	−3.23	0.0257*
s1	3.1486439	0.209717	11.75	15.01	<.0001*
s2	2.1167786	0.209717	11.75	10.09	<.0001*
s3	0.5602981	0.209717	11.75	2.67	0.0207*
s4	−3.999038	0.209717	11.75	−19.07	<.0001*
s5	−5.12726	0.209717	11.75	−24.45	<.0001*

REML Variance Component Estimates

Random Effect	Var Ratio	Var Component	Std Error	95% Lower	95% Upper	Wald p-Value	Pct of To
Whole Plots	0.894545	0.8348409	0.7640559	−0.662681	2.3323629	0.2746	47.2
Residual		0.9332576	0.3979424	0.4683308	2.6903854		52.7
Total		1.7680985	0.7895384	0.8639505	5.4337106		100.0

Fixed Effect Tests

Source	Nparm	DF	DFDen	F Ratio	Prob > F
w1	1	1	4.675	43.1976	0.0016*
w2	1	1	4.675	10.4051	0.0257*
s1	1	1	11.75	225.4126	<.0001*
s2	1	1	11.75	101.8783	<.0001*
s3	1	1	11.75	7.1379	0.0207*
s4	1	1	11.75	363.6152	<.0001*
s5	1	1	11.75	597.7254	<.0001*

FIGURE 10.2 Fixed-effect Estimates Table, Variance Estimates, and Hypothesis Tests for the Data in Table 10.4

Theoretically, in this case, it would be possible to fit the whole-plot main effect and variance if there are three whole plots. However, the estimate of the variance based on one degree of freedom is highly variable, and there is a high probability that the REML will produce a negative variance component for the whole-plot variance. A good rule of thumb is to have at least two more whole plots than whole-plot parameters. Moreover, it is important to keep in mind that the intercept is always a whole-plot parameter.

Even with two whole plots above the minimum, the hypothesis tests for the whole-plot effects will have low power. That is because the effective sample size for the whole plots is the number of whole plots. A good practice for determining whether the whole-plot factor is important is to include the two-factor interactions involving whole-plot and subplot factors in the model. Generally, these effects are tested at the subplot level, which makes the tests substantially more powerful. If any of these interaction effects is highly significant, that is an indication that the whole-plot factor is important.

When randomizing the order of the whole plots, there is a good chance that two whole plots in a row will have the same level of a whole-plot factor. When this happens, it is necessary to change the whole-plot factor away from its current setting between whole plots. Alternatively, one can choose the levels of the whole-plot factors so that they must change between whole plots. Recall the BHH experiment at the beginning of this chapter. The temperatures for the third and fourth whole plots were both 380 degrees. It would be tempting not to change the temperature between these two whole plots. However, the effect of this decision is that instead of having six whole plots, there would only be five. This would likely lead to an incorrect analysis. A safer course is to order the whole plots so that all the whole-plot factor levels change between each pair of whole plots.

Finally, even though care has been taken in design to use enough whole plots to allow for good inference, there is always the chance that the actual experimental runs will not happen as specified. That is, if a person other than you is responsible for doing the runs, he/she may sort the whole-plot runs to do all the low values first followed by all the high values (or vice versa). It is important that everyone involved in the experiment understands the importance of keeping the order of the runs as specified. If at all possible, the best practice is to observe the experiment as it is being run to make sure that all the factor settings are correctly set and each run is done in the correct order.

11 Response Surface Methods

11.1 Introduction

Response surface methodology, or **RSM**, is a collection of mathematical and statistical techniques useful for the modeling and analysis of problems in which a response of interest is influenced by several variables and the objective is to optimize this response. For example, suppose that a chemical engineer wishes to find the settings of temperature (x_1) and time (x_2) that maximize the yield (y) of a process. The process yield is a function of the settings of temperature and time, say

$$y = f(x_1, x_2) + \varepsilon$$

where ε represents the noise or error observed in the response y. If we denote the expected response by $E(y) = f(x_1, x_2) = \eta$, then the surface represented by $\eta = f(x_1, x_2)$ is called a **response surface**.

We usually represent the response surface graphically, as shown in Figure 11.1, where the response y is plotted versus the levels of x_1 and x_2. We have seen response surface plots such as these previously, particularly in the chapters on factorial designs, Chapter 7 and 8. To help visualize the shape of a response surface, we often plot the contours of the response surface as shown in Figure 11.2. In the contour plot, lines of constant response are drawn in the x_1, x_2 plane. Each contour corresponds to a particular height of the response surface. We have also previously seen the utility of contour plots. In most RSM problems, the form of the relationship between the response and the independent variables is unknown. Therefore, the first step in RSM is to find a suitable approximation for the true functional relationship between y and the independent variables. Usually, the experimenter employs a low-order polynomial in some region of the independent variables. If the response is well modeled by a linear function of the independent variables, then the appropriate approximating function is the **first-order model**

$$y = \beta_0 + \beta_1 x_1 + \beta_2 x_2 + \cdots + \beta_k x_k + \varepsilon \tag{11.1}$$

If there is curvature in the system, then a polynomial of higher degree must be used, usually the **second-order model**:

$$y = \beta_0 + \sum_{i=1}^{k} \beta_i x_1 + \sum_{i=1}^{k} \beta_{ii} x_i^2 + \sum_{i<j}\sum \beta_{ij} x_i x_j + \varepsilon \tag{11.2}$$

Almost all RSM problems use one or both of these models. Of course, it is unlikely that a polynomial model will be a reasonable approximation of the true functional relationship over the entire space of the independent variables, but for a relatively small region, they are often adequate.

The method of least squares is used to estimate the parameters in the approximating polynomials. The response surface analysis is performed using the fitted surface. If the fitted surface is

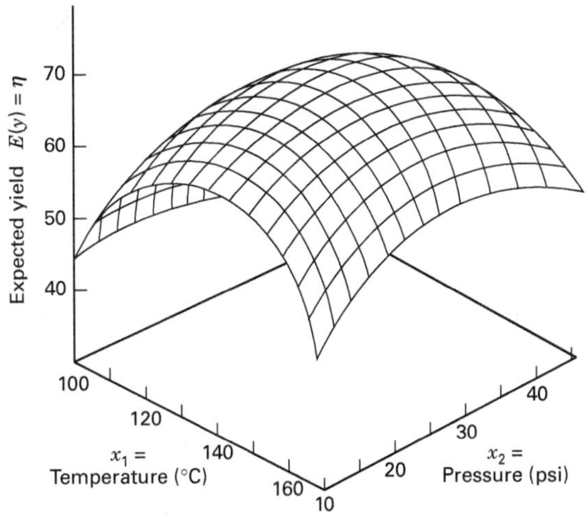

FIGURE 11.1 A Three-dimensional Response Surface Showing the Expected Yield (η) as a Function of Temperature (x_1) and Time (x_2)

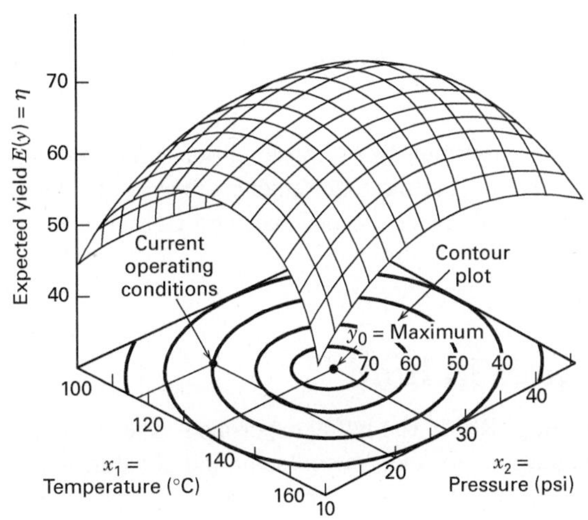

FIGURE 11.2 A Contour Plot of a Response Surface

an adequate approximation of the true response function, then analysis of the fitted surface will be approximately equivalent to analysis of the actual system. The model parameters are estimated most effectively if good experimental designs are used to collect the data. We call designs for fitting response surface models **response surface designs**. We discuss these designs in Section 11.3.

Experimenters often deploy RSM as a **sequential procedure**. The first phase of an RSM study is often a factor screening experiment to identify the active factors that affect the response. This serves to reduce the number of factors used in the RSM deployment. In the screening experiment, we often discover that we are at a point on the response surface that is far from the optimum, such as the current operating conditions shown in Figure 11.3. When this occurs, we often observe that there is little curvature in the system and the first-order model will be appropriate. Our objective here is to lead the experimenter rapidly and efficiently along a path of improvement toward the general vicinity of the **optimum**. Once we have arrived in the region of the optimum, a more elaborate model, such as the second-order model, may be employed, and an analysis is performed to locate the optimum. From Figure 11.3, we see that the analysis of a response surface can be thought of as "climbing a hill," where the top of the hill represents the point of maximum response. If the true optimum is a point of minimum response, then we may think of "descending into a valley."

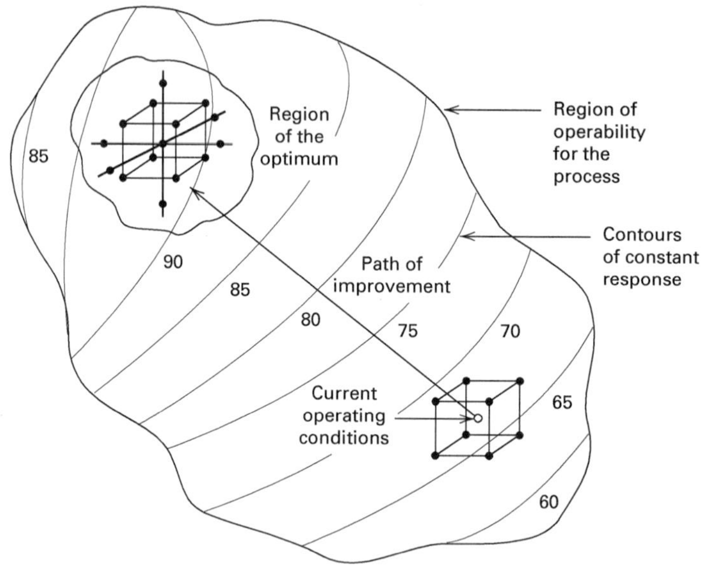

FIGURE 11.3 The Sequential Application of Response Surface Methodology

The eventual objective of RSM is to determine the optimum operating conditions for the system or to determine a region of the factor space in which operating requirements are satisfied. More extensive presentations of RSM and many useful references are in Khuri and Cornell (1996), Myers et al. (2016), and Box and Draper (2007).

11.2 Optimization Techniques in RSM

Figure 11.3 illustrates the sequential deployment of RSM. The initial experiment is usually performed around either the current operating conditions for the system, some initial estimates of these conditions that are often obtained from some smaller-scale operations such as a pilot plant or a development laboratory, or from a preliminary screening experiment. If there is an indication of curvature in the fitted response surface, the experimenter usually proceeds to fit a second-order model and estimate the optimum operating conditions. However, sometimes we find that the response values obtained at this location are disappointing; for example, a yield response may be 60% or 70%, and we desire much higher yields. In that case, the experimenter probably wants to shift the region of experimentation to an area where encountering better results are likely. This is the "path of improvement" shown in Figure 11.3.

Suppose that the experimenter has fit a first-order model in three variables, say

$$\hat{y} = 25 + 12x_1 + 8x_2 - 4x_3$$

and the objective is to maximize the response y. If we want to move as rapidly as possible toward optimum, the steps along this "path of improvement" should be in the **gradient direction**. In other words, the steps should be proportional to the model regression coefficients. In our example, this implies that we should move an amount equal to a $12/12 = 1$ coded unit step in the positive x_1 direction, $8/12 = 2/3 = 0.67$ coded units step in the positive x_2 direction, and $-4/12 = -1/3 = -0.33$ coded units step in the negative x_3 direction. Every step along this path will be proportional to these step sizes. This direction is usually called the direction of steepest ascent. If we were trying to minimize the response, then the steps would be in the negative direction (reversing the signs of the estimates of the model parameters) and this would be a direction of steepest descent. Operationally we take steps along this path and observe the response at each new point until the response changes the direction. Sometimes a second round of steepest ascent is performed, resulting in a "mid-course correction" that leads to the region that likely contains the optimum. For detailed examples of the method of steepest ascent, see Myers et al. (2016). Adding points solely on the path of steepest ascent is like a one-factor experiment. If you are not a huge distance from the optimum, it may be of value to choose points having more volume in the space of interest. In so doing, you may improve your model while moving in the direction of the desired response. See Nachtsheim and Jones (2018) for more details.

When the experimenter is relatively close to the optimum, a model that incorporates curvature is usually required to approximate the response. In most cases, the second-order model in Eq. (11.2)

$$y = \beta_0 + \sum_{i=1}^{k} \beta_i x_1 + \sum_{i=1}^{k} \beta_{ii} x_i^2 + \sum_{i<j}\sum \beta_{ij} x_i x_j + \varepsilon$$

is adequate. We now show how to use this fitted model to find the optimum set of operating conditions for the xs and to characterize the nature of the response surface.

Suppose that $\hat{\boldsymbol{\beta}}$ is the least squares estimate of the vector of model parameters $\boldsymbol{\beta}$ in Eq. (11.2). We can write the fitted model in matrix form as

$$\hat{y} = \beta_0 + \mathbf{x}'\boldsymbol{b} + \mathbf{x}'\mathbf{B}\mathbf{x} \tag{11.3}$$

where

$$\mathbf{x} = \begin{bmatrix} x_1 \\ x_2 \\ \vdots \\ x_k \end{bmatrix} \quad \boldsymbol{b} = \begin{bmatrix} \hat{\beta}_1 \\ \hat{\beta}_2 \\ \vdots \\ \hat{\beta}_k \end{bmatrix} \quad \text{and } \mathbf{B} = \begin{bmatrix} \hat{\beta}_{11}, & \hat{\beta}_{12}/2, & \cdots, & \hat{\beta}_{1k} \\ & \hat{\beta}_{22}, & \cdots, & \hat{\beta}_{2k} \\ & & \ddots & \vdots \\ & & & \hat{\beta}_{kk} \end{bmatrix}$$

That is, $\boldsymbol{b}$ is a $k \times 1$ vector of the estimates of the first-order model parameters, and $\mathbf{B}$ is a $k \times k$ symmetric matrix whose main diagonal elements are the estimates of the quadratic or second-order model parameters and whose off-diagonal elements are one-half the estimates of the interaction or mixed second-order model parameters. The optimum point on the fitted surface is the point at which all of the first derivatives of the fitted model with respect to $\mathbf{x}$ equal zero. This point is called the **stationary point** and denoted $\mathbf{x}_s$. The derivative of $\hat{y}$ with respect to the elements of the vector $\mathbf{x}$ equated to $\mathbf{0}$ is

$$\frac{\partial \hat{y}}{\partial \mathbf{x}} = \boldsymbol{b} + 2\mathbf{B}\mathbf{x} = \mathbf{0} \tag{11.4}$$

The stationary point is the solution to Eq. (11.4), or

$$\mathbf{x}_s = -\frac{1}{2}\mathbf{B}^{-1}\boldsymbol{b} \tag{11.5}$$

assuming that the matrix $\mathbf{B}$ can be inverted. Furthermore, by substituting Eq. (11.5) into Eq. (11.3), we can find the predicted response at the stationary point as

$$\hat{y}_s = \hat{\beta}_0 + \frac{1}{2}\mathbf{x}_s'\boldsymbol{b} \tag{11.6}$$

Once we have found the stationary point, it is usually necessary to characterize the response surface in the immediate vicinity of this point. By **characterize**, we mean determining whether the stationary point is a point of maximum or minimum response or a saddle point. We also usually want to study the relative sensitivity of the response to the variables $x_1, x_2, \ldots, x_k$. Figures 11.4, 11.5, and 11.6 illustrate these three types of fitted response surfaces.

A relatively straightforward way to do this is to examine a **contour plot** of the fitted model as shown in Figures 11.4, 11.5, and 11.6. If there are only two or three process variables (the *x*s), the construction and interpretation of this contour plot is relatively easy. However, even when there are relatively few variables, a more formal analysis, called the **canonical analysis**, can be useful.

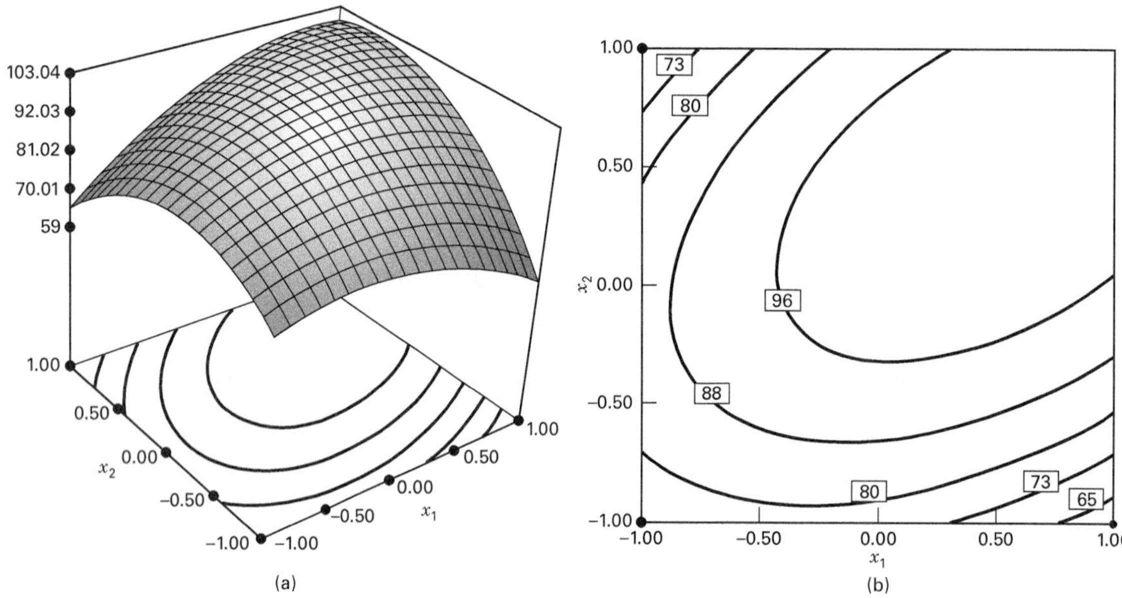

FIGURE 11.4 (a) Response Surface and (b) Contour Plot Illustrating a Surface with a Maximum

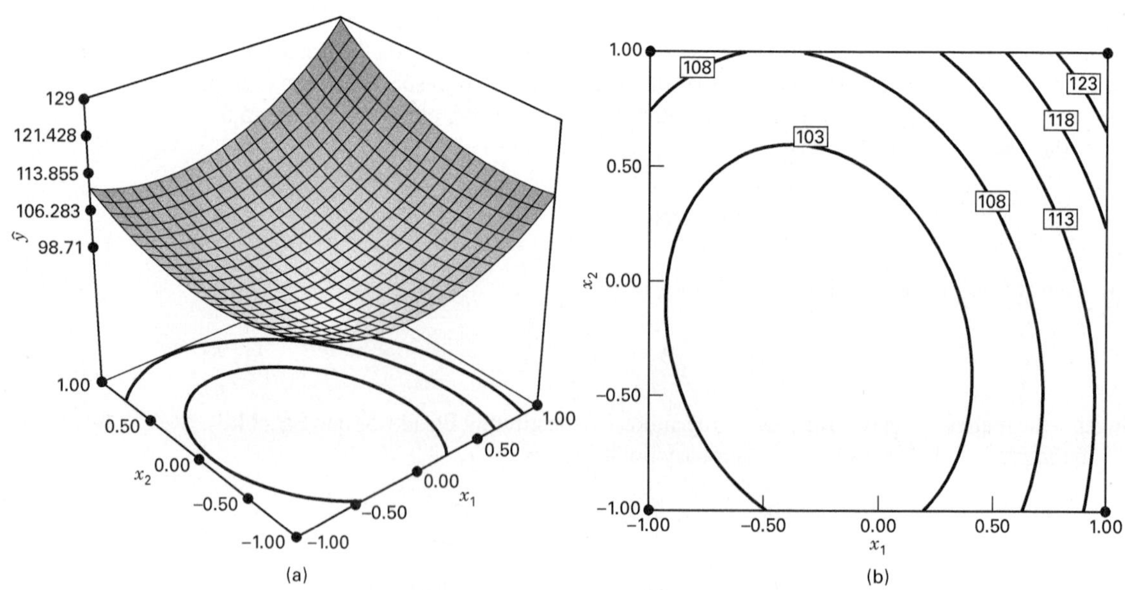

FIGURE 11.5 (a) Response Surface and (b) Contour Plot Illustrating a Surface with a Minimum

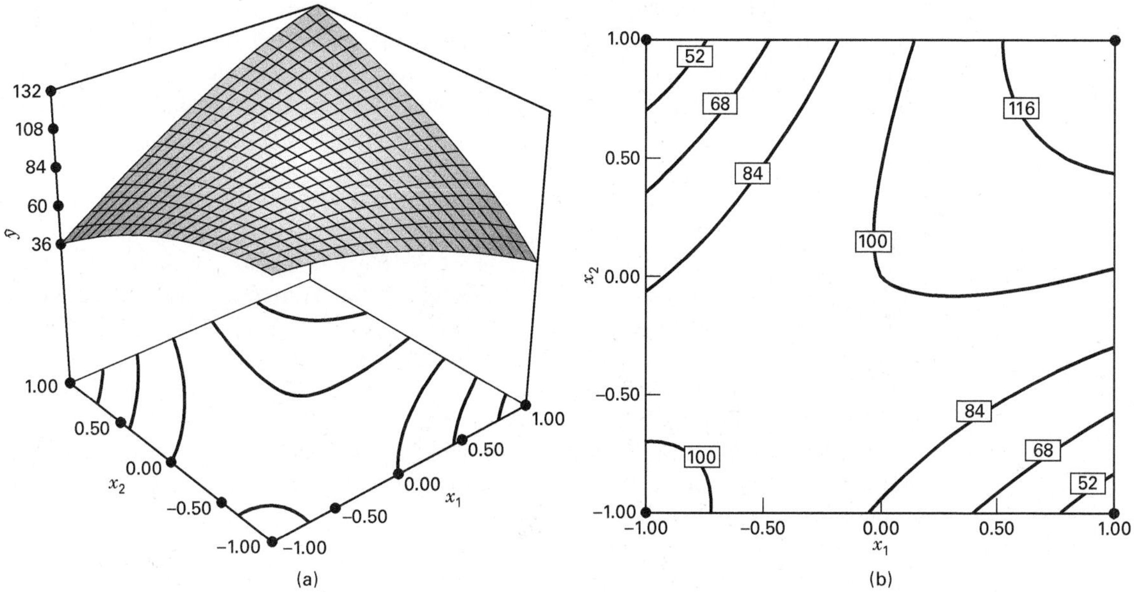

FIGURE 11.6 (a) Response Surface and (b) Contour Plot Illustrating a Saddle Point (or Minimax)

To perform the canonical analysis, we must first transform the model into a new coordinate system with the origin at the stationary point $\mathbf{x}_s$ and then rotate the axes of this system until they are parallel to the principal axes of the fitted response surface. This transformation is illustrated in Figure 11.7. We can show that this results in the fitted model

$$\hat{y} = \hat{y}_s + \lambda_1 w_1^2 + \lambda_2 w_2^2 + \cdots + \lambda_k^2 w_k^2 \tag{11.7}$$

where the $\{w_I\}$ are the transformed independent variables (usually called the **canonical variables**) and the $\{\lambda_i\}$ are constants. Equation (11.7) is called the **canonical form** of the model. Furthermore, the $\{\lambda_i\}$ are just the **eigenvalues** or **characteristic roots** of the matrix **B**.

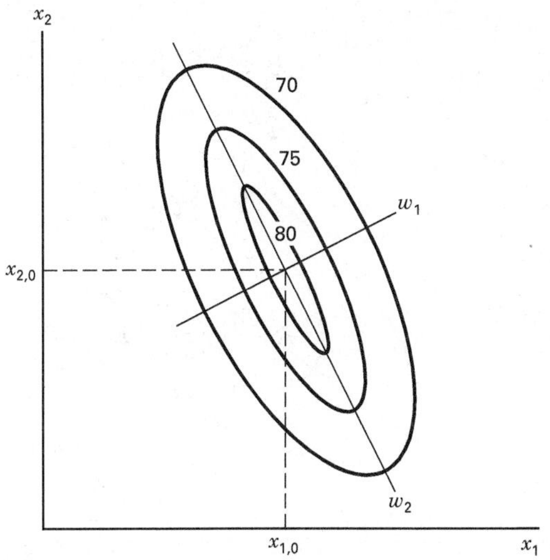

FIGURE 11.7 Canonical Form of the Second-order Model

The nature of the response surface can be determined from the stationary point and the **signs** and **magnitudes** of the eigenvalues $\{\lambda_i\}$. First, suppose that the stationary point is within the region of exploration for fitting the second-order model. If the $\{\lambda_i\}$ are all positive, the stationary point is a point of minimum response; if all of the $\{\lambda_i\}$ are negative, the stationary point is a point of maximum response; and if the $\{\lambda_i\}$ have different signs, the stationary point is a saddle point. Furthermore, the surface is steepest in the w_I direction for which $|\lambda_i|$ is the greatest. For example, Figure 11.7 depicts a system for which the stationary point is a maximum (λ_1 and λ_2 are negative) with $|\lambda_1| > |\lambda_2|$.

It is not unusual to encounter variations of the pure maximum, minimum, or saddle point response surfaces discussed in the previous section. **Ridge systems**, in particular, are fairly common. Consider the canonical form of the second-order model given in Eq. (11.7). Now suppose that the stationary point is within the region of experimentation; furthermore, let one or more of the eigenvalues be very small (e.g. $\lambda_i \simeq 0$). The response variable is then very insensitive to the variables w_i multiplied by the small eigenvalue λ_i.

A contour plot illustrating this situation is shown in Figure 11.8 for $k = 2$ variables with $\lambda_1 = 0$. (In practice, λ_1 would be close to but not exactly equal to zero.) The canonical model for this response surface is theoretically

$$\hat{y} = \hat{y}_s + \lambda_2 w_1^2$$

with λ_2 negative. Notice that the severe elongation in the w_1 direction has resulted in a line of centers at $\hat{y} = 70$ and the optimum may be taken anywhere along that line. This type of response surface is called a **stationary ridge system**.

If the stationary point is far outside the region of exploration for fitting the second-order model and one (or more) λ_i is near zero, then the surface may be a **rising ridge**. Figure 11.9 illustrates

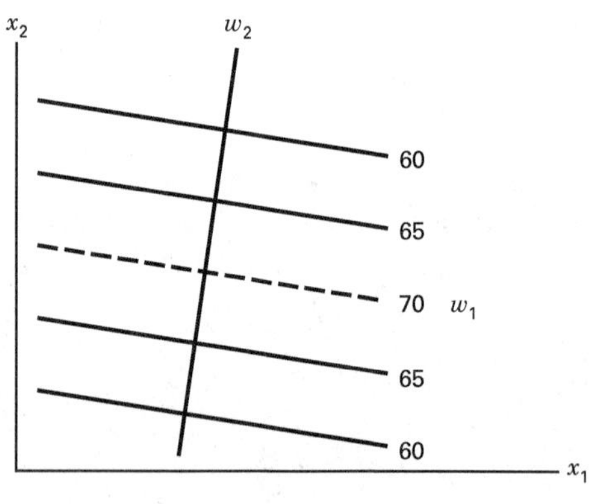

FIGURE 11.8 A Stationary Ridge System

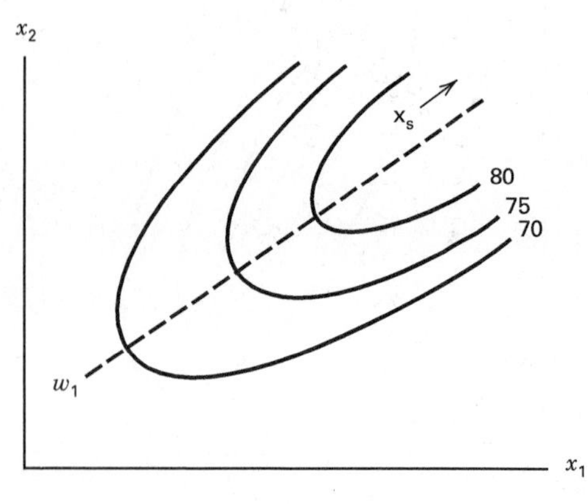

FIGURE 11.9 A Rising Ridge System

a rising ridge for $k = 2$ variables with λ_1 near zero and λ_2 negative. In this type of ridge system, we cannot draw inferences about the true surface or the stationary point because $\mathbf{x}_s$ is outside the region where we have fitted the model. However, further exploration is warranted in the w_1 direction. If λ_2 had been positive, we would call this system a falling ridge.

EXAMPLE 11.1

An engineering team conducted an experiment to improve the performance of a chemical process. Table 11.1 shows the experimental design and the results. There were two process variables, time and temperature, and two responses were observed, yield and molecular weight. The experimental design used in this study is **a central composite design (CCD)**, one of the classical RSM designs. The objective of the experiment is to find operating conditions on time and temperature that maximize yield. Table 11.2 shows the JMP RSM analysis.

The first part of the computer output in Table 11.2 shows a plot of the observed versus the predicted yield from the second-order model. The fit looks very good and the R^2 reported is 0.98, indicating that nearly all of the variability in yield is explained by the model. The Effect Summary indicated that all of the model terms are highly significant, except the interaction term, for which the *P*-value is only about 0.13. This term could possibly be removed from the model, although its removal has minimal impact on the results. There is no discernable lack of fit. The residual plots are satisfactory. The Analysis of Variance indicates that the overall model is significant and a table of the effect estimates follows.

In the Response Surface section of the output, the stationary point is given as Time = 86.8 minutes and Temperature = 176.4 degrees (°C). Both eigenvalues are negative so the response surface contains a maximum and the predicted response at the stationary point is approximately 80.2%. Furthermore, the eigenvalues are similar in magnitude, so the surface is not elongated extensively in either direction. The Prediction Profiler is positioned at the optimum conditions and the yield response surface plot shows a strong indication of a maximum.

Many response surface problems involve the analysis of several responses. For instance, in Example 11.1, the engineering team measured two responses, yield and molecular weight. In this example, we optimized the process with respect to only the yield response, say y_1. However, the objective of the experiment was to maximize both yield and molecular weight.

Table 11.1 The Chemical Process Experiment

Time, x_1	Temperature, x_2	x_1 coded	x_2 coded	Yield	Molecular weight
85	167.92893219	0	−1.414	77	57
80	170	−1	−1	76.5	62
90	170	1	−1	78	66
77.928932188	175	−1.414	0	76.6	71
85	175	0	0	79.9	72
85	175	0	0	80.3	68
85	175	0	0	80	69
85	175	0	9	79.7	70
92.071067812	175	1.414	0	78.4	68
80	180	−1	1	77	60
90	180	1	1	79.5	59
85	182.07106781	0	1.414	78.5	58

Table 11.2 JMP Output for Example 11.1

Response Yield
Actual by Predicted Plot

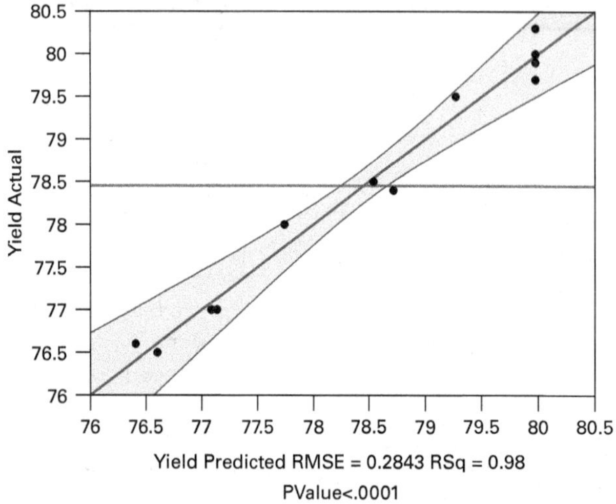

Effect Summary

Source	LogWorth	PValue
Time*Time	4.413	0.00004
Temperature*Temperature	4.142	0.00007
Time(80,90)	3.733	0.00018 ^
Temperature(170,180)	2.664	0.00217 ^
Time*Temperature	0.889	0.12914

Lack of Fit

Source	DF	Sum of Squares	Mean Square	F Ratio
Lake of Fit	3	0.29750542	0.099168	1.5867
Pure Error	3	0.18750000	0.062500	Prob > F
Total Error	6	0.48500542		0.3569
				Max RSq
				0.9916

Residual by Predicted Plot

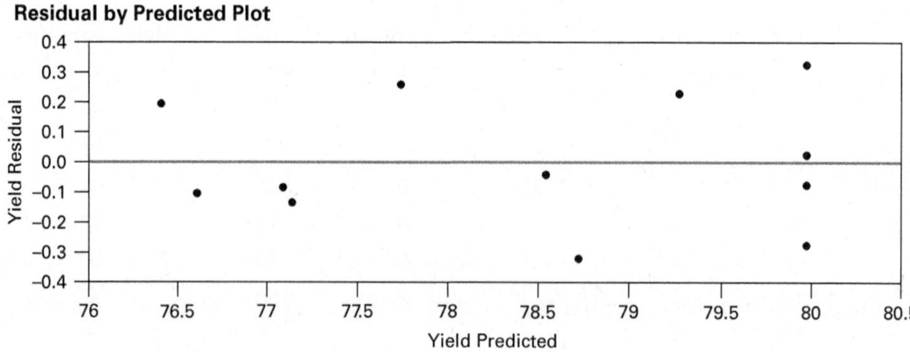

Table 11.2 ***(Continued)***

Studentized Residuals

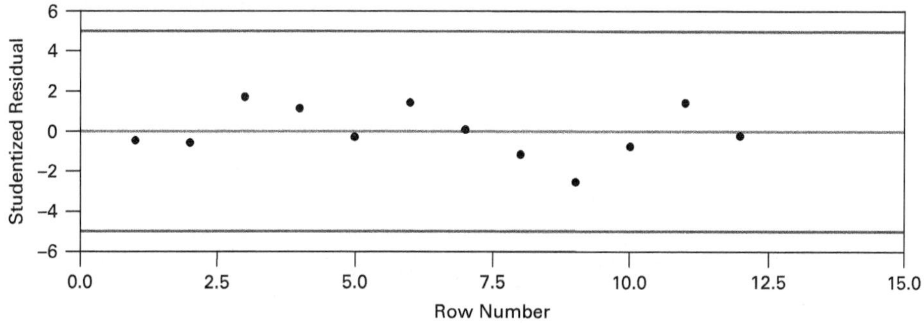

Externally Studentized Residuals with 95% Simultaneous Limits (Bonferroni)

Analysis of Variance

Source	DF	Sum of Squares	Mean Square	F Ratio
Model	5	21.744995	4.34900	53.8014
Error	6	0.485005	0.08083	**Prob > F**
C. Total	11	22.230000		<.0001*

Parameter Estimates

Term	Estimate	Std Error	t Ratio	Prob>|t|
Intercept	79.975	0.142157	562.58	<.0001*
Time(80,90)	0.8181981	0.10052	8.14	0.0002*
Temperature(170,180)	0.515165	0.10052	5.12	0.0022*
Time*Temperature	0.25	0.142157	1.76	0.1291
Time*Time	−1.20625	0.112385	−10.73	<.0001*
Temperature*Temperature	−1.08125	0.112385	−9.62	<.0001*

Response Surface

Coef

	Time(80,90)	Temperature(170,180)	Yield
Time(80,90)	−1.20625	0.25	0.8181981
Temperature(170,180)	.	−1.08125	0.515165

Solution

Variable	Critical Value
Time(80,90)	86.841239
Temperature(170,180)	176.40399

Solution is a Maximum

Predicted Value at Solution 80.197979

Canonical Curvature

Eigenvalues and Eigenvectors

Eigenvalue	−1.0040	−1.2835
Time(80,90)	0.52573	0.85065
Temperature (170,180)	0.85065	−0.52573

(continued)

Table 11.2 (*Continued*)

Prediction Profiler

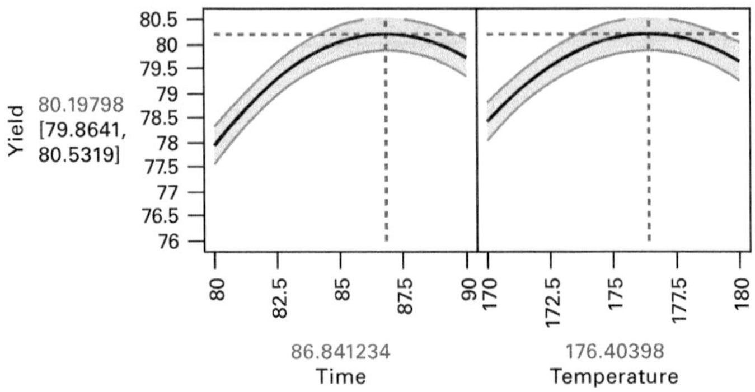

Yield

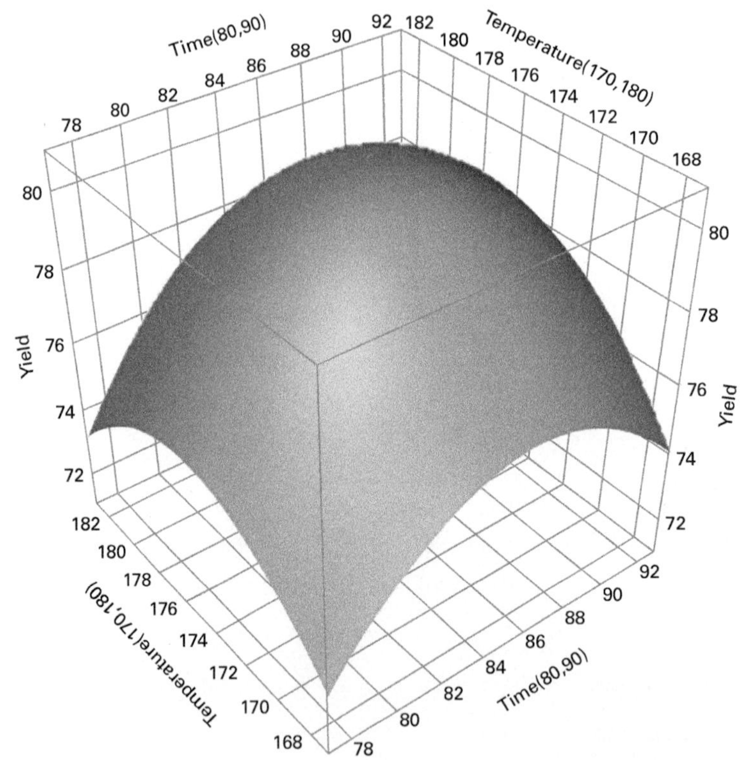

Response Grid Slider

76 [ ]

Independent Variables

X	Y		Value	Grid
(x)	( )	Time(80,90)	85	[ ]
( )	(x)	Temperature(170,180)	175	[ ]

Simultaneous consideration of multiple responses involves first building an appropriate response surface model for each response and then trying to find a set of operating conditions that in some sense optimize all responses or at least keep them in desired ranges. Myers et al. (2016) gave an extensive treatment of the multiple response optimization problem. When there are only two or three process variables in the experiment a simple way to handle the multiple response optimization is to overlay contour plots. We now show how to apply this approach to the experiment in Example 11.1.

Table 11.3 is the JMP output for the molecular weight response for the CCD in Table 11.1. The dominant term in the model is the quadratic term in temperature and the one of the eigenvalues is considerably smaller than the other. In addition, the optimum is within the region of experimentation, so the impression is that the response surface is a stationary ridge system. At the bottom of the table, an overlay contour plot of the responses yield and molecular weight is shown. From this plot, it is clear that there are several combinations of time and temperature that would produce large values of both responses. We will show subsequently that Time = 86.4 minutes and Temperature = 175.2 degrees C is a good solution.

Another useful approach to optimization of multiple responses is to use the simultaneous optimization technique popularized by Derringer and Suich (1980). Their procedure makes use of **desirability functions**. The general approach is to convert each response y_i into an individual desirability function d_i that varies over the interval $0 \le d_i \le 1$, where if the response y_i is at its goal or target, then $d_i = 1$ and if the response is outside an acceptable region, $d_i = 0$. Then the experimenter selects levels of the design variables to maximize the **overall desirability,** which is the geometric mean of the individual desirabilites as shown in Eq. (11.8):

$$D = (d_1 \times d_2 \times \cdots \times d_m)^{1/m} \tag{11.8}$$

where there are m responses. The overall desirability will be zero if any of the individual responses is undesirable.

The individual desirability functions are structured as shown in Figure 11.10. If the objective or target T for the response y is a maximum value, then

$$d = \begin{cases} 1, y < T \\ \left(\dfrac{U-y}{U-T}\right)^r, L \le y \le T \\ 0, y > U \end{cases}$$

$$d = \begin{cases} 0, y < L \\ \left(\dfrac{y-L}{T-L}\right)^{r_1}, L \le y \le T \\ \left(\dfrac{U-y}{U-T}\right)^{r_2}, T \le y \le U \\ 0, y > U \end{cases} \tag{11.9}$$

when the weight $r = 1$, the desirability function is linear. Choosing $r > 1$ places more emphasis on being close to the target value and choosing $0 < r < 1$ makes this less important. If the target for the response is a minimum value,

$$d = \begin{cases} 1, y < T \\ \left(\dfrac{U-y}{U-T}\right)^r, L \le y \le T \\ 0, y > U \end{cases} \tag{11.10}$$

Table 11.3 JMP Output for the Molecular Weight Response for the Experiment in Table 11.1 Least Squares Fit

Response Molecular Weight
Actual by Predicted Plot

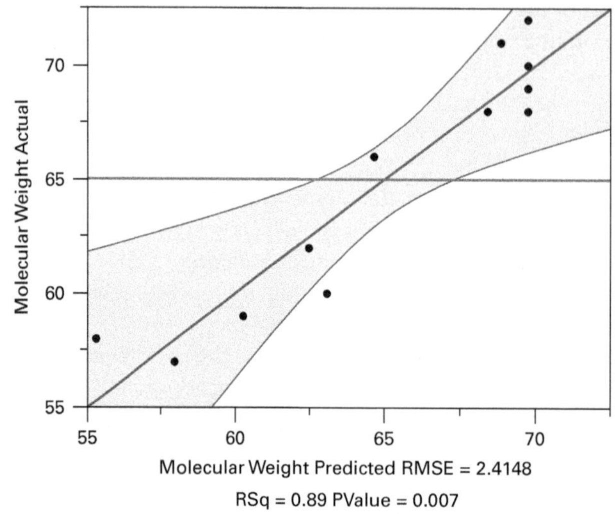

Residual by Predicted Plot

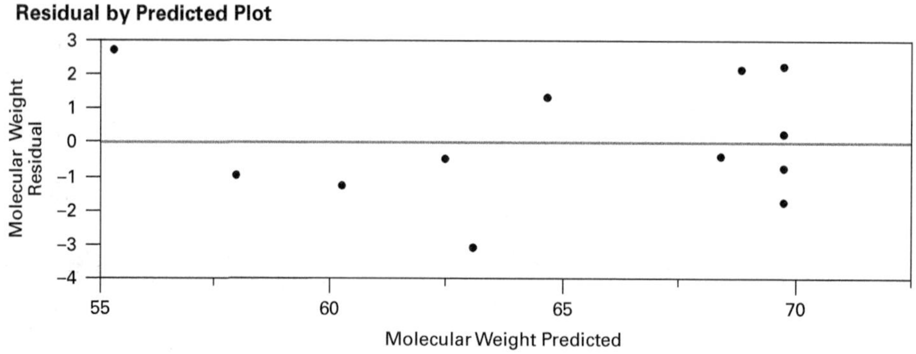

Studentized Residuals

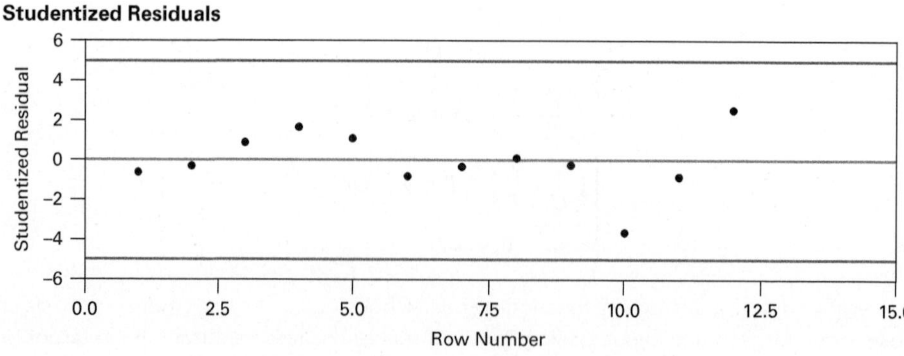

Externally Studentized Residuals with 95% Simultaneous Limits (Bonferroni)

Table 11.3 ***(Continued)***

Summary of Fit

RSquare	0.893326
RSquare Adj	0.804432
Root Mean Square Error	2.414849
Mean of Response	65
Observations (or Sum Wgts)	12

Parameter Estimates

Term	Estimate	Std Error	t Ratio	Prob>|t|
Intercept	69.75	1.207424	57.77	<.0001*
Time(80,90)	−0.15533	0.853778	−0.18	0.8616
Temperature(170,180)	−0.948223	0.853778	−1.11	0.3092
Time*Temperature	−1.25	1.207424	−1.04	0.3405
Time*Time	−0.5625	0.954553	−0.59	0.5772
Temperature*Temperature	−6.5625	0.954553	−6.87	0.0005*

Response Surface

Coef

	Time(80,90)	Temperature(170,180)	Molecular Weight
Time(80,90)	−0.5625	−1.25	−0.15533
Temperature(170,180)	.	−6.5625	−0.948223

Solution

Variable	Critical Value
Time(80,90)	84.676808
Temperature(170,180)	174.66955

Solution is a Maximum
Predicted Value at Solution 69.786354

Canonical Curvature

Eigenvalues and Eigenvectors

Eigenvalue	−0.4981	−6.6269
Time	0.99473	0.10252
Temperature	−0.10252	0.99473

Contour Profiler

Horiz	Vert	Factor		Current X
◉	○	Time		86.407442
○	◉	Temperature		175.18373

Response		Contour	Current Y	Lo Limit	Hi Limit
— Yield		78.25	80.129791	.	.
--- Molecular Weight		63.75	69.605073	.	.

(continued)

Table 11.3 (*Continued*)

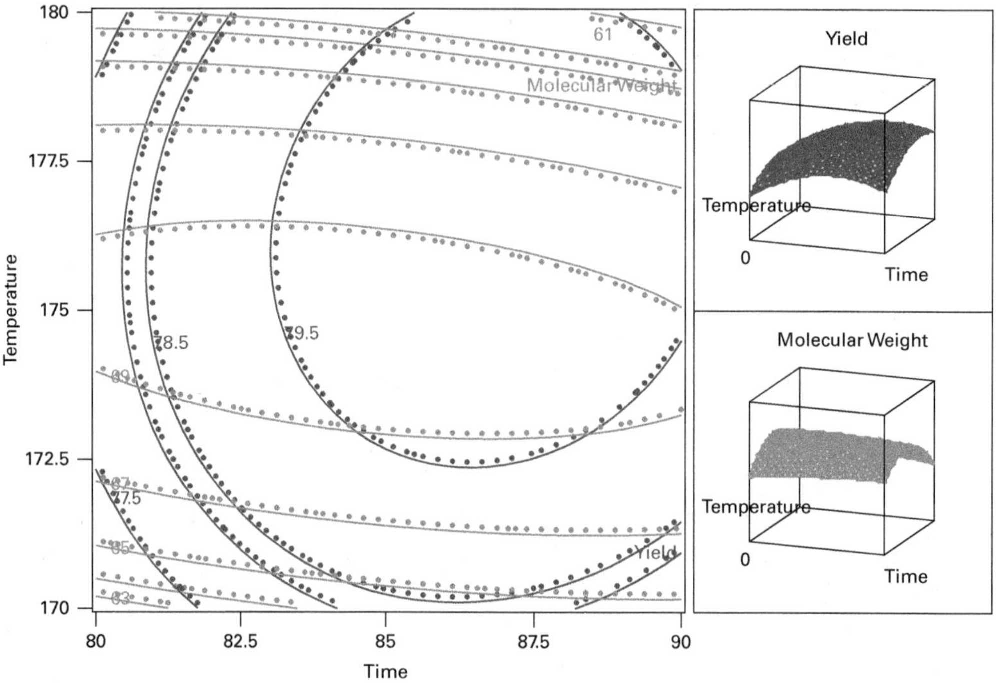

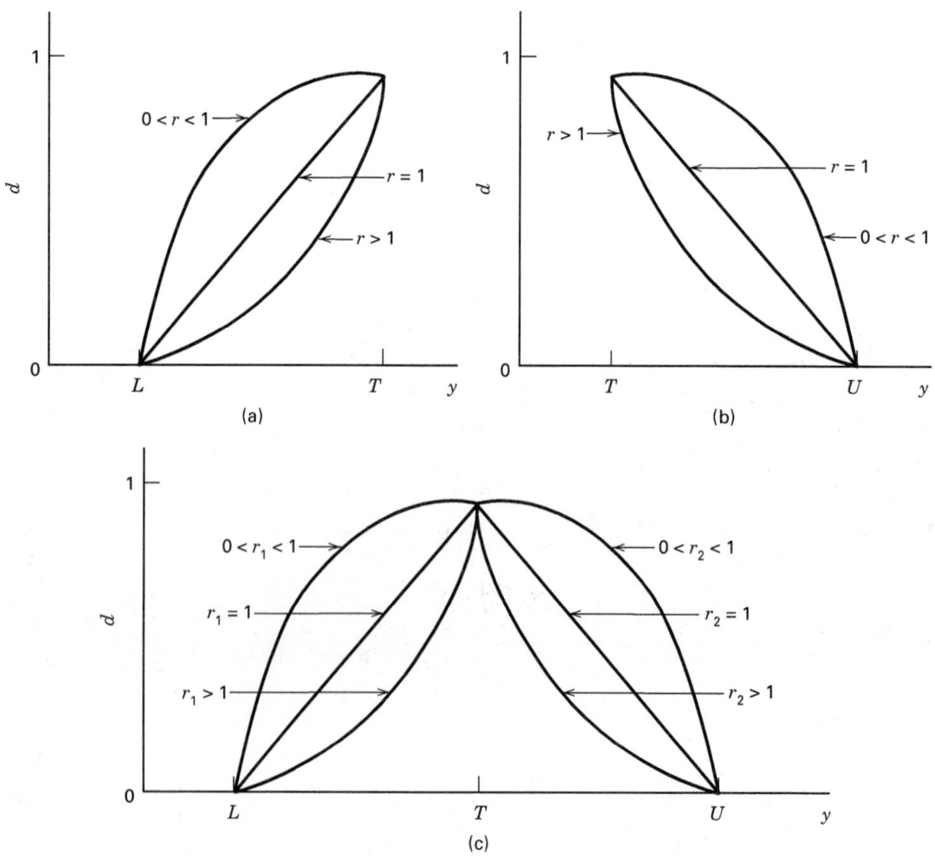

FIGURE 11.10
Individual Desirability Functions: (a) Objective (Target) Is to Maximize y (b) Objective (Target) Is to Minimize y (c) Objective is for y to Be As Close As Possible to the Target

The two-sided desirability function shown in Figure 11.10c assumes that the target is located between the lower (*L*) and upper (*U*) limits and is defined as

$$d = \begin{cases} 0, y < L \\ \left(\dfrac{y-L}{T-L}\right)^{r_1}, L \le y \le T \\ \left(\dfrac{U-y}{U-T}\right)^{r_2}, T \le y \le U \\ 0, y > U \end{cases} \tag{11.11}$$

Figure 11.11 presents the Prediction Profiler in JMP with the weights for the individual desirability functions set to unity and both responses are to be maximized. The optimal solution is shown on the horizontal axis of the time and temperature graphs.

Note that the optimal desirability is a little to the left of the maximal yield and a little to the right of the maximal molecular weight in the temperature graph. If yield were the more important response, you could choose to increase the temperature setting.

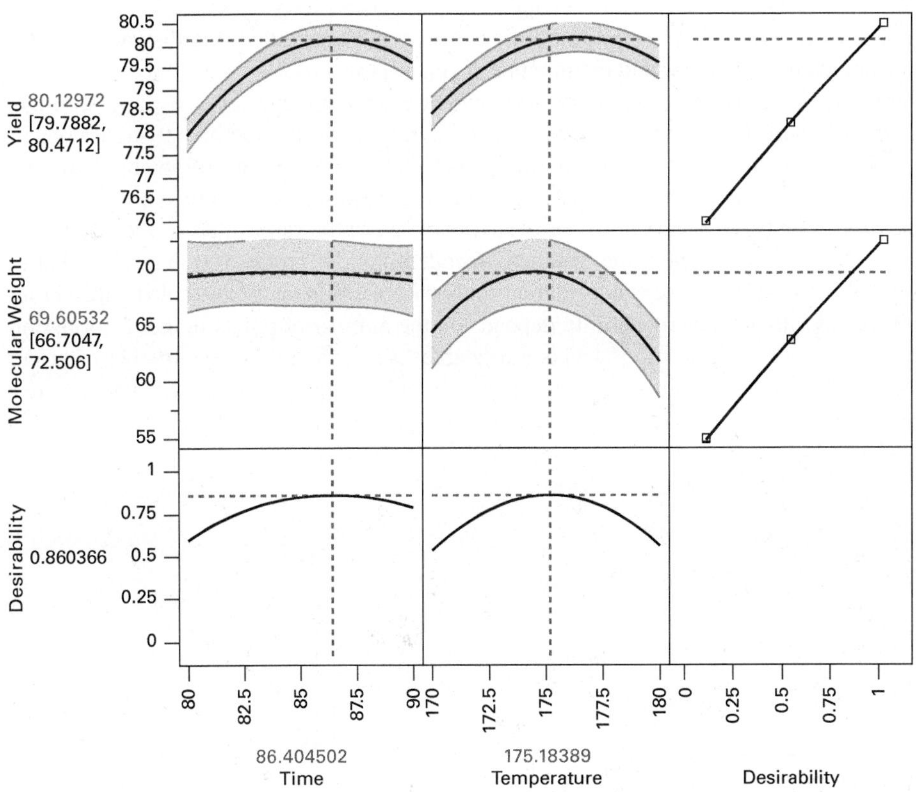

FIGURE 11.11
JMP Prediction Profiler Displaying the Optimal Solution for the Experiment in Example 11.1

11.3 Response Surface Designs

11.3.1 Classical Response Surface Designs

Suppose that we want to fit the first-order model in k variables. There is a unique class of designs that minimize the variance of the model regression coefficients. These are the **orthogonal first-order designs**. A first-order design is orthogonal if the off-diagonal elements of the $(\mathbf{X}'\mathbf{X})$ matrix are all zero. This implies that the cross products of the columns of the $\mathbf{X}$ matrix sum to zero. The class of orthogonal first-order designs includes the 2^k factorial and fractions of the 2^k series in which main effects are not aliased with each other, that is, designs of at least resolution III. In using these designs, we assume that the low and high levels of the k factors are coded to the usual ± 1 levels.

There are two designs that have been widely used in fitting second-order response surface models. The first of these is the **CCD,** which was introduced in Example 11.1. The CCD consists of a 2^k factorial (or fractional factorial of resolution V) with n_F factorial runs, 2^k axial or star runs, and n_C center runs. Figure 11.12 shows the CCD for $k = 2$ and $k = 3$ factors.

The practical deployment of a CCD often arises through **sequential experimentation**. That is, a 2^k design was used to fit a first-order model, this model has exhibited lack of fit, and the axial runs are then added to allow the quadratic terms to be incorporated into the model. The CCD is a very efficient design for fitting the second-order model in this sequential experimentation approach. There are two parameters in the design that must be specified: the distance α of the axial runs from the design center and the number of center points n_C.

It is important for the second-order model to provide good predictions of the response throughout the region of interest. One way to define "good" is to require that the model should have a reasonably consistent and stable variance of the predicted response at points of potential interest. Box and Hunter (1957) suggested that a second-order response surface design should be **rotatable**. This means that the variance of the predicted response is the same at all of the points $\mathbf{x}$ in the region of interest that are the same distance, in coded units, from the design center. That is, the variance of predicted response is constant on spheres. A CCD is made rotatable by the choice of α. The value of α to achieve rotatability depends on the number of points in the factorial portion of the design; in fact, $\alpha = (n_F)^{1/4}$ yields a rotatable CCD where n_F is the number of points used in the factorial portion of the design. For instance, in Example 11.1, the CCD had a 2^2 factorial design in the factorial portion, and since this design has four runs, the rotatable value of $\alpha = (4)^{1/4} = 1.414$.

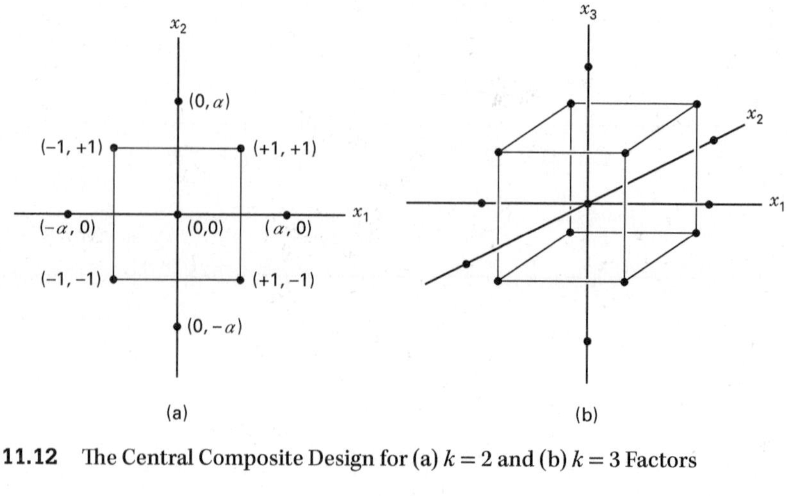

FIGURE 11.12 The Central Composite Design for (a) $k = 2$ and (b) $k = 3$ Factors

Rotatability is a **spherical property**; that is, it makes the most sense as a design criterion when the region of interest is a sphere. However, it is not important to have exact rotatability to have a good design. For a spherical region of interest, a good choice, from a prediction variance viewpoint for the CCD, is to set $\alpha = \sqrt{k}$. This design, called a **spherical CCD**, puts all the factorial and axial design points on the surface of a sphere of radius $\sqrt{k}$. There is usually a little practical difference between the rotatable and the spherical CCD. In fact, for $k = 2$ both designs are the same. For more discussion of this, see Myers et al. (2016).

In many situations, the region of interest is cuboidal rather than spherical. In these cases, a useful variation of the CCD is the **face-centered CCD** or the **face-centered cube**, in which $\alpha = 1$. This design locates the star or axial points on the centers of the faces of the cube, as shown in Figure 11.13 for $k = 3$. This variation of the CCD is also sometimes used because it requires only three levels of each factor, and in practice it is frequently difficult to change the factor levels. However, note that face-centered CCDs are not rotatable.

The CCD is not an optimal design. However, the rotatable and spherical CCDs enjoy reasonably high D-efficiencies. Generally, CCDs have D-efficiencies from about 80% to in excess of 95% depending on the number of factors and center runs. The face-centered cube design has D-efficiencies from over 85% to above 95%. Center runs generally tend to reduce the D-efficiency.

The choice of α in the CCD is dictated primarily by the region of interest. When this region is a sphere, the design must include center runs to provide reasonably stable variance of the predicted response. Generally, three to five center runs are usually adequate. The face-centered cube does not require any center runs but again in most situations at least two center runs is desirable if the experimenter wants a model-independent estimate of error. However, note that the face-centered cube with no center runs is the 14-run D-optimal design for fitting a second-order model in three factors.

Another classical RSM design is the Box–Behnken design (Box and Behnken (1960)). These are three-level designs for fitting response surfaces. These designs are formed by combining 2^k factorials with incomplete block designs. The resulting designs are usually very efficient in terms of the number of required runs, and they are either rotatable or nearly rotatable. Table 11.4 shows a three-variable Box–Behnken design. The design is shown geometrically in Figure 11.14. Notice that the three-factor Box–Behnken design is a spherical design, with all points lying on a sphere of radius $\sqrt{2}$. There are no points at the vertices of the cubic region created by the upper and lower limits for each variable. It can be shown that the Box–Behnken design with two center runs is an alias-optimal design if the model that is being protected against is the full cubic model. See Jones and Nachtsheim (2011b) for a discussion of the alias-optimal design criterion.

11.3.2 Definitive Screening Designs

Jones and Nachtsheim (2011a) introduced a useful class of response surface designs for quantitative factors whose construction can be facilitated by an optimal design algorithm. They are now

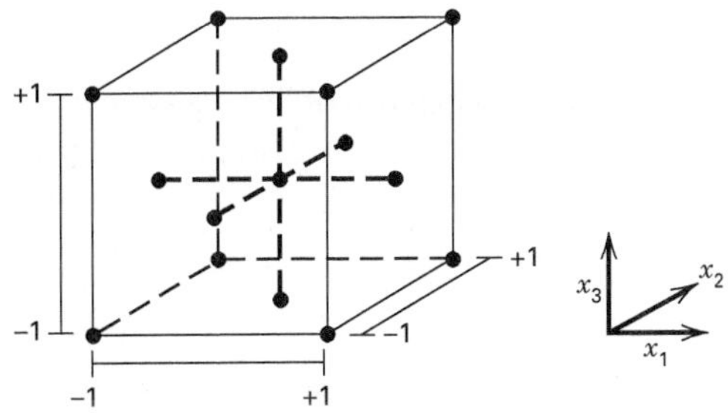

FIGURE 11.13 The Face-centered Cube for $k = 3$

Table 11.4 **A Box–Behnken Design for $k = 3$ Factors**

Run	x_1	x_2	x_3
1	−1	−1	0
2	−1	1	0
3	1	−1	0
4	1	1	0
5	−1	0	−1
6	−1	0	1
7	1	0	−1
8	1	0	1
9	0	−1	−1
10	0	−1	1
11	0	1	−1
12	0	1	1
13	0	0	0
14	0	0	0
15	0	0	0

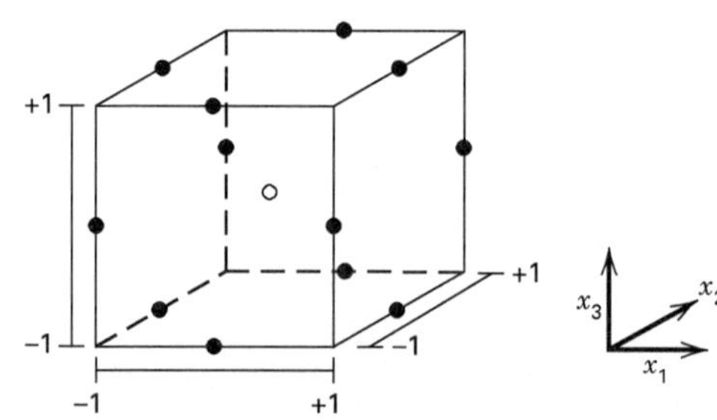

FIGURE 11.14 The Box–Behnken Design for $k = 3$ Factors

called **definitive screening designs** (or DSDs) because they are small enough to allow efficient screening of potentially many factors yet they can accommodate second-order effects in many situations without additional runs. In that regard, we can think of these designs as "one-step RSM designs". Table 11.5 shows the general structure of these designs with m factors.

Notice that for m factors, there are only $2m + 1$ runs based on m foldover pairs and an overall center point. Each run (excluding the center run) has exactly one factor level at its center point and all other factors have their levels at the extremes. These designs have the following desirable properties:

1. The number of required runs is only one more than twice the number of factors. Consequently, these are very small designs.

2. Unlike resolution III designs, main effects are completely independent of two-factor interactions. As a result, estimates of main effects are not biased by the presence of active two-factor interactions, regardless of whether the interactions are included in the model.

3. Unlike resolution IV designs, two-factor interactions are not completely confounded with other two-factor interactions, although they may be correlated.

4. Unlike resolution III, IV, and V designs with added center points, all quadratic effects can be estimated in models comprised of any number of linear and quadratic main effect terms.

Table 11.5 **General Structure of a Definitive Screening Design with *m* Factors**

		Factor levels				
Foldover pair	**Run (i)**	$x_{i,1}$	$x_{i,2}$	$x_{i,3}$	$\cdots$	$x_{i,m}$
1	1	0	±1	±1	$\cdots$	±1
	2	0	∓1	∓1	$\cdots$	∓1
2	3	±1	0	±1	$\cdots$	±1
	4	∓1	0	∓1	$\cdots$	∓1
3	5	±1	±1	0	$\cdots$	±1
	6	∓1	∓1	0	$\cdots$	∓1
$\vdots$	$\vdots$	$\vdots$	$\vdots$	$\vdots$	$\ddots$	$\vdots$
m	$2m-1$	±1	±1	±1	$\cdots$	0
	$2m$	∓1	∓1	∓1	$\cdots$	0
Center point	$2m+1$	0	0	0	$\cdots$	0

5. Quadratic effects are orthogonal to main effects and not completely confounded (though correlated) with interaction effects.
6. With six through or more factors, the designs are capable of estimating all possible full quadratic models involving three or fewer factors with very high levels of statistical efficiency.

These designs are an excellent compromise between resolution III fractions for screening and small RSM designs. They also admit the possibility of moving directly from screening to optimization using the results of a single experiment. Jones and Nachtsheim (2011a) found these designs using an optimization technique they had previously developed for finding minimum aliasing designs. Their algorithm minimizes the sum of the squares of the elements of the alias matrix subject to a constraint on the D-efficiency of the resulting design.

DSDs can also be constructed from **conference matrices** (see Xiao et al. (2012)). A conference matrix $\mathbf{C}$ is an $n \times n$ matrix that has diagonal elements equal to zero and all off-diagonal elements equal to ±1. They have the property that $\mathbf{C}'\mathbf{C}$ is a multiple of the identity matrix. For the $n \times n$ conference matrix $\mathbf{C}$, $\mathbf{C}'\mathbf{C} = (n-1)\mathbf{I}$. Conference matrices first arose in connection with a problem in telephony. They were used in constructing ideal telephone conference networks from ideal transformers. These networks were represented by conference matrices. Refer to Wikipedia on conference matrices for other applications.

The conference matrix of order 6 is

$$\begin{pmatrix} 0 & +1 & +1 & +1 & +1 & +1 \\ +1 & 0 & +1 & -1 & -1 & +1 \\ +1 & +1 & 0 & +1 & -1 & -1 \\ +1 & -1 & +1 & 0 & +1 & -1 \\ +1 & -1 & -1 & +1 & 0 & +1 \\ +1 & +1 & -1 & -1 & +1 & 0 \end{pmatrix}$$

The 13-run 6-factor DSD can be found by folding over each row of this conference matrix and adding a row of zeros at the bottom. In general, if $\mathbf{C}$ is the conference matrix of order n, the m-factor DSD matrix can be found as follows:

$$\mathbf{D} = \begin{bmatrix} \mathbf{C} \\ -\mathbf{C} \\ \mathbf{0} \end{bmatrix}$$

DSDs are intended for use with continuous factors. However, DSDs can be modified to include two-level categorical variables. The process is straightforward. Begin by constructing the usual DSD for continuous factors, except adding two rows of zeros instead of one. Change the zeros in the columns for the categorical factors to either +1 or −1. If the zeros are the added rows of zeros, make all the categorical factors −1 for the first row and +1 for the second row. If the zeros are from the conference matrix and its foldover, make the factor −1 for the first row and +1 for the second row.

As an example, consider a DSD for four continuous factors and two categorical factors. The DSD for six factors with the two added zeros is shown in the following display:

A	B	C	D	E	F
0	+1	+1	+1	+1	+1
+1	0	+1	−1	−1	+1
+1	+1	0	+1	−1	−1
+1	−1	+1	0	+1	−1
+1	−1	−1	+1	0	+1
+1	+1	−1	−1	+1	0
0	−1	−1	−1	−1	−1
−1	0	−1	+1	+1	−1
−1	−1	0	−1	+1	+1
−1	+1	−1	0	−1	+1
−1	+1	+1	−1	0	−1
−1	−1	+1	+1	−1	0
0	0	0	0	0	0
0	0	0	0	0	0

Now the zeros in the last two columns are changed to ±1 in order to produce the categorical variables E and F. The design is

A	B	C	D	E	F
0	+1	+1	+1	+1	+1
+1	0	+1	−1	−1	+1
+1	+1	0	+1	−1	−1
+1	−1	+1	0	+1	−1
+1	−1	−1	+1	−1	+1
+1	+1	−1	−1	+1	−1
0	−1	−1	−1	−1	−1
−1	0	−1	+1	+1	−1
−1	−1	0	−1	+1	+1
−1	+1	−1	0	−1	+1
−1	+1	+1	−1	+1	−1
−1	−1	+1	+1	−1	+1
0	0	0	0	−1	−1
0	0	0	0	+1	+1

When two-level categorical variables are added to a DSD as shown, the main effects of the categorical factors have some correlation with other factors, but the correlations are small. It is also possible to construct DSDs in orthogonal blocks. The procedure is as follows:

1. Start by repeating the steps for creating a design for continuous factors only except adding as many rows of zeros as there are blocks. Arrange the design in standard order.
2. Assign the first foldover pair to the first block, the second foldover pair to the second block, and so on until you get to the last block, then start over again assigning the next foldover pair to the first block.
3. Assign each center run to a separate block.

As an example, consider a DSD for six factors in two blocks. The six-factor DSD in standard order with two added zeros. Now arrange the blocks as described above, with the first foldover pair forming the first block, the second foldover pair forming the second block, and so on, and then finish by assigning the zeros to each block. The final design is as follows:

Block	A	B	C	D	E	F
1	0	+1	+1	+1	+1	+1
1	0	−1	−1	−1	−1	−1
2	+1	0	−1	+1	+1	−1
2	−1	0	+1	−1	−1	+1
1	+1	−1	0	−1	+1	+1
1	−1	+1	0	+1	−1	−1
2	+1	+1	−1	0	−1	+1
2	−1	−1	+1	0	+1	−1
1	+1	+1	+1	−1	0	−1
1	−1	−1	−1	+1	0	+1
2	+1	−1	+1	+1	−1	0
2	−1	+1	−1	−1	+1	0
1	0	0	0	0	0	0
2	0	0	0	0	0	0

11.3.3 Optimal Designs in RSM

The standard response surface designs discussed in the previous sections, such as the CCD, the Box–Behnken design, and their variations (such as the face-centered cube), are widely used in RSM applications. If the experimental region is either a cube or a sphere, often a standard response surface design will be applicable to the problem. However, quite frequently an experimenter encounters a situation where a standard response surface design may not be the obvious choice. **Optimal designs** are an alternative to use in these cases. As we have noted before, there are several very common situations where an optimal design is appropriate, for example, if a nonstandard model is required (that is something other than the usual first- or second-order polynomial), a nonstandard sample size or there is some unusual blocking requirement, or when the

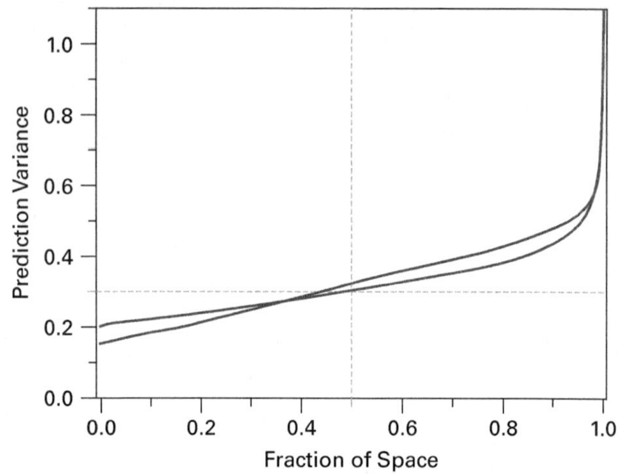

FIGURE 11.15 Fraction of Design Space Plot Comparing a Face-centered Cube and an I-optimal Design

experimental region is not the usual sphere or cube but is impacted by constraints. We recommend that I-optimal designs be used in these situations. In fact, we recommend considering I-optimal designs first because they are often superior to even very good standard designs.

For example, suppose that we are going to fit a second-order model in $k = 4$ factors and that the face-centered cube is being considered. If we use two center runs in this design it will give 26 total runs. Consider an alternative design, an I-optimal design with only 24 runs. Figure 11.15 shows the fraction of design space plot comparing these two designs. In Figure 11.15 the CCD is represented by the lower curve at the left-hand side of the graph, indicating that the CCD has better prediction performance over the initial part of the design space, but the I-optimal design begins to outperform it fairly quickly and has lower prediction variance over more than 50% of the design space. The average variance of prediction for the face-centered CCD is 0.324947 σ^2 and the average prediction variance for the I-optimal design is 0.315392 σ^2 despite the fact that it has two fewer runs.

12 Design for Models That Are Nonlinear in the Parameters

12.1 Introduction

Previously, all the models for the designs we have considered have been linear in the parameters. It means that the model for the response is the sum of an arbitrary number of terms, where each term is the product of one parameter and a function of the data. We note that a plot of the model below

$$y = \beta_0 + \beta_1 x + \beta_{11} x^2 + \varepsilon$$

will exhibit curvature from the x^2 term. However, all the unknowns, the subscripted βs, enter the model linearly. The matrix form of the above model is $\boldsymbol{y} = \mathbf{X}\boldsymbol{\beta} + \boldsymbol{\varepsilon}$. The derivative of the expected value of y with respect to the parameters is $\mathbf{X}$, which is not a function of the parameters. The efficiency of the design for a model that is a function of the parameters depends on $\mathbf{X}$, which is under our control and is not a function of the unknown parameters.

The exponential decay model is $y = e^{-\beta x} + \varepsilon$. The parameter, β, being embedded in the exponent makes the model nonlinear in the parameters. For this model, the derivative of the function with respect to β is $-1xe^{-\beta x}$. This derivative function, which is used to compute optimal designs, is a function of the unknown parameter. For nonlinear models like this, the efficiency of the design depends on the unknown parameter as well as the values we choose for x. Thus, generating an optimal design for estimating β requires that we already know what β is.

One way of dealing with this "chicken and egg" problem is to pretend that we know what the true value of the parameter β is. Then, we can plug this constant in for β in the derivative. This allows us to proceed to generate a design that is optimal if we guess correctly about what the true value of β is. This approach is called generating a **locally optimal design** because the design will be highly efficient only if our guess is close to the true value.

Another way to deal with our lack of knowledge about the true value of β is to average over a distribution of possible values of β. This averaging approach is often referred to as a **Bayesian optimal design** in that we view the distribution for β to result from our prior experience. This "Bayesian" approach tends to produce designs that perform acceptably over a broader range of possible values of the parameter(s). For the rest of this chapter, we forgo the use of quotes around the word Bayesian despite the view of some researchers that this approach is not fully Bayesian.

In this chapter, we show how to find both locally optimal and Bayesian nonlinear designs. We also consider designs where the response is not normally distributed. For example, the response might be the percentage of successes in many trials. For this case, we might consider the response to be binomially distributed. Alternatively, the response might be the observed count such as the number of defects in a linear foot of fabric. Here, it would make sense to consider the response to be a Poisson random variable. In these latter cases, the nonlinearity in the model is due to the nonconstant variance of the response as a function of the factors.

12.2 Design and Analysis of Exponential Decay

The exponential decay model mentioned in the introduction is a function of one factor and one unknown parameter, β, to be estimated. The locally optimal design puts all the experimental runs at a single value of x, which is the reciprocal of the guess for β. Figure 12.1 shows a plot of the exponential decay function for three different values of the parameter, 0.5, 1, and 2. As the parameter gets larger, the function drops more quickly.

The horizontal line on the plot is at 0.367, which is e^{-1}—the value of the function at the locally optimal design point. You can imagine that a poor guess of the parameter could result in a locally optimal design that yields a parameter estimate with a large variance.

Typically, locally optimal designs for nonlinear models have the same number of unique factor values as parameters. In the exponential decay model, there is one parameter, so the locally optimal design uses the same factor value for every run. In this case, we prefer an approach that employs more than one value of the factor. This is what Bayes optimal design does when we indicate a great deal of uncertainty through a very wide prior distribution.

For example, when the guess for β is 2 but the prior distribution is uniform from 0.1 to 100, the Bayes optimal design places design points at three values, 0.017, 0.110, and 1. These values bracket the locally optimal design value of 0.5.

A reasonable alternative would be to perform a run at the locally optimal design value and two other values on either side. Given this data, it is possible to fit the model, which will provide a data-based estimate of the parameter. Augmenting these runs, if necessary, can use this estimated parameter value.

Table 12.1 shows data obtained using this approach.

Figure 12.2 shows the analysis using the JMP Nonlinear Platform.

In this case, the guess for β is near perfect.

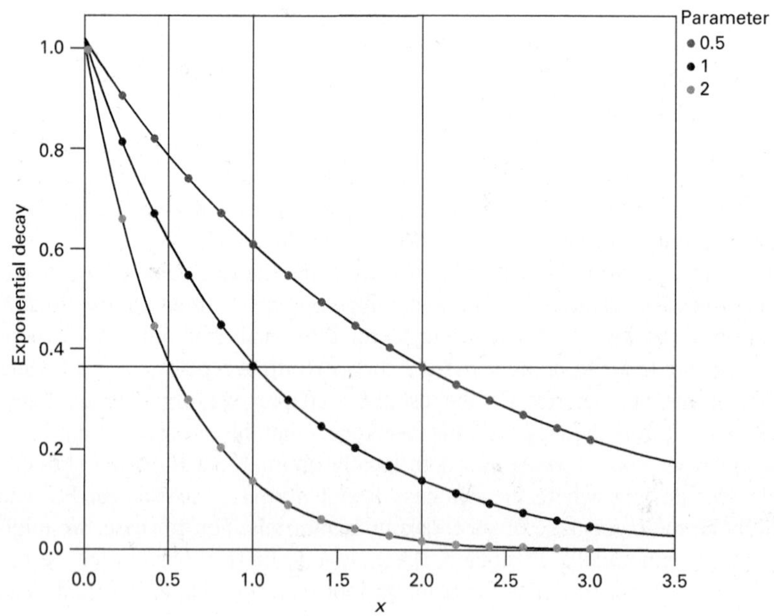

FIGURE 12.1 Plot of $y = \exp(-\beta x)$ for Three Values of β

Table 12.1 Data for Two Values of x Bracketing the Locally Optimal Design Value of 0.5 for a Guess of 2 for β

x	y
0.4	0.44
0.5	0.37
0.6	0.32

Parameter	Estimate	ApproxStdErr	Lower CL	Upper CL
theta	1.978623952	0.04450366	1.79535584	2.18046517

Plot

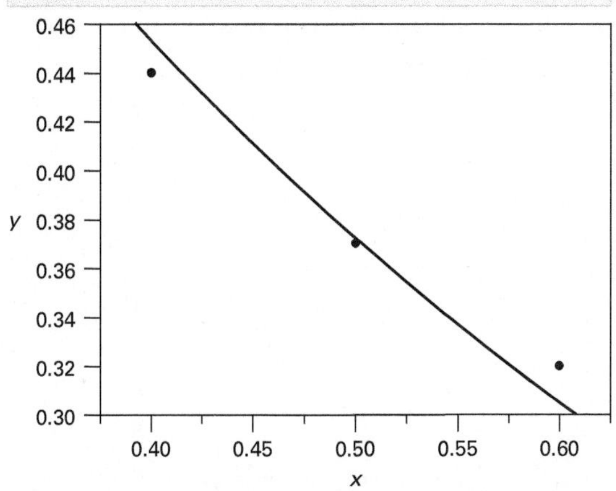

Parameter	Estimate		Low	High
theta	1.978623952		1.7116	2.24565

Prediction Profiler

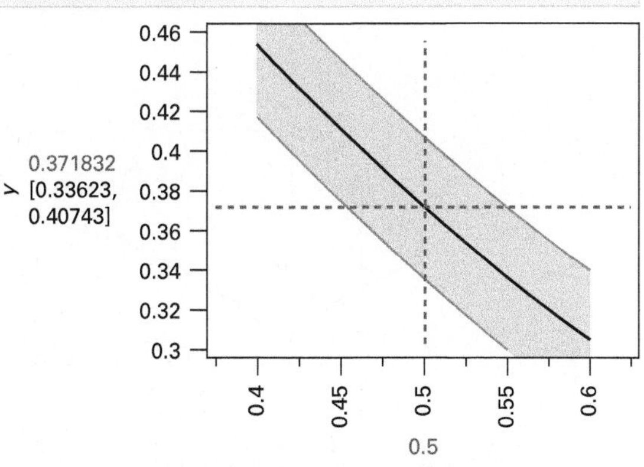

FIGURE 12.2 Nonlinear Analysis of the Exponential Decay Model

12.3 Analysis and Locally Optimal Design of the Michaelis–Menten Model

The Michaelis–Menten model is used in cases where the response to a single factor grows up to a limiting value. It has two unknown parameters. One parameter, b_1 in the following equation, is the limiting value of the response. The other parameter, b_2, is the factor value where the response is half of its limiting value.

$$y = \frac{b_1 x}{b_2 + x}$$

Reaction kinetics is one common application of the Michaelis–Menten model. Table 12.2 provides data for the response, reaction rate, as a function of enzyme concentration.

Figure 12.3 shows the analytical results from the Nonlinear Platform in JMP.

Table 12.2 Data for Fitting the Michaelis–Menten Model

Concentration	Rate
5	33.3
10	50.8
15	58.2
20	65.2
25	71.1
30	74.7
35	78.6
40	80.3
45	82.4
50	84.7
55	85.3
60	85.5
65	85.8
70	87.1

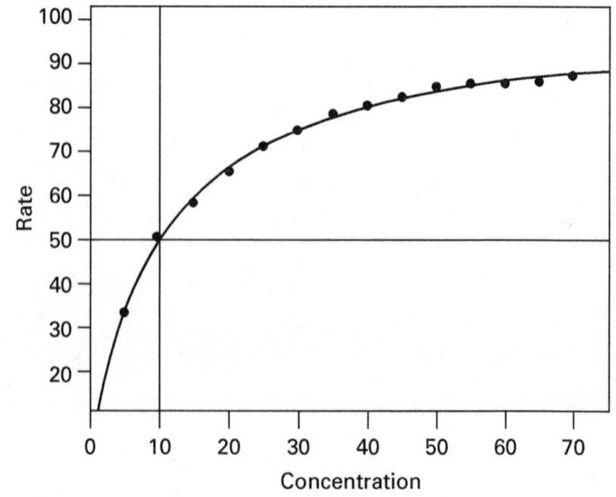

Parameter	Estimate	ApproxStdErr	Lower CL	Upper CL
b1	100.55	0.763	98.914	102.25
b2	10.275	0.320	9.595	10.996

FIGURE 12.3 Fitted Michaelis–Menten Model

In Figure 12.3, the vertical line is the concentration at the estimate for the parameter b_2, 10.275. The horizontal line is at a rate of 50.275, which is half of the estimate for b_1, 100.55.

The locally optimal design if one guessed that b_1 was 100 and b_2 was 10 and the allowable range of concentration was 5–70 puts half the runs at a concentration of 8 and the other half at a concentration of 70. The optimal design puts half of the runs at the largest allowable value to estimate the asymptote. The other half of the runs are at a concentration near the guess for b_2. If these guesses were correct, the resulting parameter estimates would have lower variances. However, running the two-point locally optimal design would not allow for assessing the appropriateness of the model. The data from the evenly spaced design in Table 12.2 provides ample evidence in the analysis that the Michaelis–Menten model provides an adequate approximation to the true relationship between concentration and reaction rate.

12.4 Yield Optimization as a Function of Reaction Temperature and Time

Consider a chemical reaction where an initial chemical reacts to form a product that decays to form an undesirable impurity. There are two reaction rates. One rate is for the initial reactant to product reaction. The other rate is for the product to impurity reaction. The goal of the experiment is twofold. First, the development team would like to estimate the two reaction rates. Armed with these estimates, the team would like to find the operating setting for the reaction temperature and the reaction time to maximize the yield of product.

Figure 12.4 shows the prediction equation for product yield as a function of temperature and time as well as two parameters, **t1** and **t3**.

The parameter **t1** is related to the first reaction, while **t3** is related to the rate of the second reaction. There are two parameters, and the locally optimal design has only two distinct points. Therefore, it would be impossible to fit an empirical model in time and temperature. Using the locally optimal design would mean that the design relies completely on the correctness of the model. The Bayes optimal design has four points, two of which are almost on top of each other as shown in Figure 12.5.

There are six runs at a temperature of 540 K and a time of 0.3 hour, three runs at a temperature of 510 K for 0.1 hour, and five runs near 540 K for 0.1 hour. A polynomial model could only just barely fit the main effects of time and temperature with this data.

Table 12.3 shows the factor settings and observed responses for the 14-run experiment.

Figure 12.6 shows the analysis of the data using the Nonlinear Analysis Platform in JMP.

The predicted optimal yield is 63.94 at a temperature of 540 K and a time of 0.19 minutes. Good practice is to run at least one verification run at this setting.

It is important to emphasize that the complexity of the fitted yield surface relies nearly completely on the correctness of the nonlinear model. If there is a lack of certainty about

$$\frac{\left(100 \cdot \left(\mathrm{Exp}\left(\mathbf{t1} - \frac{\left(1.9279 \cdot 10^{4}\right)}{\mathit{Reaction\ Temperature}}\right)\right) \cdot \left(\mathrm{Exp}\left(-\mathrm{Exp}\left(\mathbf{t1} - \frac{\left(1.9279 \cdot 10^{4}\right)}{\mathit{Reaction\ Temperature}}\right) \cdot \mathit{Reaction\ Time}\right) - \mathrm{Exp}\left(-\mathrm{Exp}\left(\mathbf{t3} - \frac{\left(1.6819 \cdot 10^{4}\right)}{\mathit{Reaction\ Temperature}}\right) \cdot \mathit{Reaction\ Time}\right)\right)\right)}{\left(\mathrm{Exp}\left(\mathbf{t3} - \frac{\left(1.6819 \cdot 10^{4}\right)}{\mathit{Reaction\ Temperature}}\right) - \mathrm{Exp}\left(\mathbf{t1} - \frac{\left(1.9279 \cdot 10^{4}\right)}{\mathit{Reaction\ Temperature}}\right)\right)}$$

FIGURE 12.4 A Function Showing the Solution of Two First-order Differential Equations for the Reaction

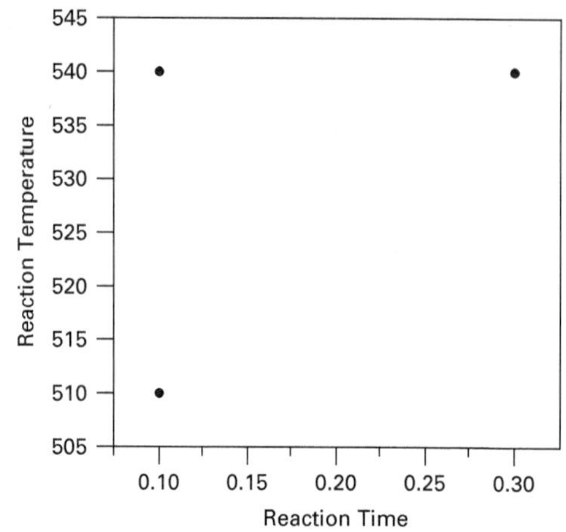

FIGURE 12.5 Scatter Plot of Bayes Optimal Design for 14 Runs

Table 12.3 **Data for the Reaction Optimization Experiment**

Reaction temperature	Reaction time	Observed yield
510.00	0.1	11.1
510.00	0.1	12.3
510.00	0.1	12.5
539.74	0.1	53.7
539.74	0.1	54.6
539.74	0.1	55.1
539.74	0.1	54.5
539.84	0.1	54.7
540.00	0.3	57.7
540.00	0.3	57.3
540.00	0.3	57
540.00	0.3	58
540.00	0.3	58.1
540.00	0.3	60

the appropriateness of this model form, the use of a response surface design as introduced in Chapter 11 would allow for fitting both a nonlinear model and a full quadratic empirical model.

12.5 Mathematical Details for Constructing Optimal Designs for Nonlinear Models

D-optimal design of linear models requires maximizing the determinant of the **information matrix, $\mathbf{X}'\mathbf{X}$.**

Similarly, design for nonlinear models also requires maximizing the determinant of an information matrix. For a nonlinear function

$$Y(\mathbf{x}) = \eta(\mathbf{x}; \theta) + \varepsilon$$

Parameter	Estimate	Standard Errro	Lower CL	Upper CL
t1	38.00	0.013	37.97	38.03
t3	32.00	0.013	31.97	32.03

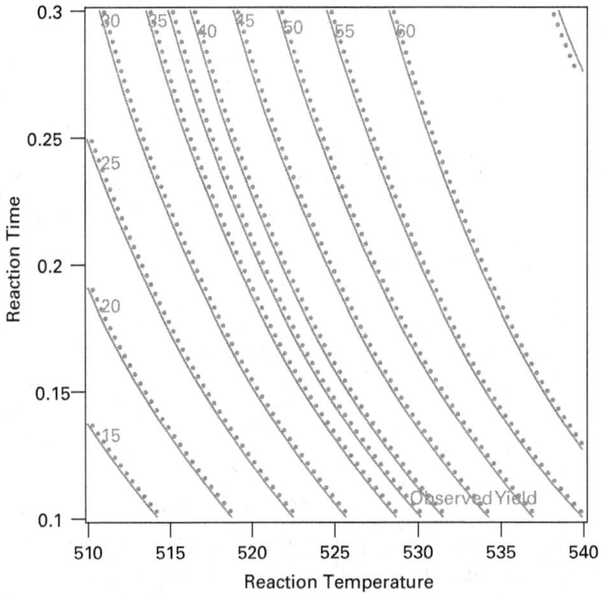

FIGURE 12.6 Parameter Estimates Predicted Yield Contours and for the Nonlinear Model

The information matrix is

$$\mathbf{I}(\theta; d) = (1/\sigma^2) \sum_{i=1}^{n} \mathbf{g}(\mathbf{x}_i; \theta)\mathbf{g}^{\mathrm{T}}(\mathbf{x}_i; \theta)$$

where the column vector

$$\mathbf{g}(\mathbf{x}_i; \theta) = \nabla\eta(\mathbf{x}_i; \theta)$$

That is, the function $\mathbf{g}$ is the gradient of the nonlinear function η with respect to θ. The information matrix is symmetric and has as many rows and columns as the number of parameters, p, in the nonlinear model, which is also the number of elements in the vector $\boldsymbol{\theta}$.

Another way to calculate the information matrix is to stack the n rows of g' on top of each other. That creates an $n \times p$ matrix, $\mathbf{G}$, that is analogous to the $\mathbf{X}$ matrix in linear design. Then, the information matrix is $\frac{1}{\sigma^2}\mathbf{G}'\mathbf{G}$.

For a model like the exponential decay model with only one parameter, the information matrix is 1×1. As mentioned in the introduction to this chapter, the derivative (gradient) of the exponential decay function is $-1xe^{-\beta x}$. For each value of x in the design, replace β with its guessed value. Figure 12.7 shows the function $xe^{-\beta x}$ for $\beta = 2$. The maximum of this function is at $\frac{1}{\beta} = \frac{1}{2}$. This illustrates why the locally optimal design requires every run to be at an x value of 0.5.

Calculating the Bayes D-optimal design involves computing the integral

$$\phi(d) = \int \log(|\mathbf{I}(\theta; d)|)d\pi(\theta)$$

In practice, this integral is intractable analytically, so we need to approximate it. Early attempts to compute this integral involved Monte Carlo sampling of many parameter vectors from the prior distribution, computing the information matrix for each vector and finding the average of their logarithms. This method was slow as well as imprecise. Gotwalt et al. (2009) provided a

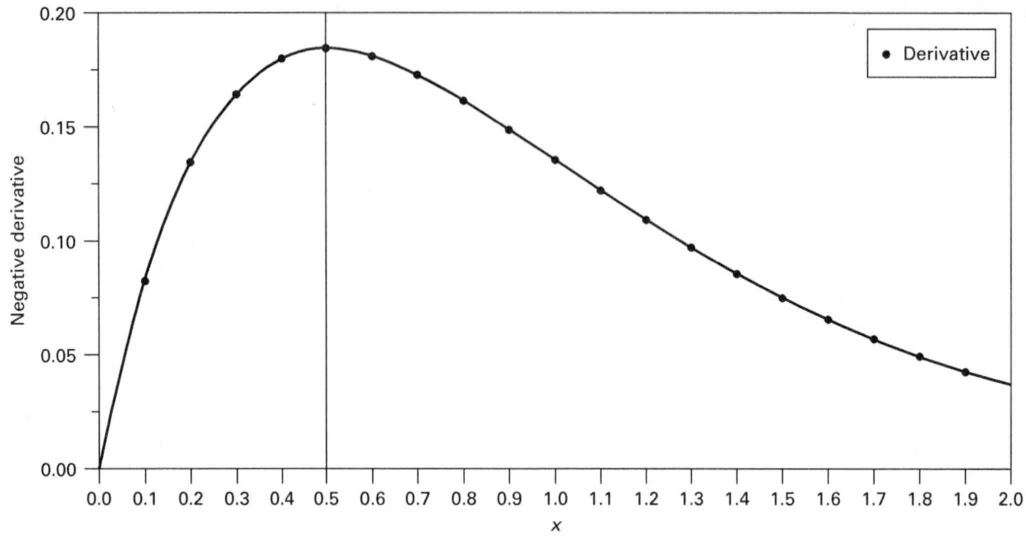

FIGURE 12.7 Negative Gradient of the Exponential Decay Function

way to approximate the integral using a wisely chosen sample of points on concentric spheres around the prior guess. Their approach dramatically reduced the number of sampled parameter vectors from the prior distribution and simultaneously provided a more precise approximation to the integral.

12.6 Optimal Design for Situations Where the Response Is Binary

For some experiments, the response is not a measurement. Rather an inspector looks at the product of an experimental run and judges it to be either good or bad. If many runs are done at the same factor settings, the responses can be combined. For example, if 10 runs are attempted and 9 of them are good, the investigator records both the number of successes and the number of trials. The **binomial distribution** is appropriate for modeling data obtained in this way.

A standard assumption made when the responses are normally distributed is that the variance of the errors is constant. For the binomial distribution, the variance is a function of the probability of a success. If p is this probability, then the variance is $p(1-p)$. When p is near zero or near one, the variance is small. The variance is maximized when $p = \frac{1}{2}$. In experiments with binomially distributed responses, the probability of success is usually assumed to be a linear function of the parameters to be estimated. The nonlinearity in optimal design for binomial responses comes from the fact that the variance of the binomial responses is a function of the parameters.

An example of a case where a binomial response makes sense is ordnance test. The US army is mandated to test ordnance to make sure that it performs according to specification. It is important that ordnance is safe under a variety of environment conditions. It is equally important that ordnance performs reliably when fired. The purpose of ordnance testing is to determine the conditions when the ordnance is virtually guaranteed not to detonate and the conditions where it is guaranteed to detonate. An example test is the response of ordnance to an increasing stimulus that at some value will cause the ordnance to fire.

Figure 12.8 shows the position of the two runs of a locally optimal design for this problem. There are two parameters, a slope and an intercept, so there are two points in the optimal design.

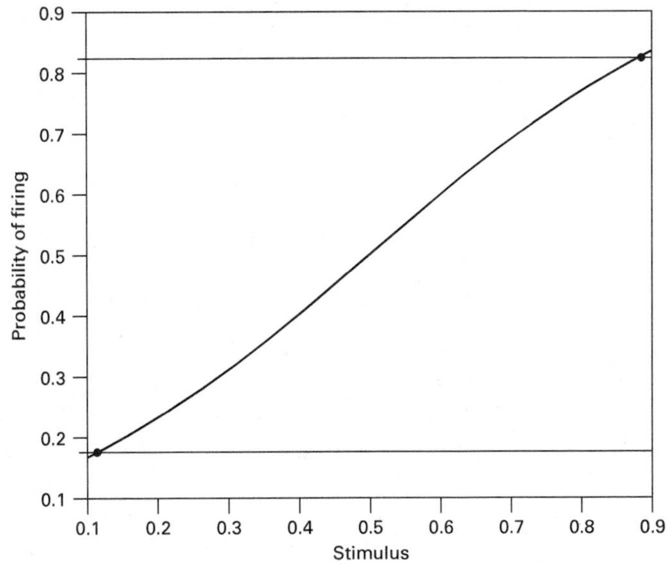

FIGURE 12.8 Probability of Firing as a Function of the Applied Stimulus

The locally optimal design puts half of the runs where the predicted probability of firing is 17.6% and the other half of the runs where the predicted probability of firing is 82.4%. Note that these probabilities are equidistant from $p = 0.5$. Recall that the variance of the binomial is $p(1 - p)$, $82.4 = 100 - 17.6$ and $17.6 = 100 - 82.4$. Therefore, the variance at each of these two settings is the same.

In optimal design for linear models, runs are placed where the variance is maximized. By locating runs at positions of maximum variance, the design reduces the variance at those points. In linear design, it is advantageous to set the levels of continuous factors far apart to increase the precision of the estimate of the main effect of a factor.

In an experiment with one continuous factor and a binomial response, setting the factors far from each other results in all failures when the stimulus is low and all successes when the stimulus is high. This is undesirable because it does not allow us to locate the function between these two extremes. Placing the runs at a high stimulus where there is a 17.6% probability that the ordnance will *not* fire, while placing the runs at a low stimulus where there is a 17.6% probability that the ordnance *will* fire represents a trade-off. The trade-off achieves some separation between the levels while bounding the variance away from zero.

The location of the low and high stimuli depends on our guesses at the unknown intercept and slope parameters. If these guesses are wrong, then the low stimulus runs might have an unacceptable number of firing events or the high stimulus runs might have an unacceptable number of duds.

Figure 12.9 shows the predicted probability for the Bayes optimal and locally optimal designs when the allowable range of x is from zero to one and the predicted parameters are -2 for the intercept and 4 for the slope.

The Bayes optimal design (diamond markers) puts four runs at zero and one and two runs at a middle value. The locally optimal design has five runs at each of its two settings. The prior distributions of the parameters for the Bayes design were independent normal distributions with their means at the guessed parameters. The variance of the intercept prior parameter was 1, while the variance of the slope prior parameter was 2.

The Bayes design puts many of its runs outside the levels of the local design but adds runs in between to help deal with uncertainty about the location of the curve.

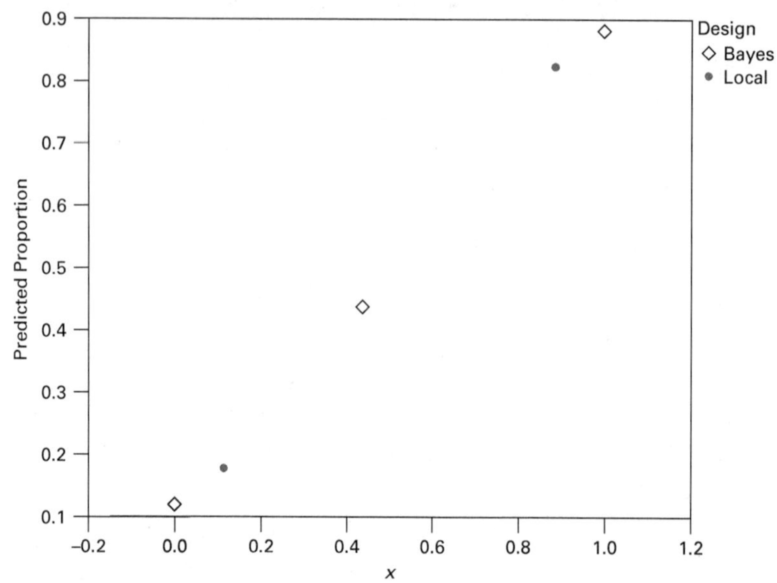

FIGURE 12.9 Bayes Optimal and Locally Optimal Designs for the Ordnance Experiment

12.7 Multifactor Binomial Model Experiments

Woods et al. (2006) described a crystallography experiment that considered the effects of four process variables (X1 = agitation rate, X2 = volume, X3 = temperature, and X4 = evaporation rate) on the probability that a desired new product would form. Their design had three replicates of a 16-run design and the response was the observed number of successes out of the three trials. Table 12.4 shows the design that JMP computes for their factor settings given their guesses for

Table 12.4 Coded Factor Settings and Responses for the Crystallography Experiment

X1	X2	X3	X4	Successes
−0.581	1.000	1.000	1.000	3
0.424	−0.500	1.000	1.000	0
1.000	−0.895	1.000	−1.000	0
−1.000	0.579	−1.000	−1.000	1
0.947	−0.386	1.000	−1.000	2
−1.000	0.699	−1.000	−1.000	3
−0.502	1.000	0.696	−1.000	3
−0.881	1.000	1.000	−1.000	1
1.000	−0.103	1.000	1.000	2
0.442	−1.000	−1.000	−1.000	0
0.759	−1.000	−1.000	1.000	3
−1.000	0.840	1.000	1.000	0
−1.000	0.177	−1.000	1.000	1
1.000	−0.043	1.000	−1.000	3
1.000	−0.590	1.000	1.000	1
0.936	−1.000	−1.000	−1.000	0

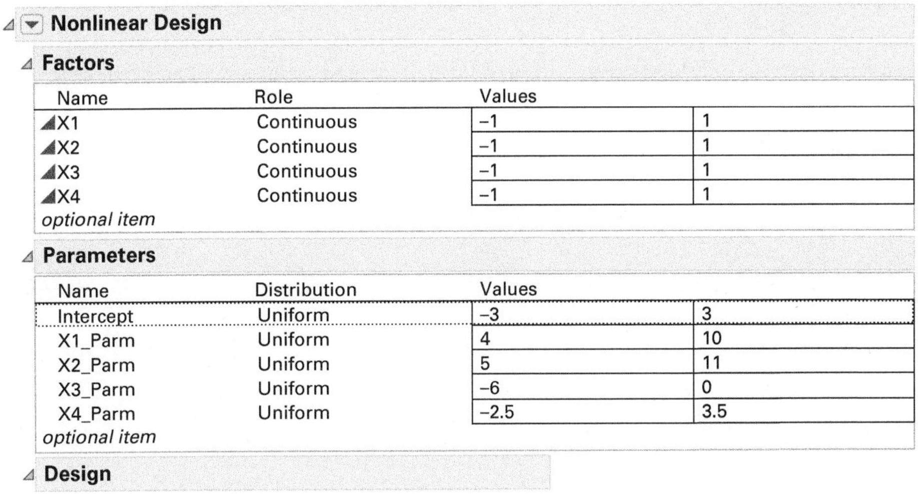

Nonlinear Design

Factors

Name	Role	Values	
X1	Continuous	–1	1
X2	Continuous	–1	1
X3	Continuous	–1	1
X4	Continuous	–1	1
optional item			

Parameters

Name	Distribution	Values	
Intercept	Uniform	–3	3
X1_Parm	Uniform	4	10
X2_Parm	Uniform	5	11
X3_Parm	Uniform	–6	0
X4_Parm	Uniform	–2.5	3.5
optional item			

Design

FIGURE 12.10 JMP User Interface for the Crystallography Design

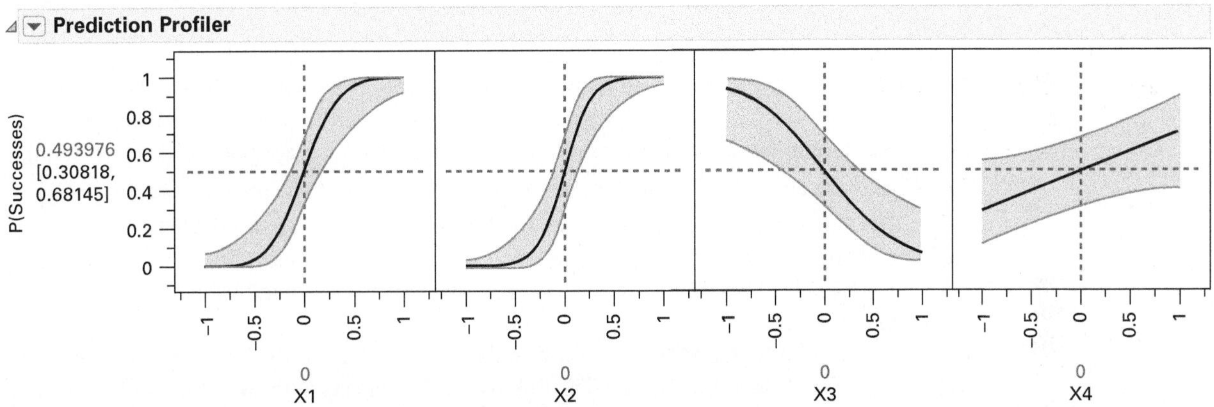

FIGURE 12.11 Success Prediction Profile at the Center Run

the parameters and uncertainties about these guesses. The Successes column is realization of the number of successes they would expect to see given their guesses for the parameters.

Figure 12.10 shows the setup for computing the Bayes optimal design for the crystallography experiment. Note that the ranges of the parameters are quite large, indicating that before running the experiment there was substantial uncertainty about the magnitude of the effects of the factors.

Figure 12.11 shows a JMP Prediction Profile plot for the predicted probabilities of success as a function of the four factors and their fitted parameter values after performing a logistic regression. In Figure 12.11, the setting of each factor is at the center of its range.

Figure 12.12 shows factor settings that are predicted to result in a very high success probability (0.99917). Factors X1 and X2 have the strongest effects and need to be set near their high values. Factors X3 and X4 have comparatively minimal effect and can be set at their default values without causing problems.

12.8 Mathematical Details for Constructing Optimal Designs for Generalized Linear Models

The experiments with a binomially distributed response are a special case of a **generalized linear model (GLM)**. These models are appropriate for distributions in the exponential family.

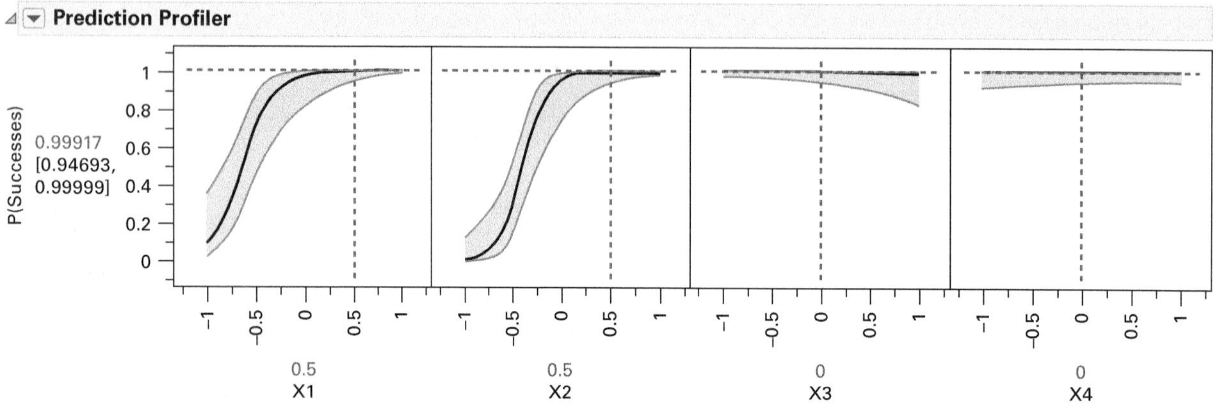

FIGURE 12.12 Factor Settings for High Success Probability

The normal, binomial, and Poisson distributions, among several others, are members of the exponential family.

The ordnance and crystallography experiments both involved a binomial response. The predicted probability of success for a binomial is between zero and one inclusive. However, this is an undesirable range for linear regression modeling because the predicted probability for some factor settings can fall outside the zero to one range so that they are not physically meaningful. The odds for success are $\frac{p}{1-p}$, which ranges from zero to infinity. The log odds are $\log\left(\frac{p}{1-p}\right)$. This metric like the normal distribution ranges from negative to positive infinity. The log odds in the parlance of GLM are called the link function. The inverse function of the log odds is the logistic function $\frac{1}{1+e^{-x\theta}}$. The logistic function allows for translating a predicted value of the log odds to a predicted probability.

Specifying a GLM requires a response in the exponential family. The expected value of the response is a function of the factors. The link function connects a linear function of the factors to the expected value of the response. Symbolically,

$h[\mu(\mathbf{x})] = \mathbf{f}^T(\mathbf{x})\boldsymbol{\theta}$, where h is the link function for the model and $\mathbf{f}^T(\mathbf{x})\boldsymbol{\theta}$ is the linear predictor.

The information matrix for a design d with observations at $\mathbf{x}_1, \ldots, \mathbf{x}_n$ is

$$\mathbf{I}(\theta; d) = \sum_{i=1}^{n} \mathbf{f}(\mathbf{x}_i)\mathbf{f}^T(\mathbf{x}_i)w_i$$

where $w_i = 1/\{\mathrm{Var}(\mu_i)[h'(\mu_i)]^2\}$ is a weight.

For binomial (logistic) regression, the linear predictor is just like the **X** matrix in linear regression. It is the weights that make the GLM nonlinear because they are functions of the parameters, which makes the information matrix also be a function of the parameters.

For our one-factor example, the linear predictor was **b0 + b1x.** The weight for each run is

$$w = \exp(-(\mathbf{b0} + \mathbf{b1x}))/(1 + \exp(-(\mathbf{b0} + \mathbf{b1x})))^2$$

The locally optimal design for GLM maximizes the determinant of the information matrix, **I** above. The Bayes design maximizes the integral of the log determinant of the information matrix with respect to a prior distribution of the parameters. These are exact analogs to the calculations described in Section 12.5.

CHAPTER 1 PROBLEMS

SS Student solution available in interactive e-text.

1.1 Suppose that you want to design an experiment to study the proportion of unpopped kernels of popcorn. Complete steps 1–3 of the guidelines for designing experiments in Section 1.3. Are there any major sources of variation that would be difficult to control?

1.2 Suppose that you want to investigate the factors that potentially affect baking a cake.

(a) What would you use as a response variable in this experiment? How would you measure the response?

(b) List all of the potential sources of variability that could impact the response.

(c) Complete the first three steps of the guidelines for designing experiments in Section 1.3.

1.3 Suppose that you want to compare the growth of flowers with different conditions of sunlight, water, fertilizer, and soil conditions. Complete steps 1–3 of the guidelines for designing experiments in Section 1.3.

1.4 Select an experiment of interest to you. Complete steps 1–3 of the guidelines for designing experiments in Section 1.3.

1.5 Search the Internet for information about Sir Ronald A. Fisher and his work on experimental design in agricultural science at the Rothamsted Experimental Station.

1.6 Find a website for a business that you are interested in. Develop a list of factors that you would use in an experiment to improve the effectiveness of this website.

1.7 Suppose that we are testing $H_0 : \mu = \mu_0$ versus $H_0 : \mu \neq \mu_0$. Calculate the *P*-value for the following observed values of the test statistic.

(a) $Z_0 = 2.25$

(b) $Z_0 = -1.78$

(c) $Z_0 = 3.09$

(d) $Z_0 = -2.03$

(e) $Z_0 = 2.85$

1.8 Consider the summary of the computer output for an experiment shown below.

One-Sample Z

Test of mu = 30 vs not = 30

The assumed standard deviation = 1.2

N	Mean	SE Mean	95% CI	Z	P
16	31.2000	0.3000	(30.6120, 31.7880)	?	?

(a) Fill in the missing values in the output. What conclusion would you draw?

(b) Is this a one-sided or two-sided test?

(c) Use the output and the normal table to find a 99% CI on the mean.

(d) What is the *P*-value if the alternative hypothesis is $H_1 : \mu > 30$?

1.9 Consider the following sample data: 9.37, 13.04, 11.69, 8.21, SS
11.18, 10.41, 13.15, 11.51, 13.21, 9.21, 10.22, and 7.75. Is there evidence to support the claim that the mean of the population is 10?

1.10 Consider the computer output summary shown below:

One-Sample T: Y

Test of mu = 91 vs. not = 91

Variable	N	Mean	Std. Dev.	SE Mean	95% CI	T	P
Y	25	92.5805	?	0.4673	(91.6160, ?)	3.38	0.002

(a) Fill in the missing values in the output. Can the null hypothesis be rejected at the 0.05 level? Why?

(b) Is this a one-sided or a two-sided test?

(c) If the hypotheses had been $H_0 : \mu = 90$ versus $H_1 : \mu \neq 90$, would you reject the null hypothesis at the 0.05 level?

(d) Find a 99% two-sided CI on the mean.

(e) What is the *P*-value if the alternative hypothesis is $H_1 : \mu > 91$?

SS **1.11** The breaking strength of a fiber is required to be at least 150 psi. Past experience has indicated that the standard deviation of breaking strength is $\sigma = 3$ psi. A random sample of five specimens is tested, and the results are 145, 153, 150, 155, and 147.

(a) State the hypotheses that you think should be tested in this experiment.

(b) Test these hypotheses using $\alpha = 0.05$ What are your conclusions?

(c) Find the *P*-value for the test in part (b).

(d) Construct a 95% CI on the mean breaking strength.

1.12 A normally distributed random variable has an unknown mean μ and a known variance $\sigma^2 = 9$. Find the sample size required to construct a 95% CI on the mean that has a total length of 1.0.

SS **1.13** The shelf life of a carbonated beverage is of interest. Ten bottles are randomly selected and tested, and the following results are obtained:

Days	
108	138
124	163
124	159
106	134
115	139

(a) We would like to demonstrate that the mean shelf life exceeds 120 days. Set up appropriate hypotheses for investigating this claim.

(b) Test these hypotheses using $\alpha = 0.01$. What are your conclusions?

(c) Find the *P*-value for the test in part (b).

(d) Construct a 99% CI on the mean shelf life.

1.14 Consider the shelf life data in Problem 1.13. Can shelf life be described or modeled adequately by a normal distribution? What effect would the violation of this assumption have on the test procedure you used in solving Problem 1.13?

1.15 Twenty observations on etch uniformity on silicon wafers are taken during a qualification experiment for a plasma etcher. The data are as follows: 5.34, 6.65, 4.76, 5.98, 7.25, 6.00, 7.55, 5.54, 5.62, 6.21, 5.97, 7.35, 5.44, 4.39, 4.98, 5.25, 6.35, 4.61, 6.00, and 5.32

(a) Test the hypothesis that the mean etch uniformity is 5.0.

(b) Find a 95% CI on mean etch uniformity.

1.16 An experiment has been performed with a single factor using a sample size of 5. The sample average is 10.12 and the standard deviation is known to be 1.1.

(a) Construct a 95% CI on the mean.

(b) Test the hypothesis that the mean equals 10. What is the *P*-value for this test?

1.17 Reconsider the situation in Problem 1.16. Suppose that the sample size was not 5. Calculate the *P*-value for the test in part (a) of Problem 1.16 assuming that the sample sizes were

(a) 10

(b) 25

(c) 50

(d) 100

(e) 1000

Do the results of these calculations give you any insight into the impact of the sample size in a statistical test on the *P*-value?

1.18 Do you think that using a significance level of 0.05 is appropriate for all experiments? In the early stages of research and development work, is there a lot of harm in identifying a factor as important when it really is not? Would that seem to justify higher levels of significance such as 0.10 or perhaps even 0.15 in some situations?

1.19 Power calculation for hypothesis testing is relatively easy to do with modern statistical software. What do you think "adequate power" should be for an experiment? What issues need to be considered in answering this question?

1.20 In the early stages of research and development experimentation, which type of error do you think is most important, type I or type II? Justify your answer.

CHAPTER 2 PROBLEMS

SS **2.1** A computer program has produced the following output for a hypothesis testing problem:

```
Difference in sample means: 2.35
Degrees of freedom: 18
Standard error of the difference in
  sample means: ?
Test statistic: t0 = 2.01
P-value: 0.0298
```

(a) What is the missing value for the standard error?

(b) Is this a two-sided or a one-sided test?

(c) If $\alpha = 0.05$, what are your conclusions?

(d) Find a 90% two-sided CI on the difference in means.

SS **2.2** Two machines are used for filling plastic bottles with a net volume of 16.0 oz. The filling processes can be assumed to be normal, with standard deviations of $\sigma_1 = .015$ and $\sigma_2 = 0.018$. The quality engineering department suspects that both machines fill to the same net volume, whether or not this volume is 16.0 oz. An experiment is performed by taking a random sample from the output of each machine. The results are as follows:

Machine 1		Machine 2	
16.03	16.01	16.02	16.03
16.04	15.96	15.97	16.04
16.05	15.98	15.96	16.02
16.05	16.02	16.01	16.01
16.02	15.99	15.99	16.00

(a) State the hypotheses that should be tested in this experiment.

(b) Test these hypotheses using $\alpha = 0.05$. What are your conclusions?

(c) Find the P-value for this test.

(d) Find a 95% CI on the difference in mean fill volume for the two machines.

2.3 Two types of plastic are suitable for use by an electronic calculator manufacturer. The breaking strength of this plastic is important. It is known that both standard deviations are 1.0 psi. From random samples of $n_1 = 10$ and $n_2 = 12$, we obtain $y_1 = 162.5$ and $y_2 = 155.0$. The company will not adopt plastic 1 unless its breaking strength exceeds that of plastic 2 by at least 10 psi. Based on the sample information, should they use plastic 1? In answering this question, set up and test appropriate hypotheses using $\alpha = 0.01$. Construct a 99% CI on the true mean difference in breaking strength. Does the CI aid in answering the question that motivated this experiment?

2.4 An article in *Solid State Technology* ("Orthogonal Design for Process Optimization and Its Application to Plasma Etching" by G. Z. Yin and D. W. Jillie (May 1987)) describes an experiment to determine the effect of the C_2F_6 flow rate on the uniformity of the etch on a silicon wafer used in integrated circuit manufacturing. All of the runs were made in random order. Data for two flow rates are as follows:

	Uniformity observation					
C_2F_6 flow (SCCM)	1	2	3	4	5	6
125	2.7	4.6	2.6	3.0	3.2	3.8
200	4.6	3.4	2.9	3.5	4.1	5.1

(a) Does the C_2F_6 flow rate affect average etch uniformity? Use $\alpha = 0.05$.

(b) What is the P-value for the test in part (a)?

(c) Does the C_2F_6 flow rate affect the wafer-to-wafer variability in etch uniformity? Use $\alpha = 0.05$.

(d) Draw box plots to assist in the interpretation of the data from this experiment.

SS **2.5** Front housings for cell phones are manufactured in an injection molding process. The time the part is allowed to cool in the mold before removal is believed to influence the occurrence of a particularly troublesome cosmetic defect, flow lines, in the finished housing. After manufacturing, the housings are inspected visually and assigned a score between 1 and 10 based on their appearance, with 10 corresponding to a perfect part and 1 to a completely defective part. An experiment was conducted using two cool-down times, 10 and 20 seconds, and 20 housings were evaluated at each level of cool-down time. All 40 observations in this experiment were run in random order. The data are as follows.

10 Seconds		**20 Seconds**	
1	3	7	6
2	6	8	9
1	5	5	5
3	3	9	7
5	2	5	4
1	1	8	6
5	6	6	8
2	8	4	5
3	2	6	8
5	3	7	7

(a) Is there evidence to support the claim that the longer cool-down time results in fewer appearance defects? Use $\alpha = 0.05$.

(b) What is the P-value for the test conducted in part (a)?

(c) Find a 95% CI on the difference in means. Provide a practical interpretation of this interval.

(d) Prepare dot diagrams to assist in interpreting the results from this experiment.

2.6 The deflection temperature under load for two different formulations of ABS plastic pipe is being studied. Two samples of 12 observations each are prepared using each formulation and the deflection temperatures (in °F) are reported below:

Formulation 1			**Formulation 2**		
206	193	192	177	176	198
188	207	210	197	185	188
205	185	194	206	200	189
187	189	178	201	197	203

(a) Do the data support the claim that the mean deflection temperature under load for formulation 1 exceeds that of formulation 2? Use $\alpha = 0.05$.

(b) What is the P-value for the test in part (a)?

2.7 Refer to the data in Problem 2.6. Do the data support the claim that the mean deflection temperature under load for formulation 1 exceeds that of formulation 2 by at least 3°F?

2.8 Two popular pain medications are being compared on the basis of the speed of absorption by the body. Specifically, a claim is made that tablet 1 is absorbed twice as fast as tablet 2. Assume that both standard deviations of the absorption speed are known. Develop a test statistic for

$$H_0 : 2\mu_1 = \mu_2$$
$$H_1 : 2\mu_1 \neq \mu_2$$

2.9 **Continuation of Problem 8.** An article in *Nature* (1972, pp. 225–226) reported on the levels of monoamine oxidase in blood platelets for a sample of 43 schizophrenic patients: $y_1 = 2.69$ and $s_1 = 2.30$ and for a sample of 45 normal patients the results were $y_2 = 6.35$ and $s_2 = 4.03$. The units are nm/mg protein/h. Use the results of the previous problem to test the claim that the mean monoamine oxidase level for normal patients is at least twice the mean level for schizophrenic patients. Assume that the sample sizes are large enough to use the sample standard deviations as the true parameter values.

2.10 Suppose that we want to construct a 95% two-sided CI on the difference in two means where the two sample standard deviations are known to be $\sigma_1 = 4$ and $\sigma_2 = 8$. The total sample size is restricted to $N = 30$. What is the length of the 95% CI if the sample sizes used by the experimenter are both 15? How much shorter would the 95% CI have been if the experimenter had used an optimal sample size allocation?

2.11 In some simple comparative experiments, we can greatly improve the precision by making comparisons within matched pairs of experimental material. For example, consider a hardness testing machine that presses a rod with a pointed tip into a metal specimen with a known force. By measuring the depth of the depression caused by the tip, the hardness of the specimen is determined. Two different tips are available for this machine, and although the precision (variability) of the measurements made by the two tips seems to be the same, we suspect that one tip produces different mean hardness readings than the other. One way to test the hypothesis of equal means would be to select some number, say N, of metal test specimens cut from available bar stock samples and assign half of them to each tip. This experimental design would produce data that could be analyzed with a two-sample t-test. This experimental design would result in potential difference in hardness between the metal specimens caused by the different samples of bar stock that might have been produced in different heats or that were not exactly homogeneous in some other way that might affect the hardness. This lack of homogeneity between specimens will contribute to the variability of the hardness measurements and will tend to inflate the experimental error, thus making a true difference between tips harder to detect. An alternative experimental design would test each tip on each coupon and use the difference in hardness measurements on each specimen as a measure of the difference in hardness reading for the two tips. Does this alternative design solve the problem of variability between metal coupons inflating the experimental error? Which of these two experimental designs would you prefer to use and why?

2.12 **Continuation of Problem 2.11.** Suppose that the hardness testing experiment described in Problem 2.11 was conducted by pairing measurements from both tips on a single specimen. The data that results from this experiment with 10 specimens are as follows.

Specimen	Tip 1	Tip 2
1	7	6
2	3	3
3	3	5
4	4	3
5	8	8
6	3	2
7	2	4
8	9	9
9	5	4
10	4	5

(a) Develop a test procedure for testing the equality of the mean hardness readings that result from the two types of tips.

Hint: Testing the equality of the two means is equivalent to testing that the mean difference in hardness readings from each specimen is zero.

(b) Apply the procedure that you developed in part (a) to the data from this experiment. This test procedure is called the **paired t-test**, and the experimental design employed in which the observations from both tips are paired on each specimen to eliminate the "nuisance" variability that results from the different hardness between the specimens is an example of the blocking principle, discussed in detail in Chapter 6.

2.13 **Continuation of Problems 2.11 and 2.12.** Use the data from Problem 2.12 to construct two-sided 95% CIs on the difference in mean hardness. Construct one interval assuming that the data were collected by randomly assigning 10 specimens to each tip. Then construct another interval assuming that the paired design was used. Compare the lengths of the two CIs. What does this tell you about the relative advantages or disadvantages of the two experimental designs?

2.14 An article in the journal *Neurology* (1998, Vol. 50, pp. 1246–1252) observed that monozygotic twins share numerous physical, psychological, and pathological traits. The investigators measured an intelligence score of 10 pairs of twins. The data obtained are as follows: SS

Pair	Birth order: 1	Birth order: 2
1	6.08	5.73
2	6.22	5.80
3	7.99	8.42
4	7.44	6.84
5	6.48	6.43
6	7.99	8.76
7	6.32	6.32
8	7.60	7.62
9	6.03	6.59
10	7.52	7.67

(a) Find a 95% CI on the difference in mean score. Is there any evidence that mean score depends on birth order?

(b) Test an appropriate set of hypotheses indicating that the mean score does not depend on birth order.

2.15 A new filtering device is installed in a chemical unit. Before its installation, a random sample yielded the following information about the percentage of impurity: $\bar{y}_1 = 12.5$, $S_1^2 = 101.25, n_1 = 12$. After installation, a random sample yields $\bar{y}_1 = 10.8, S_1^2 = 98.7, n_1 = 8$. Has the new filtering device reduced the percentage of impurity significantly? Use $\alpha = 0.05$.

CHAPTER 3 PROBLEMS

3.1 An experimenter has conducted a single-factor experiment with four levels of the factor, and each factor level was replicated six times. The computed value of the F-statistic is $F_0 = 3.26$. Find bounds on the P-value.

3.2 A computer ANOVA output is shown below. Fill in the blanks. You may give bounds on the P-value.

Source	DF	SS	MS	F	P
Factor	3	36.15	?	?	?
Error	?	?	?		
Total	19	196.04			

3.3 An article appeared in *The Wall Street Journal* on Tuesday, April 27, 2010, with the title "Eating Chocolate Is Linked to Depression." The article reported on a study funded by the National Heart, Lung and Blood Institute (part of the National Institutes of Health) and conducted by faculty at the University of California, San Diego, and the University of California, Davis. The research was also published in the *Archives of Internal Medicine* (2010, pp. 699–703). The study examined 931 adults who were not taking antidepressants and did not have known cardiovascular disease or diabetes. The group was about 70% men and the average age of the group was reported to be about 58. The participants were asked about chocolate consumption and then screened for depression using a questionnaire. People who score less than 16 on the questionnaire were not considered depressed, while those with scores above 16 and less than or equal to 22 were considered possibly depressed, while those with scores above 22 were considered likely to be depressed. The survey found that people who were not depressed ate an average 5.4 servings of chocolate per month, possibly depressed individuals ate an average of 8.4 servings of chocolate per month, while those individuals who scored above 22 and were likely to be depressed ate the most chocolate, an average of 11.8 servings per month. No differentiation was made between dark and milk chocolate. Other foods were also examined, but no pattern emerged between other foods and depression. Is this study really a designed experiment? Does it establish a cause-and-effect link between chocolate consumption and depression? Should gender and age have been considered explicitly in the study? How should the study be conducted to establish such a cause-and effect link?

3.4 An article in *Bioelectromagnetics* ("Electromagnetic Effects on Forearm Disuse Osteopenia: A Randomized, Double-Blind, Sham-Controlled Study," Vol. 32, 2011, pp. 273–282) described a randomized, double-blind, sham-controlled, feasibility, and dosing study to determine if a common pulsing electromagnetic field (PEMF) treatment could moderate the substantial osteopenia that occurs after forearm disuse. Subjects were randomized into four groups after a distal radius fracture or carpal surgery requiring immobilization in a cast. Active or identical sham PEMF transducers were worn on the distal forearm for 1, 2, or 4 h/day for eight weeks starting after cast removal ("baseline") when bone density continues to decline. Bone mineral density (BMD) and bone geometry were measured in the distal forearm by dual-energy X-ray absorptiometry (DXA) and peripheral quantitative computed tomography (pQCT). The data below are the percent losses in BMD measurements on the radius after 16 weeks for patients wearing the active or sham PEMF transducers for 1, 2, or 4 h/day (data were constructed to match the means and standard deviations read from a graph in the paper).

(a) Is there evidence to support a claim that PEMF usage affects BMD loss? If so, analyze the data to determine which specific treatments produce the differences.

(b) Analyze the residuals from this experiment and comment on the underlying assumptions and model adequacy.

Sham	PEMF 1 h/day	PEMF 2 h/day	PEMF 4 h/day
4.51	5.32	4.73	7.03
7.95	6.00	5.81	4.65
4.97	5.12	5.69	6.65
3.00	7.08	3.86	5.49
7.97	5.48	4.06	6.98
2.23	6.52	6.56	4.85
3.95	4.09	8.34	7.26
5.64	6.28	3.01	5.92
9.35	7.77	6.71	5.58
6.52	5.68	6.51	7.91
4.96	8.47	1.70	4.90
6.10	4.58	5.89	4.54
7.19	4.11	6.55	8.18
4.03	5.72	5.34	5.42
2.72	5.91	5.88	6.03
9.19	6.89	7.50	7.04
5.17	6.99	3.28	5.17
5.70	4.98	5.38	7.60
5.85	9.94	7.30	7.90
6.45	6.38	5.46	7.91

SS **3.5** The tensile strength of Portland cement is being studied. Four different mixing techniques are available that can be used economically. A completely randomized experiment was conducted and the following data were collected:

Mixing technique	Tensile strength (lb/in.^{2})			
1	3129	3000	2865	2890
2	3200	3300	2975	3150
3	2800	2900	2985	3050
4	2600	2700	2600	2765

(a) Test the hypothesis that mixing techniques affect the strength of the cement. Use $\alpha = 0.05$.

(b) Construct a graphical display to compare the mean tensile strengths for the four mixing techniques. What are your conclusions?

(c) Use the Fisher LSD method with $\alpha = 0.05$ to make comparisons between pairs of means.

(d) Construct a normal probability plot of the residuals.

(e) What conclusion would you draw about the validity of the normality assumption?

(f) Plot the residuals versus the predicted tensile strength. Comment on the plot.

(g) Prepare a scatter plot of the results to aid the interpretation of the results of this experiment.

3.6 Refer to Problem 3.5.

(a) Rework part (c) using Tukey's test with $\alpha = 0.05$. Do you get the same conclusions from Tukey's test that you did from the graphical procedure and/or the Fisher LSD method?

(b) Explain the difference between the Tukey and Fisher procedures.

3.7 Reconsider the experiment in Problem 3.4. Find a 95% confidence interval on the mean tensile strength of the Portland cement produced by each of the four mixing techniques. Also, find a 95% confidence interval on the difference in means for techniques 1 and 3. Does this aid you in interpreting the results of the experiment?

3.8 A product developer is investigating the tensile strength of a new synthetic fiber that will be used to make cloth for men's shirts. Strength is usually affected by the percentage of cotton used in the blend of materials for the fiber. The engineer conducts a completely randomized experiment with five levels of cotton content and replicates the experiment five times. The data are shown in the following table.

Cotton weight percent	Observations				
15	7	7	15	11	9
20	12	17	12	18	18
25	14	19	19	18	18
30	19	25	22	19	23
35	7	10	11	15	11

(a) Is there evidence to support the claim that cotton content affects the mean tensile strength? Use $\alpha = 0.05$.

(b) Use the Fisher LSD method to make comparisons between the pairs of means. What conclusions can you draw?

(c) Analyze the residuals from this experiment and comment on model adequacy.

3.9 Reconsider the experiment described in Problem 3.8. Suppose that 30% cotton content is a control. Use Dunnett's test with $\alpha = 0.05$ to compare all of the other means with the control.

3.10 Suppose that an experimenter is interested in comparing the means of five treatments. If any treatment mean differs from the others by two standard deviations, the experimenter would like to detect this with high power.

(a) How many replicates would you recommend that the experimenter use if "high power" is defined as 0.70?

(b) How many replicates would you recommend that the experimenter use if "high power" is defined as 0.90?

3.11 Suppose that an experimenter is interested in comparing the means of three treatments. If any treatment mean differs from the others by one standard deviation, the experimenter would like to detect this with high power.

(a) How many replicates would you recommend that the experimenter use?

(b) Suppose that the experimenter is concerned about larger differences, say two or more standard deviations. How many replicates would you recommend?

SS **3.12** Consider testing the equality of the means of two normal populations, where the variances are unknown but can be assumed equal. The appropriate test procedure is the pooled t-test. Show that the pooled t-test is equivalent to the single-factor analysis of variance.

SS **3.13** Three brands of batteries are under study. We suspect that the lives (in weeks) of the three brands are different. Five randomly selected batteries of each brand are tested with the following results:

Weeks of life		
Brand 1	**Brand 2**	**Brand 3**
100	76	108
96	80	100
92	75	96
96	84	98
92	82	100

(a) Are the lives of these brands of batteries different?

(b) Analyze the residuals from this experiment.

(c) Construct a 95% confidence interval estimate on the mean life of battery brand 2. Construct a 99% confidence interval estimate on the mean difference between the lives of battery brands 2 and 3.

3.14 Referring to Problem 3.13, which brand would you select for use? If the manufacturer will replace without charge any battery that fails in less than 85 weeks, what percentage would the company expect to replace?

3.15 Four catalysts that may affect the concentration of one component in a three-component liquid mixture are being investigated. The following concentrations are obtained from a completely randomized experiment: SS

Catalyst			
1	**2**	**3**	**4**
58.2	56.3	50.1	52.9
57.2	54.5	54.2	49.9
58.4	57.0	55.4	50.0
55.8	55.3		51.7
54.9			

(a) Do the four catalysts have the same effect on the concentration?

(b) Analyze the residuals from this experiment.

(c) Construct a 99% confidence interval estimate of the mean response for catalyst 1.

CHAPTER 4 PROBLEMS

4.1 An engineering team experimented with changing the pressure in a reaction chamber from 10 to 20 torr. The average yield at 10 torr was 50%, while the average yield at 20 torr was 70%. Assuming the relationship between pressure and yield is linear, what is the slope of the relationship in engineering units? What is the slope in coded units?

SS **4.2** Consider the arbitrary 2 × 2 matrix $\mathbf{A} = \begin{bmatrix} a & b \\ c & d \end{bmatrix}$. Its determinant, det(**A**) = $ad - bc$.

(a) Show that the inverse of **A**, $\mathbf{A}^{-1} = \frac{1}{(ad-bc)}\begin{bmatrix} d & -b \\ -c & a \end{bmatrix}$.

(b) Show that the determinant of the inverse of **A** is the inverse of the determinant of **A**.

4.3 The 10-run D-optimal design for a linear model in one continuous factor places five runs each at the coded values of −1 and 1. Show that $\mathbf{X^tX} = \begin{bmatrix} 10 & 0 \\ 0 & 10 \end{bmatrix}$.

(a) What is the determinant of $\mathbf{X^tX}$?

(b) What is the determinant of $\mathbf{X^tX}^{-1}$?

(c) Recall that for the design in coded units in Table 4.2, $(\mathbf{X'X})^{-1} = \begin{bmatrix} 0.092 & 0.013 \\ 0.013 & 0.209 \end{bmatrix}$. What is the determinant of this matrix?

(d) Compare the determinant from your answer in part (b) to the determinant of your answer for part (c).

4.4 Consider the arbitrary 3×3 matrix $\mathbf{A} = \begin{matrix} a_{11} & a_{12} & a_{13} \\ a_{21} & a_{22} & a_{23} \\ a_{31} & a_{32} & a_{32} \end{matrix}$. Its determinant is det(**A**) $= a_{11}a_{22}a_{33} + a_{12}a_{23}a_{31} + a_{13}a_{21}a_{32} - (a_{11}a_{23}a_{32} + a_{12}a_{21}a_{33} + a_{13}a_{22}a_{31})$.

(a) The nine-run D-optimal design for the quadratic model in one continuous factor places three runs each at the coded values −1, 0, and 1. Show that $\mathbf{X^tX} = \begin{matrix} 9 & 0 & 6 \\ 0 & 6 & 0 \\ 6 & 0 & 6 \end{matrix}$.

(b) Compute the determinant of the matrix above.

(c) Compute **X′X** for the coded data in Table 4.2.

(d) What is the relative efficiency of the 11-run design in Table 4.2 to the 9-run D-optimal design?

4.5 Consider the *x*/*y* data in the table below originally published by Frank Anscombe. SS

x	10	8	13	9	11	14	6	4	12	7	5
y	8.04	6.95	7.58	8.81	8.33	9.96	7.24	4.26	10.84	4.82	5.68

(a) Make a scatter plot of the data.

(b) What are the coded values of *x*?

(c) Fit a line to the data. What are the estimates of the slope, intercept, and error variance?

(d) Plot the residuals versus the predicted values of *y*. Does the line seem like a good model for these data?

4.6 Consider the *x*/*y* data published by Anscombe in the same paper.

x	10	8	13	9	11	14	6	4	12	7	5
y	9.14	8.14	8.74	8.77	9.26	8.1	6.13	3.1	9.13	7.26	4.74

(a) Fit a line to the data. What are the estimates of the slope, intercept, and error variance?

(b) Plot the residuals versus the predicted values of *y*. Does the line seem like a good model for these data?

(c) Fit a quadratic model to the data. What are your parameter estimates? Has your estimate of the error variance decreased?

SS **4.7** Consider the x/y data below

x	−1	−1	−1	−1	−1	−1	−1	−1
y	1.68	1.98	2.07	2.44	3.28	3.44	4.09	4.53
x	1	1	1	1	1	1	1	1
y	4.98	5.7	7.77	9.07	9.43	9.97	11.75	16.3

(a) Fit a line to the data. What are the estimates of the slope, intercept, and error variance?

(b) Plot the residuals versus the predicted values of y. Does the variation in y at the high values of x seem larger than for low values of x?

(c) Make a table of x and $\log(y)$.

(d) Fit a line to these data and plot the residuals versus the predicted values of $\log(y)$. Which model do you prefer and why?

4.8 Consider the temperature/time data below

Temperature (°C)	163	149	163	163	149	163	135
Time	13.6	13.9	89	3.7	6.1	17	4
Temperature (°C)	135	135	149	149	149	149	135
Time	4.6	15.4	10.4	18.1	3.1	18.2	6.6

(a) Fit a line to the data. What are the estimates of the slope, intercept, and error variance?

(b) Plot the residuals versus the predicted values of y. Does one of the values of y seem unusual? Which one?

(c) Suppose you exclude that value from the data set. Fit a line to the remaining data. What are your new estimates of the slope, intercept, and error variance?

(d) Suppose you talked to the operator who collected the data and found out that this value was a typo and should have been 10 times smaller. Fit a line to the resulting data. What are your conclusions?

4.9 Consider the binary dose response data below.

Dose	1	1	1	1	1	2	2	2	2	2	2	3	3	3	3
Response	1	1	1	0	1	1	1	1	1	1	0	1	1	0	0
Dose	5	5	5	5	5	5	8	8	8	8	12	12	12	12	12
Response	0	0	1	1	0	0	0	0	1	0	0	0	0	0	0

(a) Fit a line to the data using least squares regression. What are the estimates of the slope, intercept, and error variance?

(b) What is the predicted value of response when the dose is 12?

(c) Does your predicted value make sense? See Chapter 12 for an alternative modeling approach.

4.10 Consider the x/y data below.

x	0.1667	0.5500	0.3300	0.6333	0.2483	0.2500	0.2083
y	4.0222	3.5526	3.3097	2.3626	3.6719	2.6066	3.2916

x	0.5167	0.3250	0.4000	0.3883	0.4400	0.5917
y	2.2972	1.6451	1.5888	3.2020	3.8526	2.8745

(a) Fit a line to the data. What are the estimates of the slope, intercept, and error variance?

(b) Compute the covariance matrix of the parameters. Compare the square root of the result in the second row and second column (this value is called the standard error of the slope) to the estimate of the slope.

(c) What do you think is the true relationship between x and y? Defend your point of view.

CHAPTER 5 PROBLEMS

SS **5.1** Consider the following incomplete ANOVA table:

Source	SS	DF	MS	F
A		1	50.00	
B	80.00		40.00	
AB	30.00	2	15.00	
Error		12		
Total	172.00	17		

(a) Complete the ANOVA calculations.

(b) Provide an interpretation of this experiment.

(c) The pure error estimate of the standard deviation of the sample observations is 1. **True / False**

5.2 Physics PhD student Laura Van Ertia has conducted a two-factor factorial experiment. From her experimental results, Laura has constructed the following incomplete ANOVA display:

Source	SS	DF	MS	F
A	350.00	2		
B	300.00		150	
AB	200.00		50	
Error	150.00	18		
Total	1000.00			

(a) How many levels of factor B did she use in the experiment?

(b) How many degrees of freedom are associated with interaction?

(c) The error mean square is ________.

(d) The mean square for factor A is ________.

(e) How many replicates of the experiment were conducted?

(f) What are your conclusions about interaction and the two main effects?

(g) An estimate of the standard deviation of the response variable is ________.

5.3 The C. F. Eye Care company manufactures lenses for transplantation into the eye following cataract surgery. An engineering group has conducted an experiment involving two factors to determine their effect on the lens polishing process. The results of this experiment are summarized in the following ANOVA display:

Source	DF	SS	MS	F	P-value
Factor A	—	—	0.0833	0.05	0.952
Factor B	—	96.333	96.3333	57.80	<0.001
Interaction	2	12.167	6.0833	3.65	—
Error	6	10.000	—		
Total	11	118.667			

(a) The sum of squares for factor A is ________.

(b) The number of degrees of freedom for factor A in the experiment is ________.

(c) The number of degrees of freedom for factor B is ________.

(d) The mean square for error is ________.

(e) An upper bound for the P-value for the interaction test statistic is ________.

(f) The engineers used ________ levels of factor A in this experiment.

(g) The engineers used ___________ levels of factor B in this experiment.

(h) There are ___________ replicates of this experiment.

(i) Would you conclude that the effect of factor B depends on the level of factor A? **Yes / No**

(j) An estimate of the standard deviation of the response variable is ___________.

5.4 Suppose that you are designing an experiment with two categorical factors A and B. Factor A has two levels and factor B has three levels. You decide to use a full-factorial experiment but now must determine the number of replicates. After considering the budget available, you think that two replicates (12 runs) would be affordable. If the size of all of the model parameters is equal to one standard deviation of the experimental error, what is the power of this design? If you could afford an additional replicate, how much would the power increase? Do you think that the increased power is worth the additional expense of conducting the experiment?

5.5 Suppose that you are designing an experiment with two factors A and B. Factor A is continuous and factor B is categorical with four levels. You think it is necessary to account for quadratic effects in the continuous factor and to consider possible interactions between the linear and quadratic effects of A and the levels of the categorical factor. Construct an appropriate design and determine the number of replicates to achieve adequate power.

5.6 An article in *Biotechnology Progress* (2001, Vol. 17, pp. 366–368) described an experiment to investigate nisin extraction in aqueous two-phase solutions. A two-factor factorial experiment was conducted using factors A = concentration of PEG and B = concentration of Na_2SO_4. Data similar to that reported in the article are presented below.

A	B	Extraction (%)
13	11	62.9
13	11	65.4
15	11	76.1
15	11	72.3
13	13	87.5
13	13	84.2
15	13	102.3
15	13	105.6

(a) Analyze the extraction response. Draw appropriate conclusions about the effects of the significant factors on the response.

(b) Prepare appropriate residual plots and comment on model adequacy.

(c) Construct contour plots to aid in practical interpretation of the response.

5.7 An experiment was performed to investigate the keyboard feel on a computer (crisp or mushy) and the size of the keys (small, medium, or large). The response variable is typing speed. Three replicates of the experiment were performed. The experimental design and the data are as follows. SS

	Keyboard feel	
Key size	**Mushy**	**Crisp**
Small	31, 33, 35	36, 40, 41
Medium	36, 35, 33	40, 41, 42
Large	37, 34, 33	38, 36, 39

(a) Analyze the data from this experiment.

(b) Investigate model adequacy by constructing appropriate residual plots.

(c) What conclusions can you draw?

5.8 Reconsider the experiment in Problem 5.7. Suppose that the experimenter was interested in detecting a difference between the keyboard feel factor levels of one error standard deviation. SS

(a) Do you think that three replicates are sufficient to achieve adequate power in this experiment?

(b) Suppose that due to time constraints the experimenter asked you to consider reducing the number of replications. What power would result if you decided to use two replicates? Are you comfortable with this decision?

(c) Suppose the effect size of interest was two standard deviations. Find the power associated with the use of either two or three replicates. What would you recommend to the experimenter in this situation?

5.9 Johnson and Leone (*Statistics and Experimental Design in Engineering and the Physical Sciences*, Wiley, 1977) describe an experiment to investigate warping of copper plates. The two factors studied were the temperature and the copper content of the plates. The response variable was a measure of the amount of warping. The data are as follows:

	Copper content (%)			
Temperature (°C)	**40**	**60**	**80**	**100**
50	17, 20	16, 21	24, 22	28, 27
75	12, 9	18, 13	17, 12	27, 31
100	16, 12	18, 21	25, 23	30, 23
125	21, 17	23, 21	23, 22	29, 31

(a) Is there any indication that either factor affects the amount of warping? Is there any interaction between the factors? Use $\alpha = 0.05$.

(b) Analyze the residuals from this experiment.

(c) Fit a second-order model to the data from this experiment and use the JMP prediction profiler to draw conclusions about the appropriate choices of temperature and copper content. Also, construct a contour plot to aid in interpretation.

5.10 **Continuation of Problem 5.9**. Suppose that temperature factor cannot be easily controlled in the environment in which the copper plates are to be used. What impact does this have on your decision about the best level for copper content?

5.11 You have been asked to assist in designing an experiment for two continuous factors. The experimenter thinks it will be necessary to model the results with a second-order polynomial. However, the experiments are very expensive and take a long time to run, and the budget for this experiment is only eight tests. Find a D-optimal design for this problem. Evaluate the prediction performance of this design.

5.12 **Continuation of Problem 5.11.** Suppose that the experimenter in Problem 5.11 has found additional resources and can now afford to use 11 tests in his experiment. Find a new D-optimal design with this increased sample size. How does it compare in performance to the original eight-run design you found in Problem 5.11? SS

5.13 **Continuation of Problem 5.11.** Reconsider the situation in Problem 5.11. Find an eight-run I-optimal design for this problem. How does it compare to the original D-optimal design you found in that problem.

5.14 You have been asked to help design an experiment for two continuous factors. The experimenter thinks it will be necessary to model the results with at least a second-order polynomial. Based on her experience, she has some concern about stronger curvature in the x_1 direction, so asks about the possibility of a design for a model that in addition to the usual quadratic terms, also includes the terms x_1^3 and $x_1^2x_2$. She has a budget of 12 runs. Find a D-optimal design for the problem.

5.15 **Continuation of Problem 5.14.** Reconsider the situation in Problem 5.14. Find an I-optimal design for this problem. Compare this design to the original design from Problem 5.14. If prediction is the goal of the experimenter, which design would you recommend that she should use?

CHAPTER 6 PROBLEMS

SS **6.1** Suppose that a single-factor experiment with four levels of the factor has been conducted. There are six replicates and the experiment has been conducted in blocks. The error sum of squares is 500 and the block sum of squares is 250. If the experiment had been conducted as a completely randomized designs the estimate of the error variance σ^2 would be

(a) 25.0

(b) 25.5

(c) 35.0

(d) 37.5

(e) None of the above

6.2 The key concept of the phrase "Block if you can, randomize if you can't." is that:

(a) It is usually better not to randomize within blocks.

(b) Blocking violates the assumption of constant variance.

(c) Create blocks by using each level of the nuisance factor as a block and randomize within blocks.

(d) Randomizing the runs is preferable to randomizing blocks.

6.3 The ANOVA from a randomized complete block experiment output is shown below.

```
Source      DF       SS       MS       F   P
Treatment    4  1010.56        ?   29.84   ?
Block        ?        ?   64.765       ?   ?
Error       20   169.33        ?
Total       29  1503.71
```

(a) Fill in the blanks. You may give bounds on the P-value.

(b) How many blocks were used in this experiment?

(c) What conclusions can you draw?

6.4 A consumer products company relies on direct mail marketing pieces as a major component of its advertising campaigns. The company has three different designs for a new brochure and wants to evaluate their effectiveness, as there are substantial differences in costs between the three designs. SS

The company decides to test the three designs by mailing 5000 samples of each to potential customers in four different regions of the country. Since there are known regional differences in the customer base, regions are considered as blocks. The number of responses to each mailing is as follows.

	Region			
Design	**NE**	**NW**	**SE**	**SW**
1	250	350	219	375
2	400	525	390	580
3	275	340	200	310

(a) Analyze the data from this experiment.

(b) Use the Fisher LSD method to make comparisons among the three designs to determine specifically which designs differ in the mean response rate.

(c) Analyze the residuals from this experiment.

6.5 Consider the direct mail marketing experiment in Problem 6.4. Suppose that this experiment had been run as a completely randomized design, ignoring potential regional differences, but that exactly the same data was obtained. Reanalyze the experiment under this new assumption. What difference would ignoring blocking have on the results and conclusions?

6.6 An aluminum master alloy manufacturer produces grain refiners in ingot form. The company produces the product in four furnaces. Each furnace is known to have its own unique operating characteristics, so any experiment run in SS the foundry that involves more than one furnace will consider furnaces as a nuisance variable. The process engineers suspect that stirring rate affects the grain size of the product. Each furnace can be run at four different stirring rates. A randomized block design is run for a particular refiner, and the resulting grain size data is as follows.

	Furnace			
Stirring rate (rpm)	**1**	**2**	**3**	**4**
5	8	4	5	6
10	14	5	6	9
15	14	6	9	2
20	17	9	3	6

(a) Is there any evidence that stirring rate affects the grain size?

(b) Graph the residuals from this experiment on a normal probability plot. Interpret this plot.

(c) Plot the residuals versus furnace and stirring rate. Does this plot convey any useful information?

(d) What should the process engineers recommend concerning the choice of stirring rate and furnace for this particular grain refiner if small grain size is desirable?

6.7 Consider the RCBD for the vascular graft experiment in Table 6.1. This design and analysis treats the pressure factor as categorical. This is the usual way the ANOVA is performed—treating all factors as categorical. However, pressure is continuous. Suppose that you wanted to conduct this experiment treating pressure as a continuous factor that could potentially have a quadratic effect. You still have six batches of raw material that can be used. Construct an appropriate design for this version of the experiment.

SS **6.8** The effect of five different ingredients (A, B, C, D, E) on the reaction time of a chemical process is being studied. Each batch of new material is only large enough to permit five runs to be made. Furthermore, each run requires approximately 1.5 hours, so only five runs can be made in one day. The experimenter decides to run the experiment as a Latin square so that day and batch effects may be systematically controlled. She obtains the data that follow. Analyze the data from this experiment (use $\alpha = 0.05$) and draw conclusions.

	Day				
Batch	**1**	**2**	**3**	**4**	**5**
1	$A = 8$	$B = 7$	$D = 1$	$C = 7$	$E = 3$
2	$C = 11$	$E = 2$	$A = 7$	$D = 3$	$B = 8$
3	$B = 4$	$A = 9$	$C = 10$	$E = 1$	$D = 5$
4	$D = 6$	$C = 8$	$E = 6$	$B = 6$	$A = 10$
5	$E = 4$	$D = 2$	$B = 3$	$A = 8$	$C = 8$

6.9 Reconsider the situation in Problem 6.8. Suppose that each batch of material can only produce four runs. The experimenter still wants to perform the experiment over a five-day period.

(a) Construct an appropriate design for this problem.

(b) Compare this design to the Latin square design employed in Problem 6.8. Are the effects of the five ingredients orthogonal?

(c) Suppose that an additional day or two could be used to conduct the experiment. Would this be a better alternative to the design you found in part (a)?

6.10 An experimenter wants to compare eight treatments in blocks of size four. How many blocks would you recommend? What is the minimum number of blocks to use for a BIBD? How does you recommended design compare to the minimal design?

6.11 An engineer is studying the mileage performance characteristics of five types of gasoline additives. In the road test, he wishes to use cars as blocks; however, because of a time constraint, he must use an incomplete block design. He runs the balanced design with the five blocks that follow. Analyze the data from this experiment (use $\alpha = 0.05$) and draw conclusions.

	Car				
Additive	**1**	**2**	**3**	**4**	**5**
1		17	14	13	12
2	14	14		13	10
3	12		13	12	9
4	13	11	11	12	
5	11	12	10		8

6.12 An experimenter wishes to compare four treatments in blocks of two runs. Find an optimal design for this experiment with six blocks. Is this a BIBD?

6.13 Reconsider the gasoline additive experiment in Problem 6.11. Suppose that only four cars are available for the experiment. Construct an appropriate design. How does this design compare to the one used in Problem 6.11? Would this allow you to make a convincing argument for the use of a fifth car so that the original design could be used?

6.14 **A missing observation in the BCRD**. Occasionally an observation in a randomized block design is missing. Let x denote the missing observation.

(a) Show that the estimated value of the missing observation that minimizes the error sum of squares is

$$x = \frac{ay'_{i.} + by'_{.j} + y'_{..}}{(a-1)(b-1)}$$

where the totals with the primes (′) are the treatment and block totals containing the missing observations and grand total.

(b) Consider the design in Problem 6.6. Suppose that the observation for stirring rate 15 on Furnace 3 is missing. Replace the missing value of this observation with an estimate computed using the equation from part (a). Analyze the data, and compare your results to the original analysis of the complete experiment from Problem 6.6.

6.15 **Continuation of Problem 6.14**. Consider the missing observation situation from Problem 6.14. Analyze the data from this experiment ignoring the missing value. How does this "exact analysis" compare to the "approximate analysis" from Problem 6.14 using an estimate of the missing observation?

CHAPTER 7 PROBLEMS

SS **7.1** An engineer is interested in the effects of cutting speed (A), tool geometry (B), and cutting angle (C) on the life (in hours) of a machine tool. Two levels of each factor are chosen, and three replicates of a 2^3 factorial design are run. The results are as follows:

A	*B*	*C*	Treatment combination	Replicate 1	2	3
–	–	–	(1)	22	31	25
+	–	–	*a*	32	43	29
–	+	–	*b*	35	34	50
+	+	–	*ab*	55	47	46
–	–	+	*c*	44	45	38
+	–	+	*ac*	40	37	36
–	+	+	*bc*	60	50	54
+	+	+	*abc*	39	41	47

(a) Estimate the factor effects. Which effects appear to be large?

(b) Use the analysis of variance to confirm your conclusions for part (a).

(c) Write down a regression model for predicting tool life (in hours) based on the results of this experiment.

(d) Analyze the residuals. Are there any obvious problems?

(e) On the basis of an analysis of main effect and interaction plots, what coded factor levels of A, B, and C would you recommend using?

7.2 Reconsider part (c) of Problem 7.1. Use the regression model to generate response surface and contour plots of the tool life response. Interpret these plots. Do they provide insight into the desirable operating conditions for this process?

7.3 Suppose that the experiment in Problem 7.1 was conducted in blocks, and each replicate was treated as a block. Analyze the data from this version of the experiment. Is the block effect important? What impact does this have on your conclusions about desirable operating conditions?

7.4 A router is used to cut locating notches on a printed circuit board. The vibration level at the surface of the board as it is cut is considered to be a major source of dimensional variation in the notches. Two factors potentially influence vibration: bit size (A) and cutting speed (B). Two bit sizes (1/16 and 1/8 in.) and two speeds (40 and 90 rpm) are selected, and four boards are cut at each set of conditions shown below. The response variable is vibration measured as the resultant vector of three accelerometers (x, y, and z) on each test circuit board. SS

A	*B*	Treatment combination	Replicate 1	2	3	4
–	–	(1)	18.2	18.9	12.9	14.4
+	–	*a*	27.2	24.0	22.4	22.5
–	+	*b*	15.9	14.5	15.1	14.2
+	+	*ab*	41.0	43.9	36.3	39.9

(a) Analyze the data from this experiment.

(b) Construct a normal probability plot of the residuals, and plot the residuals versus the predicted vibration level. Interpret these plots.

(c) Draw the AB interaction plot. Interpret this plot. What levels of bit size and speed would you recommend for routine operation?

7.5 Reconsider the experiment described in Problem 7.1. Suppose that the experimenter only performed the eight trials from Replicate 1. In addition, he ran four center points and obtained the following response values: 36, 40, 43, 45.

(a) Estimate the factor effects. Which effects are large?

(b) Perform an analysis of variance, including a check for pure quadratic curvature. What are your conclusions?

(c) Write down an appropriate model for predicting tool life, based on the results of this experiment. Does this model differ in any substantial way from the model in Problem 7.1, part (c)?

(d) Analyze the residuals. Are there any indications of model inadequacy or violations of the assumptions?

(e) What conclusions would you draw about the appropriate operating conditions for this process?

SS **7.6** An experiment was performed to improve the yield of a chemical process. Four factors were selected, and two replicates of a completely randomized experiment were run. The results are shown in the following table:

(a) Estimate the factor effects. Prepare an analysis of variance table and determine which factors are important in explaining yield.

Treatment combination	Replicate 1	Replicate 2	Treatment combination	Replicate 1	Replicate 2
(1)	90	93	*d*	98	95
a	74	78	*ad*	72	76
b	81	85	*bd*	87	83
ab	83	80	*abd*	85	86
c	77	78	*cd*	99	90
ac	81	80	*acd*	79	75
bc	88	82	*bcd*	87	84
abc	73	70	*abcd*	80	80

(b) Write down a regression model for predicting yield, assuming that all four factors were varied over the range from -1 to $+1$ (in coded units).

(c) Plot the residuals versus the predicted yield and on a normal probability scale. Does the residual analysis appear satisfactory?

(d) Two three-factor interactions, *ABC* and *ABD*, apparently have large effects. Draw a cube plot in the factors *A*, *B*, and *C* with the average yields shown at each corner. Repeat using the factors *A*, *B*, and *D*. Do these two plots aid in data interpretation? Where would you recommend running the process with respect to the four variables?

7.7 A nickel–titanium alloy is used to make components for jet turbine aircraft engines. Cracking is a potentially serious problem in the final part because it can lead to nonrecoverable failure. A test is run at the parts producer to determine the effect of four factors on cracks. The four factors are pouring temperature (*A*), titanium content (*B*), heat treatment method (*C*), and amount of grain refiner used (*D*). Two replicates of a 2^4 design are run, and the length of crack (in mm $\times 10^{-2}$) induced in a sample coupon subjected to a standard test is measured. The data are shown in the next column.

(a) Estimate the factor effects. Which factor effects appear large?

(b) Conduct an analysis of variance. Do any of the factors affect cracking? Use $\alpha = 0.05$.

(c) Write down a regression model that can be used to predict the crack length as a function of the significant main effects and interactions you have identified in part (b).

(d) Analyze the residuals from this experiment.

(e) Is there an indication that any of the factors affect the variability in cracking? What recommendations would you make regarding process operations? Use interaction and/or main effect plots to assist in drawing conclusions.

A	*B*	*C*	*D*	Treatment combination	Replicate 1	Replicate 2
−	−	−	−	(1)	7.037	6.376
+	−	−	−	*a*	14.707	15.219
−	+	−	−	*b*	11.635	12.089
+	+	−	−	*ab*	17.273	17.815
−	−	+	−	*c*	10.403	10.151
+	−	+	−	*ac*	4.368	4.098
−	+	+	−	*bc*	9.360	9.253
+	+	+	−	*abc*	13.440	12.923
−	−	−	+	*d*	8.561	8.951
+	−	−	+	*ad*	16.867	17.052
−	+	−	+	*bd*	13.876	13.658
+	+	−	+	*abd*	19.824	19.639
−	−	+	+	*cd*	11.846	12.337
+	−	+	+	*acd*	6.125	5.904
−	+	+	+	*bcd*	11.190	10.935
+	+	+	+	*abcd*	15.653	15.053

7.8 Continuation of Problem 7.7. One of the variables in the experiment described in Problem 7.7, heat treatment method (C), is a categorical variable. Assume that the remaining factors are continuous.

(a) Write two regression models for predicting crack length, one for each level of the heat treatment method variable. What differences, if any, do you notice in these two equations?

(b) Generate appropriate response surface contour plots for the two regression models in part (a).

(c) What set of conditions would you recommend for the factors A, B, and D if you use heat treatment method $C = +$?

(d) Repeat part (c) assuming that you wish to use heat treatment method $C = -$.

7.9 Consider the 2^4 factorial experiment in Problem 7.7. Suppose that each replicate had to be run on a different day. Would it be appropriate to consider days as blocks? Analyze the experiment assuming that replicates are blocks. Are there any differences in conclusions from this analysis and the original analysis from Problem 7.7?

7.10 Suppose that only the data from the first replicate of the experiment in Problem 7.7 were available. Analyze the data and draw conclusions. Do these conclusions differ from those in Problem 7.7 where both replicates were analyzed?

7.11 An experiment was run in a semiconductor fabrication plant in an effort to increase yield. Five factors, each at two levels, were studied. The factors (and levels) were A = aperture setting (small, large), B = exposure time (20% below nominal, 20% above nominal), C = development time (30 and 45 seconds), D = mask dimension (small, large), and E = etch time (14.5 and 15.5 minutes). The unreplicated 2^5 design shown below was run.

$(1) = 7$	$d = 8$	$e = 8$	$de = 6$
$a = 9$	$ad = 10$	$ae = 12$	$ade = 10$
$b = 34$	$bd = 32$	$be = 35$	$bde = 30$
$ab = 55$	$abd = 50$	$abe = 52$	$abde = 53$
$c = 16$	$cd = 18$	$ce = 15$	$cde = 15$
$ac = 20$	$acd = 21$	$ace = 22$	$acde = 20$
$bc = 40$	$bcd = 44$	$bce = 45$	$bcde = 41$
$abc = 60$	$abcd = 61$	$abce = 65$	$abcde = 63$

(a) Construct a normal probability plot of the effect estimates. Which effects appear to be large?

(b) Conduct an analysis of variance to confirm your findings for part (a).

(c) Write down the regression model relating yield to the significant process variables.

(d) Construct a normal probability plot of the residuals. Is the plot satisfactory?

(e) Plot the residuals versus the predicted yields and versus each of the five factors. Comment on the plots.

(f) Interpret any significant interactions.

(g) What are your recommendations regarding process operating conditions?

(h) Project the 2^5 design in this problem into a 2^k design in the important factors. Sketch the design and show the average and range of yields at each run. Does this sketch aid in interpreting the results of this experiment?

7.12 **Continuation of Problem 7.11**. Suppose that the experimenter had run four center points in addition to the 32 trials in the original experiment. The yields obtained at the center point runs were 68, 74, 76, 70.

(a) Reanalyze the experiment, including a test for quadratic curvature.

(b) Discuss what your next step would be.

7.13 Suppose that you want to add four runs to an unreplicated 2^4 factorial design to obtain an estimate of pure error. The design in Table 7.5 shows the way to do this. Another strategy is to add four center runs to the original 16-run 2^4. Construct this design and compare it to the design in Table 7.5. Which design has the higher D-efficiency? Discuss the circumstances that might lead you to select one design over the other.

7.14 In a process development study on yield, four factors were studied, each at two levels: time (A), concentration (B), pressure (C), and temperature (D). A single replicate of a 2^4 design was run, and the resulting data are shown below. SS

Run number	Actual run order	A	B	C	D	Yield (lb)	Factor levels	Low (−)	High (+)
1	5	−	−	−	−	12	A (h)	2.5	3
2	9	+	−	−	−	18	B (%)	14	18
3	8	−	+	−	−	13	C (psi)	60	80
4	13	+	+	−	−	16	D (°C)	225	250
5	3	−	−	+	−	17			
6	7	+	−	+	−	15			
7	14	−	+	+	−	20			
8	1	+	+	+	−	15			

(continued)

Run number	Actual run order	A	B	C	D	Yield (lb)	Factor levels Low (−)	High (+)
9	6	−	−	−	+	10		
10	11	+	−	−	+	25		
11	2	−	+	−	+	13		
12	15	+	+	−	+	24		
13	4	−	−	+	+	19		
14	16	+	−	+	+	21		
15	10	−	+	+	+	17		
16	12	+	+	+	+	23		

(a) Construct a normal probability plot of the effect estimates. Which factors appear to have large effects?

(b) Conduct an analysis of variance using the normal probability plot in part (a) for guidance in forming an error term. What are your conclusions?

(c) Write down a regression model relating yield to the important process variables.

(d) Analyze the residuals from this experiment. Does your analysis indicate any potential problems?

(e) Can this design be collapsed into a 2^3 design with two replicates? If so, sketch the design with the average and range of yield shown at each point in the cube. Interpret the results.

7.15 **Continuation of Problem 7.14**. Use the regression model in part (c) of Problem 7.14 to generate a response surface contour plot of yield. Discuss the practical value of this response surface plot.

7.16 Reconsider the 2^4 design in Problem 7.14. Suppose that it was necessary to run this experiment in blocks, but due to material availability, the blocks can only contain three runs. Use experimental design software to construct an appropriate design. Evaluate the design. If the block size could be 4, a much different design could be used. Construct and evaluate this design. What does the restriction on the block size do to design performance?

7.17 **A missing value in a 2^k factorial**. It is not unusual to find that one of the observations in a 2^k design is missing due to faulty measuring equipment, a spoiled test, or some other reason. If the design is an unreplicated factorial ($n = 1$) one commonly used approach is to estimate the missing value with a number that makes the highest-order interaction effect estimate zero.

(a) Apply this technique to the experiment in Problem 7.14 assuming that standard order run number 5 is missing. Compare the results with the results found in Problem 7.14 using the full data set.

(b) Analyze the data using only the 15 available runs. Compare your results with those from part (a) and with those from the original analysis using the full data set.

7.18 Consider the 2^3 design below.

Run	Process variables: Temp (°C)	Pressure (psig)	Conc (g/l)	Coded variables: x_1	x_2	x_3	Yield, y
1	120	40	15	−1	−1	−1	32
2	160	40	15	1	−1	−1	46
3	120	80	15	−1	1	−1	57
4	160	80	15	1	1	−1	65
5	120	40	30	−1	−1	1	36
6	160	40	30	1	−1	1	48
7	120	80	30	−1	1	1	57
8	160	80	30	1	1	1	68
9	140	60	22.5	0	0	0	50
10	140	60	22.5	0	0	0	44
11	140	60	22.5	0	0	0	53
12	140	60	22.5	0	0	0	56

$$x_1 = \frac{\text{Temp} - 140}{20}, \quad x_2 = \frac{\text{Pressure} - 60}{20}, \quad x_3 = \frac{\text{Conc} - 22.5}{7.5}$$

When running a designed experiment, it is sometimes difficult to reach and hold the precise factor levels required by the design. Small discrepancies are not important, but large ones are potentially of more concern. To illustrate, the experiment presented below shows a variation of the original 2^3 design, where many of the test combinations are not exactly the ones specified in the design. Most of the difficulty seems to have occurred with the temperature variable. Fit a first-order model to both the original data and the data below. Compare the inference from the two models. What conclusions can you draw from this simple example?

	Process variables			Coded variables			
Run	Temp (°C)	Pressure (psig)	Conc (g/l)	x_1	x_2	x_3	Yield, y
1	125	41	14	−0.75	−0.95	−1.133	32
2	158	40	15	0.90	−1	−1	46
3	121	82	15	−0.95	1.1	−1	57
4	160	80	15	1	1	−1	65
5	118	39	33	−1.10	−1.05	1.14	36
6	163	40	30	1.15	−1	1	48
7	122	80	30	−0.90	1	1	57
8	165	83	30	1.25	1.15	1	68
9	140	60	22.5	0	0	0	50
10	140	60	22.5	0	0	0	44
11	140	60	22.5	0	0	0	53
12	140	60	22.5	0	0	0	56

SS **7.19** The display below summarizes the results of analyzing a 2^4 factorial design.

Term intercept	Effect estimate	Sum of squares	% Contribution
A		6.25	3.25945
B	5.25	110.25	57.4967
C	3.5	49	25.5541
D	0.75		1.1734
AB	0.75	2.25	1.1734
AC	−0.5	1	0.521512
AD	0.75	2.25	1.1734
BC	1.5	9	
BD	0.25	0.25	0.130378
CD	0.5	1	0.521512
ABC	−1	4	2.08605
ABD		2.25	1.1734
ACD	−0.5		0.521512
BCD	0	0	0
ABCD	−0.5	1	

(a) Fill in the missing information in this table.

(b) Analyze the data. Which factors seem to be active?

7.20 Reconsider the 2^4 factorial design in Problem 7.6. Suppose that this design had to be run in blocks of size 4. Set up a design with *ABCD* confounded in Replicate 1 and *ABC* confounded in Replicate 2. What factorial effects can be estimated from this experiment and what are their relative standard errors? What advantage does this design have over one that confounds *ABCD* in both replicates? (Note: This type of blocking scheme is often called partial confounding.)

7.21 Suppose that you need to plan an experiment with six factors, each at two levels. You need to estimate all 6 main effects and all 15 two-factor interactions. The smallest regular fractional factorial that can do this is the 2^{6-1} design, with 32 runs. However, the experimental budget is not sufficient for this many runs. The minimum number of runs that can accomplish the objectives is 22.

(a) Using software, find a D-optimal design for this problem with 22 runs.

(b) Suppose that the budget will allow two more runs. Using software, find a D-optimal design with 24 runs. Compare this design to the one you found in part (a). Explain how you might justify the additional expense associated with the larger experiment.

CHAPTER 8 PROBLEMS

SS **8.1** R. D. Snee ("Experimenting with a Large Number of Variables," in *Experiments in Industry: Design, Analysis and Interpretation of Results,* by R. D. Snee, L. B. Hare, and J. B. Trout, Editors, ASQC, 1985) describes an experiment in which a 2^{5-1} design with I = ABCDE was used to investigate the effects of five factors on the color of a chemical product. The factors are A = solvent/reactant, B = catalyst/reactant, C = temperature, D = reactant purity, and E = reactant pH. The responses obtained are as follows:

$e = -0.63$	$d = 6.79$
$a = 2.51$	$ade = 5.47$
$b = -2.68$	$bde = 3.45$
$abe = 1.66$	$abd = 5.68$
$c = 2.06$	$cde = 5.22$
$ace = 1.22$	$acd = 4.38$
$bce = -2.09$	$bcd = 4.30$
$abc = 1.93$	$abcde = 4.05$

(a) Which effects seem active?

(b) Calculate the residuals. Construct a normal probability plot of the residuals and plot the residuals versus the fitted values. Comment on the plots.

(c) If any factors are negligible, collapse the fractional design into a full factorial in the active factors. Comment on the resulting design, and interpret the results.

SS **8.2** An article in *Industrial and Engineering Chemistry* ("More on Planning Experiments to Increase Research Efficiency," 1970, pp. 60–65) uses a 2^{5-2} design to investigate the effect of A = condensation temperature, B = amount of material 1, C = solvent volume, D = condensation time, and E = amount of material 2 on yield. The results obtained are as follows:

$e = 23.2$ $ad = 16.9$ $cd = 23.8$ $bde = 16.8$ $ab = 15.5$ $bc = 16.2$ $ace = 23.4$ $abcde = 18.1$

(a) Verify that the design generators used were I = ACE and I = BDE.

(b) Write down the complete defining relation and the aliases for this design.

(c) Estimate the main effects.

(d) Prepare an analysis of variance table. Verify that the AB and AD interactions are available to use as error.

8.3 Construct a 2^{7-2} design by choosing two four-factor interactions as the independent generators. Write down the complete alias structure for this design. Outline the analysis of variance table. What is the resolution of this design?

8.4 Construct a 2^{5-1} design. Show how the design may be run in two blocks of eight observations each. Are any main effects or two-factor interactions confounded with blocks?

8.5 Consider a 2^4 design. We must estimate the four main effects and the six two-factor interactions, but the full 16-run factorial cannot be run. The largest possible block size contains 12 runs. These 12 runs can be obtained from the four one-quarter replicates defined by I = ±AB = ±ACD = ±BCD by omitting the principal fraction. Show how the remaining three fractions can be combined to estimate the required effects, assuming three-factor and higher interactions are negligible. This design could be thought of as a three-quarter fraction.

8.6 Reconsider the design situation in Problem 8.5. Construct a D-optimal design for this situation. Compare your design to the one from Problem 8.5.

8.7 A 16-run experiment was performed in a semiconductor manufacturing plant to study the effects of six factors on the curvature or camber of the substrate devices produced. The six variables and their levels are shown in the following tables. SS

(a) What type of design did the experimenters use?

(b) What are the alias relationships in this design?

(c) Do any of the process variables affect average camber?

(d) Do any of the process variables affect the variability in camber measurements?

(e) If it is important to reduce camber as much as possible, what recommendations would you make?

Run	Lamination temperature (°C)	Lamination time (s)	Lamination pressure (tn)	Firing temperature (°C)	Firing cycle time (h)	Firing dew point (°C)
1	55	10	5	1580	17.5	20
2	75	10	5	1580	29	26
3	55	25	5	1580	29	20
4	75	25	5	1580	17.5	26
5	55	10	10	1580	29	26
6	75	10	10	1580	17.5	20
7	55	25	10	1580	17.5	26
8	75	25	10	1580	29	20
9	55	10	5	1620	17.5	26
10	75	10	5	1620	29	20
11	55	25	5	1620	29	26
12	75	25	5	1620	17.5	20
13	55	10	10	1620	29	20
14	75	10	10	1620	17.5	26
15	55	25	10	1620	17.5	20
16	75	25	10	1620	29	26

Run	Camber for replicate (in./in.)				Total (10^{-4}in./in.)	Mean (10^{-4}in./in.)	Standard deviation
	1	2	3	4			
1	0.0167	0.0128	0.0149	0.0185	629	157.25	24.418
2	0.0062	0.0066	0.0044	0.0020	192	48.00	20.976
3	0.0041	0.0043	0.0042	0.0050	176	44.00	4.083
4	0.0073	0.0081	0.0039	0.0030	223	55.75	25.025
5	0.0047	0.0047	0.0040	0.0089	223	55.75	22.410
6	0.0219	0.0258	0.0147	0.0296	920	230.00	63.639
7	0.0121	0.0090	0.0092	0.0086	389	97.25	16.029
8	0.0255	0.0250	0.0226	0.0169	900	225.00	39.42
9	0.0032	0.0023	0.0077	0.0069	201	50.25	26.725
10	0.0078	0.0158	0.0060	0.0045	341	85.25	50.341
11	0.0043	0.0027	0.0028	0.0028	126	31.50	7.681
12	0.0186	0.0137	0.0158	0.0159	640	160.00	20.083
13	0.0110	0.0086	0.0101	0.0158	455	113.75	31.12
14	0.0065	0.0109	0.0126	0.0071	371	92.75	29.51
15	0.0155	0.0158	0.0145	0.0145	603	150.75	6.75
16	0.0093	0.0124	0.0110	0.0133	460	115.00	17.45

8.8 A spin coater is used to apply photoresist to a bare silicon wafer. This operation usually occurs early in the semiconductor manufacturing process, and the average coating thickness and the variability in the coating thickness have an important impact on downstream manufacturing steps. Six variables are used in the experiment. The variables and their high and low levels are as follows:

Factor	Low level	High level
Final spin speed	7350 rpm	6650 rpm
Acceleration rate	5	20
Volume of resist applied	3 cc	5 cc
Time of spin	14 s	6 s
Resist batch variation	Batch 1	Batch 2
Exhaust pressure	Cover off	Cover on

The experimenter decides to use a 2^{6-1} design and to make three readings on resist thickness on each test wafer. The design and the data are shown in the following table.

							Resist thickness				
Run	*A* Volume	*B* Batch	*C* Time (sec)	*D* Speed	*E* Acc.	*F* Cover	Left	Center	Right	Avg.	Range
1	5	Batch 2	14	7350	5	Off	4531	4531	4515	4525.7	16
2	5	Batch 1	6	7350	5	Off	4446	4464	4428	4446	36
3	3	Batch 1	6	6650	5	Off	4452	4490	4452	4464.7	38
4	3	Batch 2	14	7350	20	Off	4316	4328	4308	4317.3	20
5	3	Batch 1	14	7350	5	Off	4307	4295	4289	4297	18
6	5	Batch 1	6	6650	20	Off	4470	4492	4495	4485.7	25
7	3	Batch 1	6	7350	5	On	4496	4502	4482	4493.3	20
8	5	Batch 2	14	6650	20	Off	4542	4547	4538	4542.3	9
9	5	Batch 1	14	6650	5	Off	4621	4643	4613	4625.7	30
10	3	Batch 1	14	6650	5	On	4653	4670	4645	4656	25
11	3	Batch 2	14	6650	20	On	4480	4486	4470	4478.7	16
12	3	Batch 1	6	7350	20	Off	4221	4233	4217	4223.7	16
13	5	Batch 1	6	6650	5	On	4620	4641	4619	4626.7	22
14	3	Batch 1	6	6650	20	On	4455	4480	4466	4467	25
15	5	Batch 2	14	7350	20	On	4255	4288	4243	4262	45
16	5	Batch 2	6	7350	5	On	4490	4534	4523	4515.7	44
17	3	Batch 2	14	7350	5	On	4514	4551	4540	4535	37
18	3	Batch 1	14	6650	20	Off	4494	4503	4496	4497.7	9
19	5	Batch 2	6	7350	20	Off	4293	4306	4302	4300.3	13
20	3	Batch 2	6	7350	5	Off	4534	4545	4512	4530.3	33
21	5	Batch 1	14	6650	20	On	4460	4457	4436	4451	24
22	3	Batch 2	6	6650	5	On	4650	4688	4656	4664.7	38
23	5	Batch 1	14	7350	20	Off	4231	4244	4230	4235	14
24	3	Batch 2	6	7350	20	On	4225	4228	4208	4220.3	20
25	5	Batch 1	14	7350	5	On	4381	4391	4376	4382.7	15
26	3	Batch 2	6	6650	20	Off	4533	4521	4511	4521.7	22
27	3	Batch 1	14	7350	20	On	4194	4230	4172	4198.7	58
28	5	Batch 2	6	6650	5	Off	4666	4695	4672	4677.7	29
29	5	Batch 1	6	7350	20	On	4180	4213	4197	4196.7	33
30	5	Batch 2	6	6650	20	On	4465	4496	4463	4474.7	33
31	5	Batch 2	14	6650	5	On	4653	4685	4665	4667.7	32
32	3	Batch 2	14	6650	5	Off	4683	4712	4677	4690.7	35

(a) Verify that this is a 2^{6-1} design. Discuss the alias relationships in this design.

(b) What factors appear to affect average resist thickness?

(c) Because the volume of resist applied has little effect on average thickness, does this have any important practical implications for the process engineers?

(d) Project this design into a smaller design involving only the significant factors. Graphically display the results. Does this aid in interpretation?

(e) Use the range of resist thickness as a response variable. Is there any indication that any of these factors affect the variability in resist thickness?

(f) Where would you recommend that the engineers run the process?

SS **8.9** An article in *Soldering & Surface Mount Technology* ("Characterization of a Solder Paste Printing Process and Its Optimization," 1999, Vol. 11, No. 3, pp. 23–26) describes the use of a 2^{8-3} fractional factorial experiment to study the effect of eight factors on two responses: percentage volume matching (PVM)—the ratio of the actual printed solder paste volume to the designed volume; and nonconformities per unit (NPU)—the number of solder paste printing defects determined by visual inspection (20 × magnification) after printing according to an industry workmanship standard. The test matrix and response data are in the following table.

	Parameters									
Run order	*A*	*B*	*C*	*D*	*E*	*F*	*G*	*H*	PVM	NPU (%)
4	−	−	−	−	−	−	−	+	1.00	5
13	+	−	−	−	−	+	+	+	1.04	13
6	−	+	−	−	−	+	+	−	1.02	16
3	+	+	−	−	−	−	−	−	0.99	12
19	−	−	+	−	−	+	−	−	1.02	15
25	+	−	+	−	−	−	+	−	1.01	9
21	−	+	+	−	−	−	+	+	1.01	12
14	+	+	+	−	−	+	−	+	1.03	17
10	−	−	−	+	−	−	+	−	1.04	21
22	+	−	−	+	−	+	−	−	1.14	20
1	−	+	−	+	−	+	−	+	1.20	25
2	+	+	−	+	−	−	+	+	1.13	21
30	−	−	+	+	−	+	+	+	1.14	25
8	+	−	+	+	−	−	−	+	1.07	13
9	−	+	+	+	−	−	−	−	1.06	20
20	+	+	+	+	−	+	+	−	1.13	26
17	−	−	−	−	+	−	−	−	1.02	10
18	+	−	−	−	+	+	+	−	1.10	13
5	−	+	−	−	+	+	+	+	1.09	17
26	+	+	−	−	+	−	−	+	0.96	13
31	−	−	+	−	+	+	−	+	1.02	14
11	+	−	+	−	+	−	+	+	1.07	11
29	−	+	+	−	+	−	+	−	0.98	10
23	+	+	+	−	+	+	−	−	0.95	14
32	−	−	−	+	+	−	+	+	1.10	28
7	+	−	−	+	+	+	−	+	1.12	24
15	−	+	−	+	+	+	−	−	1.19	22
27	+	+	−	+	+	−	+	−	1.13	15
12	−	−	+	+	+	+	+	−	1.20	21
28	+	−	+	+	+	−	−	−	1.07	19
24	−	+	+	+	+	−	−	+	1.12	21
16	+	+	+	+	+	+	+	+	1.21	27

(a) Verify that the generators are I = ABCF, I = ABDG, and I = BCDEH for this design.

(b) What are the aliases for the main effects and two-factor interactions? You can ignore all interactions of order three and higher.

(c) Analyze both PVM and NPU responses.

(d) Analyze the residual for both responses. Are there any problems with model adequacy?

(e) The ideal value of PVM is unity and the NPU response should be as small as possible. Recommend suitable operating conditions for the process based on the experimental results.

8.10 You need to design a screening experiment for six factors. Four of the factors are continuous, but two of them are categorical with three levels. You only want to estimate the main effects of these factors.

(a) What is the smallest design that you can use, assuming that all interactions are negligible? Construct a D-optimal design with this number of runs.

(b) Find the alias relationships between the main effects and two-factor interactions for this design.

(c) Suppose that you can afford up to four additional runs. Find this design using the D-criterion. Compare the alias relationships for the design to the minimal run design from part (a). What did you gain from the additional runs?

8.11 A DSD with six or more factors can fit a full quadratic model in any three of the factors. Suppose the responses for the DSD in Table 8.16 are as in the table below.

Run #	y
1	44.5
2	39.3
3	46.2
4	46.1
5	34.9
6	51.2
7	51.8
8	42.1
9	43.2
10	40.3
11	38.0
12	34.8
13	49.7

There are 20 ways to choose three factors from six possible factors. For each of these 20 combinations, fit a full quadratic model to the responses above and record the root mean squared error for each model.

(a) Which model fits best?

(b) Which factors are present in this model?

(c) What would you advise management to do if the goal were to maximize the response y?

(d) What would you advise management to do if the goal were to minimize the response y?

(e) What would you advise management to do if the goal were to match a target value of 40 for y?

8.12 Critique the analytical method above. How could this approach fail?

8.13 For the design in Table 8.16, there are 15 possible two-factor interactions. Suppose that no quadratic effects are important. What is the maximum number of two-factor interactions that can be estimated?

CHAPTER 9 PROBLEMS

SS **9.1** A consumer products company relies on direct mail marketing pieces as a major component of its advertising campaigns. The company has three different designs for a new brochure and wants to evaluate their effectiveness, as there are substantial differences in costs between the three designs.

The company decides to test the three designs by mailing 5000 samples of each to potential customers in four different regions of the country. Since there are known regional differences in the customer base, regions are considered as blocks. The number of responses to each mailing is as follows.

	Region			
Design	**NE**	**NW**	**SE**	**SW**
1	250	350	219	375
2	400	525	390	580
3	275	340	200	310

(a) Analyze the data from this experiment treating the four regions as a random sample rather than fixed blocks.

(b) Compare your results to a fixed-block analysis.

SS **9.2** What is the **Z** matrix for the design in Problem 9.1?

SS **9.3** Assume that the error variance and the block variance are both equal to 1.

(a) Compute the **V** matrix

(b) Compute $\mathbf{V}^{-1}$.

(c) Effects coding for design 1 is [1 0], [0 1] for design 2, and [−1 −1] for design 3. What is the information matrix, $\mathbf{X}'\mathbf{V}^{-1}\mathbf{X}$?

SS **9.4** Box, Hunter, and Hunter (1978) presented results of a blocked experiment investigating the effects of four different treatments on the yield of penicillin manufacture. The data are shown below.

Block	A	B	C	D
1	89	88	97	94
2	84	77	92	79
3	81	87	87	85
4	87	92	89	84
5	79	81	80	88

(a) Analyze this experiment as a fixed-block experiment. Is one treatment better than the others?

(b) Analyze this experiment treating the blocks as random. Are your conclusions different?

9.5 Generate a D-optimal design with eight random blocks of three runs each where there are two continuous factors and one categorical factor at two levels for a main effects model.

(a) Compute the $\mathbf{V}^{-1}$ matrix assuming the error variance and block variance are both equal to 1.

(b) What is the information matrix, $\mathbf{X}'\mathbf{V}^{-1}\mathbf{X}$?

(c) Generate the D-optimal design assuming the blocks are fixed instead of random. Compare the designs.

9.6 Generate a 28-run D-optimal design for six continuous factors and a model containing all the main effects and two-factor interactions.

(a) Suppose you wanted to block this design with four runs per fixed block. What is the minimum possible number of runs?

(b) Now suppose the blocks are random. How many runs are necessary?

9.7 A specialty coffee producer wants to measure the strength of coffee measured using a colorimeter as a function of the grind (medium or coarse), temperature (195, 205), charge (1.6, 2.4) at three different coffee stations. The data are shown below.

Grind	Temperature	Time	Charge	Station	Strength
Medium	205	4	2.4	1	1.78
Coarse	195	3	1.6	1	1.25
Medium	205	3	1.6	1	1.10
Coarse	195	4	2.4	1	1.63
Coarse	205	4	1.6	2	1.26
Medium	195	4	2.4	2	1.63
Medium	195	3	1.6	2	1.22
Coarse	205	3	2.4	2	1.51
Coarse	205	4	1.6	3	1.07
Coarse	195	3	2.4	3	1.26
Medium	195	4	1.6	3	1.13
Medium	205	3	2.4	3	1.25

(a) Fit a main effects model treating station as a fixed blocking factor.

(b) Repeat your analysis treating station as a random factor.

(c) What is the most important factor?

9.8 There is substantial folk lore about what factors affect the taste of wine. At Chehalem Winery, the owners decided to run an experiment to determine what factors are most beneficial for good tasting wines. The table of factors is shown below.

Factors and Levels for the Winemaking Experiment

Variable	Low level (−)	High level (+)
A = Pinot Noir clone	Pommard	Wadenswil
B = Oak type	Allier	Troncais
C = Age of barrel	Old	New
D = Yeast/skin contact	Champagne	Montrachet
E = Stems	None	All
F = Barrel toast	Light	Medium
G = Whole cluster	None	10%
H = Fermentation temperature	Low (75°F max)	High (92°F max)

Each wine was tasted by five raters designated HPN, JPN, CAL, DCM, and RGB. Each rater ranked the 16 wines in the experiment from 1 (best) to 16 (worst).

Design and Results for Wine Tasting Experiment

	Variable								Panel rankings				
Run	A	B	C	D	E	F	G	H	HPN	JPN	CAL	DCM	RGB
1	−	−	−	−	−	−	−	−	12	6	13	10	7
2	+	−	−	−	−	+	+	+	10	7	14	14	9
3	−	+	−	−	+	−	+	+	14	13	10	11	15
4	+	+	−	−	+	+	−	−	9	9	7	9	12
5	−	−	+	−	+	+	+	−	8	8	11	8	10
6	+	−	+	−	+	−	−	+	16	12	15	16	16
7	−	+	+	−	−	+	−	+	6	5	6	5	3
8	+	+	+	−	−	−	+	−	15	16	16	15	14
9	−	−	−	+	+	+	−	+	1	2	3	3	2
10	+	−	−	+	+	−	+	−	7	11	4	7	6
11	−	+	−	+	−	+	+	−	13	3	8	12	8
12	+	+	−	+	−	−	−	+	3	1	5	1	4
13	−	−	+	+	−	−	+	+	2	10	2	4	5
14	+	−	+	+	−	+	−	−	4	4	1	2	1
15	−	+	+	+	+	−	−	−	5	15	9	6	11
16	+	+	+	+	+	+	+	+	11	14	12	13	13

(a) Analyze the experiment using the average rating as the response.

(b) Reanalyze the data treating each rater as one level of a fixed blocking factor with five levels.

(c) Reanalyze the data treating the blocks as random.

9.9 Pignatiello and Ramberg (1985) used a replicated fractional factorial design to investigate the effect of five factors on the free height of leaf springs used as shock absorbers. The factors were A = temperature, B = heating time, C = transfer time, D = hold down time, E = quench oil temperature.

The data are shown below.

Leaf Spring Experiment

A	B	C	D	E	Free height		
−	−	−	−	−	7.78	7.78	7.81
+	−	−	+	−	8.15	8.18	7.88
−	+	−	+	−	7.50	7.56	7.50
+	+	−	−	−	7.59	7.56	7.75
−	−	+	+	−	7.54	8.00	7.88
+	−	+	−	−	7.69	8.09	8.06
−	+	+	−	−	7.56	7.52	7.44
+	+	+	+	−	7.56	7.81	7.69
−	−	−	−	+	7.50	7.25	7.12
+	−	−	+	+	7.88	7.88	7.44
−	+	−	+	+	7.50	7.56	7.50
+	+	−	−	+	7.63	7.75	7.56
−	−	+	+	+	7.32	7.44	7.44
+	−	+	−	+	7.56	7.69	7.62
−	+	+	−	+	7.18	7.18	7.25
+	+	+	+	+	7.81	7.50	7.59

Imagine that each replicate of the experiment was run on a separate day.

(a) Analyze the data using the average free height as the response.

(b) Reanalyze the data treating Day as a fixed blocking factor. Is there a significant block effect?

(c) Reanalyze the data treating Day as random.

9.10 Consider the design in Problem 9.9.

(a) Compute the D-optimal design with 48 runs for a main effects plus two-factor interactions model treating Day as a fixed blocking factor.

(b) For the same scenario, treat Day as random and generate the D-optimal random block design.

(c) Compare these designs.

CHAPTER 10 PROBLEMS

10.1 Suppose that, in the Filtration experiment introduced in Chapter 9, the temperature was hard to change. The engineering team decided to field the experiment as a split-plot experiment with four whole plots. The data are shown below.

Whole plots	Temperature	Pressure	Concentration	Stirring rate	Filtration rate
1	1	−1	1	−1	55
1	1	1	−1	−1	68
1	1	−1	−1	1	101
1	1	1	1	1	84
2	−1	−1	1	−1	79
2	−1	−1	−1	1	46
2	−1	1	1	1	76
2	−1	1	−1	−1	53
3	1	−1	−1	−1	80
3	1	1	1	−1	64
3	1	−1	1	1	99
3	1	1	−1	1	102
4	−1	−1	−1	−1	51
4	−1	1	1	−1	75
4	−1	−1	1	1	75
4	−1	1	−1	1	52

(a) Fit the data as if the design were completely randomized.

(b) Reanalyze the data as a split-plot design.

(c) How are the fitted models alike?

(d) How are the fitted models different?

10.2 Construct the 16×4 element $\mathbf{Z}$ matrix.

10.3 Assuming both the error variance and the whole-plot variance are 1, compute the $\mathbf{V}$ matrix. Compute the inverse of the $\mathbf{V}$ matrix. Compute $|\mathbf{X}'\mathbf{V}^{-1}\mathbf{X}|$ given the same assumptions about the error variance and the whole-plot variance.

10.4 Suppose an operator noticed that whole plots 1 and 3 had the same temperature setting as well as whole plots 2 and 4. The operator then combined whole plots 1 and 3 into one whole plot having eight runs and did the same thing with whole plots 2 and 4. The resulting design has only two whole plots.

(a) Construct the 16×2 $\mathbf{Z}$ matrix.

(b) Repeat the steps of Problem 10.3 for this design.

(c) Compute the D-efficiency of this design to the experiment with four whole plots.

(d) Is it possible to estimate the fixed effect of temperature as well as the whole-plot variance with this design?

10.5 Using software, generate a split-plot design for six factors, *A*–*F*, where *A*–*C* are hard to change. Use a model with all the main effects and two-factor interactions. Your design should have eight whole plots of four runs each. Assume the error variance and the whole-plot variance are both equal to 1.

10.6 Using software, generate a D-optimal split-plot design for four factors, *A*–*D*, where *A* and *B* are hard to change. Use a main effects model. Your design should have 7 whole plots of 4 runs each for a total of 28 runs. Assume the error variance and the whole-plot variance are both equal to 1. Are the columns for the hard-to-change factors correlated? If so, why?

10.7 Repeat Problem 10.6 with 24 runs and 4 whole plots of 6 runs. Are the columns for the hard-to-change factors correlated? What is the relative D-efficiency of the 24-run design with respect to the 28-run design?

10.8 Montgomery (2020) reports the results of a split-plot experiment to find the relationship of the tensile strength of paper to the factors, temperature and pulp preparation method. The data are below.

Whole plot	Temperature	Pulp preparation	Tensile strength
1	200	1	30
1	225	1	35
1	250	1	37
1	275	1	36
2	200	2	34
2	225	2	41
2	250	2	38
2	275	2	42
3	200	3	29
3	225	3	26
3	250	3	33
3	275	3	36
4	200	1	28
4	225	1	32
4	250	1	40
4	275	1	41
5	200	2	31
5	225	2	36
5	250	2	42
5	275	2	40
6	200	3	31
6	225	3	30
6	250	3	32
6	275	3	40
7	200	1	31
7	225	1	37
7	250	1	41
7	275	1	40
8	200	2	35
8	225	2	40
8	250	2	39
8	275	2	44
9	200	3	32
9	225	3	34
9	250	3	39
9	275	3	45

(a) Analyze the experiment as if it were a completely randomized design treating temperature as a categorical factor.

(b) The pulp preparation method is a whole-plot factor. Reanalyze the experiment as a split-plot design with a model having both main effects and the two-factor interaction.

(c) Compare the two analyses.

(d) If you wanted to maximize the tensile strength of the paper, what settings of the factors would you use?

10.9 Reanalyze the design in Problem 10.8 treating temperature as continuous and adding a quadratic term for temperature. Does this analytical approach change your conclusions?

10.10 Compute two D-optimal designs treating temperature as continuous. In the first design, assume that the effect of temperature on tensile strength is linear. In the second design, assume that the effect of temperature on tensile strength is quadratic. Each design should have 36 runs with 9 whole plots of 4 runs. Compare the D-efficiencies of the two designs for the model having the main effect of temperature, the main effect of pulp preparation, and the two-factor interaction involving these two factors.

10.11 Steel is normalized by heating above the critical temperature, soaking, and then air cooling. This process increases the strength of the steel, refines the grain, and homogenizes the structure. An experiment is performed to determine the effect of temperature and heat treatment time on the strength of normalized steel. Two temperatures and three times are selected. The experiment is performed by heating the oven to a randomly selected temperature and inserting three specimens. After 10 minutes one specimen is removed, after 20 minutes the second is removed, and after 30 minutes the final specimen is removed. Then the temperature is changed to the other level and the process is repeated. Four shifts are required to collect the data, which are shown below. Analyze the data and draw conclusions. SS

		Temperature (°F)	
Shift	Time (min)	1500	1600
	10	63	89
1	20	54	91
	30	61	62
	10	50	80
2	20	52	72
	30	59	69

Shift	Time (min)	Temperature (°F) 1500	1600
3	10	48	73
	20	74	81
	30	71	69
4	10	54	88
	20	48	92
	30	59	64

10.12 An experiment was performed to study pigment dispersion in paint. Four different mixes of a particular pigment are studied. The procedure consists of preparing a particular mix and then applying that mix to a panel by three application methods (brushing, spraying, and rolling). The response measured is the percentage reflectance of pigment. Three days are required to run the experiment, and the data obtained given below. Analyze the data and draw conclusions.

Day	Application method	Mix 1	2	3	4
1	1	64.5	66.3	74.1	66.5
	2	68.3	69.5	73.8	70.0
	3	70.3	73.1	78.0	72.3
2	1	65.2	65.0	73.8	64.8
	2	69.2	70.3	74.5	68.3
	3	71.2	72.8	79.1	71.5
3	1	66.2	66.5	72.3	67.7
	2	69.0	69.0	75.4	68.6
	3	70.8	74.2	80.1	72.4

10.13 An article in *Quality Engineering* ("Quality Quandaries: Two-Level Factorials Run as Split-Plot Experiments," Bisgaard et al., Vol. 8, No. 4, pp. 705–708, 1996) describes a 2^5 factorial experiment in a plasma process focused on making paper more susceptible to ink. Four of the factors (A–D) are difficult to change from run to run, so the experimenters set up the reactor at the sets of conditions specified by the low and high levels of those factors and then processed the two paper types (factor E) together. The placement of the paper specimens in the reactor (right versus left) was randomized. This produces a split-plot design with A–D as the whole-plot factors and factor E as the subplot factor. The data from this experiment are shown below. Analyze the data from this experiment and draw conclusions. SS

Standard order	Run number	A = Pressure	B = Power	C = Gas flow	D = Gas type	E = Paper type	y Contact angle
5	1	−1	−1	+1	Oxygen	$E1$	37.6
21	2	−1	−1	+1	Oxygen	$E2$	43.5
2	3	+1	−1	−1	Oxygen	$E1$	41.2
18	4	+1	−1	−1	Oxygen	$E2$	38.2
10	5	+1	−1	−1	$SiCl_4$	$E1$	56.8
26	6	+1	−1	−1	$SiCl_4$	$E2$	56.2
14	7	+1	−1	+1	$SiCl_4$	$E1$	47.5
30	8	+1	−1	+1	$SiCl_4$	$E2$	43.2
11	9	−1	+1	−1	$SiCl_4$	$E1$	25.6
27	10	−1	+1	−1	$SiCl_4$	$E2$	33.0
3	11	−1	+1	−1	Oxygen	$E1$	55.8
19	12	−1	+1	−1	Oxygen	$E2$	62.9

Standard order	Run number	A = Pressure	B = Power	C = Gas flow	D = Gas type	E = Paper type	y Contact angle
13	13	−1	−1	+1	$SiCl_4$	$E1$	13.3
29	14	−1	−1	+1	$SiCl_4$	$E2$	23.7
6	15	+1	−1	+1	Oxygen	$E1$	47.2
22	16	+1	−1	+1	Oxygen	$E2$	44.8
16	17	+1	+1	+1	$SiCl_4$	$E1$	49.5
32	18	+1	+1	+1	$SiCl_4$	$E2$	48.2
9	19	−1	−1	−1	$SiCl_4$	$E1$	5.0
25	20	−1	−1	−1	$SiCl_4$	$E2$	18.1
15	21	−1	+1	+1	$SiCl_4$	$E1$	11.3
31	22	−1	+1	+1	$SiCl_4$	$E2$	23.9
1	23	−1	−1	−1	Oxygen	$E1$	48.6
17	24	−1	−1	−1	Oxygen	$E2$	57.0
8	25	+1	+1	+1	Oxygen	$E1$	48.7
24	26	+1	+1	+1	Oxygen	$E2$	44.4
7	27	−1	+1	+1	Oxygen	$E1$	47.2
23	28	−1	+1	+1	Oxygen	$E2$	54.6
4	29	+1	+1	−1	Oxygen	$E1$	53.5
20	30	+1	+1	−1	Oxygen	$E2$	51.3
12	31	+1	+1	−1	$SiCl_4$	$E1$	41.8
28	32	+1	+1	−1	$SiCl_4$	$E2$	37.8

SS **10.14** Reconsider the experiment in Problem 10.13. This is a rather large experiment, so suppose that the experimenter had used a 2^{5-1} design instead. Set up the 2^{5-1} design in a split-plot, using the principal fraction. Then select the response data using the information from the full factorial. Analyze the data and draw conclusions. Do they agree with the results of Problem 10.13?

CHAPTER 11 PROBLEMS

SS **11.1** The data shown in the table below were collected in an experiment to optimize crystal growth as a function of three variables x_1, x_2, and x_3. Large values of y (yield in grams) are desirable. Fit a second-order model and analyze the fitted surface. Under what set of conditions is maximum growth achieved?

x_1	x_2	x_3	y
−1	−1	−1	66
−1	−1	1	70
−1	1	−1	78
−1	1	1	60
1	−1	−1	80
1	−1	1	70
1	1	−1	100
1	1	1	75
−1.682	0	0	100
1.682	0	0	80
0	−1.682	0	68
0	1.682	0	63
0	0	−1.682	65
0	0	1.682	82
0	0	0	113
0	0	0	100
0	0	0	118
0	0	0	88
0	0	0	100
0	0	0	85

11.2 The data in the following table were collected by a chemical engineer. The response y is filtration time, x_1 is temperature, and x_2 is pressure. Fit a second-order model. SS

(a) What operating conditions would you recommend if the objective is to minimize the filtration time?

(b) What operating conditions would you recommend if the objective is to operate the process at a mean filtration rate very close to 46?

x_1	x_2	y
−1	−1	54
−1	1	45
1	−1	32
1	1	47
−1.414	0	50
1.414	0	53
0	−1.414	47
0	1.414	51
0	0	41
0	0	39
0	0	44
0	0	42
0	0	40

11.3 The design in the table below is used in an experiment that has the objective of fitting a second-order model:

x_1	x_2	y
1	0	68
0.5	$\sqrt{0.75}$	74
−0.5	$\sqrt{0.75}$	65
−1	0	60
−0.5	$-\sqrt{0.75}$	63
0.5	$-\sqrt{0.75}$	70
0	0	58
0	0	60
0	0	57
0	0	55
0	0	69

(a) Plot the design points in a two-dimensional graph.

(b) Construct a contour plot of the variance of predicted response. Is the design rotatable?

(c) Fit a second-order model. What operating conditions on x_1 and x_2 lead to the stationary point? What kind of surface have you found?

(d) Where would you run this process if the objective is to obtain a response that is as close to 65 as possible?

11.4 Verify that an orthogonal first-order design is also first-order rotatable.

SS **11.5** A CCD is run in a chemical vapor deposition process, resulting in the experimental data shown in the table below. Four experimental units were processed simultaneously on each run of the design, and the responses are the mean and the variance of thickness, computed across the four units.

x_1	x_2	$\bar{y}$	s^2
−1	−1	360.6	6.689
1	−1	445.2	14.230
−1	1	412.1	7.088
1	1	601.7	8.586
1.414	0	518.0	13.130
−1.414	0	411.4	6.644
0	1.414	497.6	7.649
0	−1.414	397.6	11.740
0	0	530.6	7.836
0	0	495.4	9.306
0	0	510.2	7.956
0	0	487.3	9.127

(a) Fit a model to the mean response. Analyze the residuals.

(b) Fit a model to the variance response. Analyze the residuals.

(c) Fit a model to $\ln(s^2)$. Is this model superior to the one you found in part (b).

(d) Suppose that you want the mean thickness to be in the interval 450 ± 25. Find a set of operating conditions that achieves this objective and simultaneously minimizes the variance.

11.6 Suppose that you want to fit a second-order model in $k = 5$ factors. You cannot afford more than 25 runs. Construct both a *D*-optimal and an I-optimal design for this situation. Compare the prediction variance properties of the two designs. Which design would you prefer?

11.7 The table below presents a six-variable DSD from Jones and Nachtsheim (2011). Analyze the response data from this experiment.

Run (i)	$x_{i.1}$	$x_{i.2}$	$x_{i.3}$	$x_{i.4}$	$x_{i.5}$	$x_{i.6}$	y_i
1	0	1	−1	−1	−1	−1	21.04
2	0	−1	1	1	1	1	10.48
3	1	0	−1	1	1	−1	17.89
4	−1	0	1	−1	−1	1	10.07
5	−1	−1	0	1	−1	−1	7.74
6	1	1	0	−1	1	1	21.01
7	−1	1	1	0	1	−1	16.53
8	1	−1	−1	0	−1	1	20.38
9	1	−1	1	−1	0	−1	8.62
10	−1	1	−1	1	0	1	7.80
11	1	1	1	1	−1	0	23.56
12	−1	−1	−1	−1	1	0	15.24
13	0	0	0	0	0	0	19.91

SS **11.8** An article in *Quality Progress* ("For Starbucks, It's in the Bag," March 2011, pp. 18–23) describes using a CCD to improve the packaging of one-pound coffee. The objective is to produce an airtight seal that is easy to open without damaging the top of the coffee bag. The experimenters studied three factors—x_1 = plastic viscosity (300–400 centipoise), x_2 = clamp pressure (170–190 psi), and x_3 = plate gap (−3,+3 mm) and two responses tear and leakage. The design is shown below. The tear response was measured on a scale from 0 to 9 (good to bad) and leakage was proportion failing. Each run used a sample of 20 bags for response measurement.

Run	Viscosity	Pressure	Plate gap	Tear	Leakage
Center	350	180	0	0	0.15
Axial	350	170	0	0	0.5
Factorial	319	186	1.8	0.45	0.15
Factorial	380	174	1.8	0.85	0.05
Center	350	180	0	0.35	0.15
Axial	300	180	0	0.3	0.45
Axial	400	180	0	0.7	0.25
Axial	350	190	0	1.9	0
Center	350	180	0	0.25	0.05
Factorial	319	186	−1.8	0.1	0.35
Factorial	380	186	−1.8	0.15	0.4
Axial	350	180	3	3.9	0
Factorial	380	174	−1.8	0	0.45
Center	350	180	0	0.55	0.2
Axial	350	180	−3	0	1
Factorial	319	174	−1.8	0.05	0.2
Factorial	319	174	1.8	0.4	0.25
Factorial	380	186	1.8	4.3	0.05
Center	350	180	0	0	0

(a) Build a second-order model for the tear response.

(b) Build a second-order model for the leakage response.

(c) Analyze the residuals for both models. Do transformations seem necessary for either response? If so, refit the models in the transformed metric.

(d) Construct response surface plots and contour plots for both responses. Provide interpretations for the fitted surfaces.

(e) What conditions would you recommend for process operation to minimize leakage and keep tear below 0.75?

11.9 Suppose that you need to design an experiment to fit a quadratic model over the region $-1 \leq x_i \leq 1$, $i = 1, 2$ subject to the constraint $x_1 + x_2 \leq 1$. If the constraint is violated, the process will not work properly. You can afford to make no more than $n = 12$ runs. Find a D-optimal design for this problem.

11.10 Reconsider the situation in Problem 11.9. Find an I-optimal design for this problem. Compare this design to the D-optimal design from Problem 11.9. Which design would you prefer? Justify your choice.

11.11 Suppose that you want to fit a second-order model in $k = 5$ factors. You cannot afford more than 25 runs. Construct both a *D*-optimal and an *I*-optimal design for this situation. Compare the prediction variance properties of the designs. Which design would you prefer?

11.12 Suppose that you want to fit a second-order response surface model in a situation where there are $k = 4$ factors; however, one of the factors is categorical with three levels. What model should you consider for this experiment? Suggest an appropriate design for this situation.

11.13 An article in the *Journal of Chromatography A* ("Optimization of the Capillary Electrophoresis Separation of Ranitidine and Related Compounds," Vol. 766, pp. 245–254) describes an experiment to optimize the production of ranitidine, a compound that is the primary active ingredient of Zantac. This is a pharmaceutical product used to treat ulcers, gastroesophageal reflux disease (a condition in which backward flow of acid from the stomach causes heartburn and injury of the esophagus), and other conditions where the stomach produces too much acid, such as Zollinger–Ellison syndrome. The authors used three factors (x_1 = pH of the buffer solution, x_2 = the electrophoresis voltage, and x_3 = the concentration of one component of the buffer solution) in a CCD. The response is chromatographic exponential function (CEF), which should be minimized. The table below shows the design.

(a) Fit a second-order model to the CEF response. Analyze the residuals from this model. Does it seem that all model terms are necessary?

(b) Reduce the model from part (a) as necessary. Did model reduction improve the fit?

(c) Does transformation of the CEF response seem like a useful idea? What aspect of either the data or the residual analysis suggests that transformation would be helpful?

(d) Fit a second-order model to the transformed CEF response. Analyze the residuals from this model. Does it seem that all model terms are necessary? What would you choose as the final model?

(e) Suppose that you had some information that suggests that the separation process malfunctioned during run 7. Delete this run and analyze the data from this experiment again.

(f) What conditions would you recommend to minimize CEF?

Standard order	X1	X2	X3	CEF
1	−1	−1	−1	17.3
2	1	−1	−1	45.5
3	−1	1	−1	10.3
4	1	1	−1	11757.1
5	−1	−1	1	16.942
6	1	−1	1	25.4
7	−1	1	1	31697.2
8	1	1	1	12039.2
9	−1.68	0	0	7.5
10	1.68	0	0	6.3
11	0	−1.68	0	11.1
12	0	1.68	0	6.664
13	0	0	−1.68	16548.7
14	0	0	1.68	26351.8
15	0	0	0	9.9
16	0	0	0	9.6
17	0	0	0	8.9
18	0	0	0	8.8
19	0	0	0	8.013
20	0	0	0	8.059

11.14 Set up a DSD for seven factors. Show how this design could be run in blocks of size 2. Find the correlation structure among the factor effects for this design.

11.15 Set up a DSD for 6 continuous factors and 2 two-level categorical factors. Find the correlation structure among the factor effects for this design.

11.16 Reconsider the situation described in Problem 11.9. Suppose that there is an additional constraint, $0.25 \leq x_1$. You can afford to make no more than $n = 12$ runs. Find a D-optimal design for this problem. Find an I-optimal design for this problem. Evaluate and compare the two designs. Which design would you choose? Justify your choice.

CHAPTER 12 PROBLEMS

SS **12.1** Consider the exponential decay model $y = e^{-\beta x} + \varepsilon$. Suppose the true value of β is 1.

(a) What value of x makes the design locally optimal?

(b) What is the value of y for this value of x?

(c) Show that the maximum of the absolute value of the derivative function, $-1xe^{-x}$, is e^{-1}.

12.2 Consider the following data on enzyme kinetics.

Concentration	Rate
0.2	32
0.5	67
1	97
1	103
4	158
4	162

(a) Fit the Michaelis–Menten model to these data. What are the estimates of the two parameters?

(b) Suppose that the true parameters were 200 for the asymptote and 1 for the concentration value where the predicted rate is 100. What is the predicted rate for a concentration of 4?

(c) Given the true parameters above, and suppose you have no data. What is the locally optimal two-run design if the concentration is only allowed to range between 0.2 and 4?

12.3 Consider the following data on demand for oxygen from algae added to bottles of water and allowed to sit for one to seven days.

Time	Demand
1	6.5
1	7.5
2	11
2	12
3	14
3	16
4	15
4	17
5	17
5	18
7	18.5
7	20

Fit the Michaelis–Menten model to this data. Fit this alternative model to the data.

$$y = \beta_1(1 - e^{\beta_2 x})$$

Compare the two model fits. Which model has the lower MSE? At what values of time do the two models disagree most?

12.4 For models that are nonlinear in the parameters, the analog SS to the **X** matrix introduced in Chapter 4 has a column for the derivative of the model with respect to each parameter. This analog of **X** is sometimes called **J** (for Jacobian). In the case of the exponential decay model, there is only one parameter, so this matrix has only one column formed by plugging in the guessed value of the parameter in each row. As mentioned in the introduction of this chapter, the derivative function for the exponential decay model is $-1xe^{-\beta x}$. We obtain the single column of **J**, row by row by plugging in the x value of the row and the guessed value of β into the expression for the derivative. For example, if $x = 1/2$ and our guess for β is 2, the value of the derivative function is $-1/2e$. The information matrix, which in this case has only a single value, is proportional to $\mathbf{J}'\mathbf{J}$.

(a) Compute $\mathbf{J}'\mathbf{J}$ for the design consisting of three runs at $x = 1/2$.

(b) Compute $\mathbf{J}'\mathbf{J}$ for the design consisting of three runs at x equals 0.4, 0.5, and 0.6.

(c) The relative D-efficiency of the designs is the ratio of the two numbers you just computed. What is the relative efficiency of the design in part (b) to the design in part (a)?

12.5 The derivative of the Michaelis–Menten model with respect to b_1 is $\frac{x}{b_2+x}$. The derivative of the Michaelis–Menten model with respect to b_2 is $\frac{-1b_1x}{(b_2+x)^2}$.

(a) Compute $\mathbf{I}_1 = \mathbf{J}'\mathbf{J}$ for the design in Table 12.2 assuming the true parameters are $b_1 = 100$ and $b_2 = 10$.

(b) What is the determinant of $\mathbf{I}_1$, written $|\mathbf{I}_1|$?

(c) Compute $\mathbf{I}_2 = \mathbf{J}'\mathbf{J}$ for the design with seven runs at a setting of eight and seven runs at a setting of 70.

(d) Compute $|\mathbf{I}_2|$.

(e) The relative efficiency of the design in part (a) versus the design in part (b) is $\left(\frac{|I_1|}{|I_2|}\right)^{1/2}$. Compute this relative efficiency. Note that the ratio is raised to the power of ½. To compute the relative efficiencies of two designs each having p parameters, we raise the ratio of their determinants to the $1/p$ power. In this case, there are two parameters, so the exponent is ½.

SS **12.6** Consider the data in the following table:

X1	X2	Y	n Trials
−1	−1	6	20
1	1	20	20
−1	1	13	20
1	−1	18	20

(a) Assume that Y is distributed binomial. Fit the GLM assuming the main effects model and the link function is the logistic function. What is the estimate of the intercept? What are the estimates of the two slope parameters?

(b) What is the predicted probability of a success at the point [1 1]?

12.7 Using the parameters estimates from Problem 12.6, find the locally optimal design with four runs assuming each factor ranges from −1 to 1.

(a) Rounding to three decimal places, how many unique rows are there?

(b) Explain why the locally optimal design is not the same as the 2 × 2 factorial.

12.8 Using the locally optimal design from Problem 12.7,

(a) Compute the information matrix of the design.

(b) Compute the information matrix of the 2 × 2 factorial design.

(c) Compute the relative D-efficiency of the 2 × 2 design to the locally optimal design.

12.9 For the table in Problem 12.6, assume that Y is distributed Poisson instead of binomial.

(a) What are the parameter estimates?

(b) What are the predicted values at each design point?

(c) Compare the Poisson predictions to the binomial predictions.

(d) Given that you know that there were only 20 trials, which model makes more sense?

12.10 Using the parameter estimates from Problem 12.9, find the locally optimal design with four runs assuming each factor ranges from −1 to 1.

(a) Rounding to three decimal places, how many unique rows are there?

(b) Explain why the locally optimal design is not the same as the 2 × 2 factorial.

JMP Scripting Commands for Computing Distribution Probabilities and Quantiles

A

For creating confidence intervals and doing hypothesis tests, many DOE textbooks provide tables of distributions. JMP Scripting Language (JSL) has functions to compute probabilities and quantiles (or critical values) for many distributions. This appendix provides examples of all the necessary commands.

First, it is necessary to know how to create a script in JMP. On Windows, holding down the Ctrl key and simultaneously pressing the letter "T" is the keyboard shortcut that creates a script window. On the Mac, the keyboard shortcut is to hold down the Shift and Command keys and simultaneously type the letter "N".

The results of scripts appear in the JMP Log. On Windows, you can open the Log by choosing Log off the View menu. On the Mac, the Log is found on the Windows menu.

An alternative to using a script is to make a formula column in a JMP table. On Windows, you can create a new table using the keyboard shortcut holding down the Ctrl and "N" keys simultaneously. You can open the formula editor for a column by clicking on the column header to select it and then choosing Formula … from the Cols menu. On the Mac, you create a new table by holding down the Command and "N" keys simultaneously. Creating a formula column is accomplished the same way as on Windows.

Also, note that spaces are not delimiters in JSL, so you can put spaces between words in a command and you can leave the spaces out as well.

EXAMPLE A.1 The Standard Normal Distribution

```
Distribution function p = Normal Distribution(x).
```

"p" is the probability that the standard normal distribution is less than or equal to "x."

```
p = Normal Distribution(1.645);
```

Given the command above, the JMP Log shows the value below.

```
0.950015094460879
```

```
Quantile function x = Normal Quantile(p)
```

This is the inverse of the distribution function. Providing the desired probability, p, JMP returns the quantile or critical value, x.

```
x = Normal Quantile(0.95);
```

Given the command above, the JMP Log shows the value below.

```
1.64485362695147
```

EXAMPLE A.2 The Normal Distribution Given μ and σ

This command makes it unnecessary to compute the quantity (x-μ)/σ.

```
Distribution function p = Normal Distribution(x,mu,sigma);

p = Normal Distribution(12,10,2);
```

Given the command above, where x =12, mu = 2, and sigma = 2, the JMP Log shows the value below.

```
0.841344746068543

Quantile function x = Normal Quantile(p,mu,sigma);

x = Normal Quantile(0.975,10,2);
```

Given the command above, where p = 0.975 =12, mu = 2, and sigma = 2, the JMP Log shows the value below.

```
13.9199279690801
```

EXAMPLE A.3 95% Confidence Interval

```
Show(Normal Quantile(0.025,10,2) || Normal Quantile(0.975,10,2);
```

The command above produces the output below in the JMP Log:

```
Normal Quantile(0.025, 10, 2) || Normal Quantile(0.975, 10, 2) =
   [6.08007203091989 13.9199279690801];
The symbol, ||, is a horizontal concatenation operator.
```

EXAMPLE A.4 The *t*-Distribution

```
Distribution function p = t Distribution(x, df)
```

"p" is the probability that the t distribution is less than or equal to "x" where the number of degrees of freedom is df.

```
p = t Distribution(2,5);
```

Given the command above, the JMP Log shows the value below.

```
0.949030260585071

Quantile function x = t Quantile(p,df);
```

This is the inverse of the distribution function. Providing the desired probability, p, and degrees of freedom, df, JMP returns the quantile or critical value, x.

```
x = t Quantile(0.95,5);
```

Given the command above, the JMP Log shows the value below.

```
2.01504837333302
```

EXAMPLE A.5 The χ^2 Distribution

```
Distribution function p = ChiSquare Distribution(x, df)
```

"p" is the probability that the chi-square distribution is less than or equal to "x" where the number of degrees of freedom is df.

```
p = ChiSquare Distribution(12,6);
```

Given the command above, the JMP Log shows the value below.

```
0.938031195583341
```

```
Quantile function x = ChiSquare Quantile(p,df)
```

This is the inverse of the distribution function. Providing the desired probability, p, and degrees of freedom, df, JMP returns the quantile or critical value, x.

```
x = ChiSquare Quantile(0.95, 6);
```

Given the command above, the JMP Log shows the value below.

```
12.591587243744
```

EXAMPLE A.6 The Noncentral χ^2 Distribution

```
Distribution function p = ChiSquare Distribution(x, df, nc);
```

"p" is the probability that the noncentral chi-square distribution is less than or equal to "x" where the number of degrees of freedom is df and the noncentrality parameter is nc.

```
p = ChiSquare Distribution(12,4,2);
```

Given the command above, the JMP Log shows the value below.

```
0.916387992935805
```

```
Quantile function x = ChiSquare Quantile(p, df, nc);
```

This is the inverse of the distribution function. Providing the desired probability, p, and degrees of freedom, df, and noncentrality parameter, nc, JMP returns the quantile or critical value, x.

```
x = ChiSquare Quantile(0.95, 4, 2);
```

Given the command above, the JMP Log shows the value below.

```
13.7009921451443
```

EXAMPLE A.7 The *F* Distribution

`Distribution function p = F Distribution(x, dfn, dfd);`

"p" is the probability that the F distribution is less than or equal to "x" where the numerator degrees of freedom is dfn and the denominator degrees of freedom is dfd.

```
p = F Distribution(8,5,5);
```

Given the command above, the JMP Log shows the value below.

```
0.980247095162329
```

`Quantile function x = F Quantile(p, dfn, dfd);`

This is the inverse of the distribution function. Providing the desired probability, p, where the numerator degrees of freedom is dfn and the denominator degrees of freedom is dfd, JMP returns the quantile or critical value, x.

```
x = F Quantile(0.95, 5, 5);
```

Given the command above, the JMP Log shows the value below.

```
5.05032905763264
```

EXAMPLE A.8 The noncentral *F* distribution

`Distribution function p = F Distribution(x, dfn, dfd, nc);`

"p" is the probability that the noncentral F distribution is less than or equal to "x" where the numerator degrees of freedom is dfn, the denominator degrees of freedom is dfd, and the noncentrality parameter is nc.

```
p = F Distribution(12,5,5,3);
```

Given the command above, the JMP Log shows the value below.

```
0.97831131532358
```

`Quantile function x = F Quantile(p, dfn, dfd, nc);`

This is the inverse of the distribution function. Providing the desired probability, p, where the numerator degrees of freedom is dfn, the denominator degrees of freedom is dfd, and noncentrality parameter, nc, JMP returns the quantile or critical value, x.

```
x = F Quantile(0.95, 5, 5, 3);
```

Given the command above, the JMP Log shows the value below.

```
7.95522217132703
```

REFERENCES

Barnett, V., and T. Lewis (1994). *Outliers in Statistical Data*. 3rd edition. Wiley, New York.

Bartlett, M. S. (1947). "The Use of Transformations." *Biometrics*, Vol. 3, pp. 39–52.

Bergquist, B., Vanhatalo, E. and Wordenvaad, M.L. (2011) "A Bayesian Analysis of Unreplicated Two-Level Factorials Using Effect Sparsity, Hierarchy and Heredity", *Quality Engineering*, Vol. 23, No. 2, pp. 152–166.

Bingham, D., and R. R. Sitter (1999). "Minimum Aberration Two-Level Fractional Factorial Split-Plot Designs." *Technometrics*, Vol. 41, pp. 62–70.

Box, G. E. P. (1999). "Statistics as a Catalyst to Learning by Scientific Method Part II—A Discussion" (with discussion). *Journal of Quality Technology*, 31, pp. 16–29.

Box, G. E. P., and D. R. Cox (1964). "An Analysis of Transformations." *Journal of the Royal Statistical Society B*, Vol. 26, pp. 211–243.

Box, G. E. P., and D. W. Behnken (1960). "Some New Three Level Designs for the Study of Quantitative Variables." *Technometrics*, 2, pp. 455–476.

Box, G. E. P., S. Bisgaard, and C. A. Fung (1988). "An Explanation and Critique of Taguchi's Contributions to Quality Engineering." *Quality and Reliability Engineering International*, 4, pp. 123–131.

Box, G. E. P., and N. R. Draper (2007). *Response Surfaces, Mixtures, and Ridge Analysis*. Wiley, New York.

Box, G. E. P., and J. S. Hunter (1957). "Multifactor Experimental Designs for Exploring Response Surfaces." *Annals of Mathematical Statistics*, 28, pp. 195–242.

Box, G. E. P., and K. G. Wilson (1951). "On the Experimental Attainment of Optimum Conditions." *Journal of the Royal Statistical Society B*, 13, pp. 1–45.

Box, G. E. P., W. G. Hunter, and J. S. Hunter (1978). *Statistics for Experimenters*. Wiley, New York.

Box, G. E. P., W. G. Hunter, and J. S. Hunter (2005). *Statistics for Experimenters*. 2nd edition Wiley, New York.

Box, J. F. (1978). *R. A. Fisher: The Life of a Scientist*. Wiley, New York.

Carmer, S. G., and M. R. Swanson (1973). "Evaluation of Ten Pairwise Multiple Comparison Procedures by Monte Carlo Methods." *Journal of the American Statistical Association*, Vol. 68, No. 314, pp. 66–74.

Coleman, D. E., and D. C. Montgomery (1993). "A Systematic Approach to Planning for a Designed Industrial Experiment" (with discussion). *Technometrics*, 35(1), pp. 1–27.

Daniel, C. (1959). "Use of Half-Normal Plots in Interpreting Two-Level Experiments", *Technometrics*, Vol. 1, pp. 311–342.

Derringer, G., and R. Suich (1980). "Simultaneous Optimization of Several Response Variables." *Journal of Quality Technology*, 12, pp. 214–219.

Dolby, J. L. (1963). "A Quick Method for Choosing a Transformation." *Technometrics*, Vol. 5, pp. 317–326.

Draper, N. R., and W. G. Hunter (1969). "Transformations: Some Examples Revisited." *Technometrics*, Vol. 11, pp. 23–40.

Dunnett, C. W. (1964). "New Tables for Multiple Comparisons with a Control." *Biometrics*, Vol. 20, pp. 482–491.

Errore, A., B. Jones, W. Li, and C. J. Nachtsheim (2017). "Benefits and Fast Construction of Efficient Two-Level Foldover Designs." *Technometrics*, 59(1), pp. 48–57.

Gotwalt, C., B. Jones, and D. Steinberg (2009). "Fast Computation of Designs Robust to Parameter Uncertainty for Nonlinear Settings." *Technometrics*, 51, pp. 88–95.

Hunter, J. S. (1985). "Statistical Design Applied to Product Design." *Journal of Quality Technology*, 17, pp. 210–221.

Hunter, J. S. (1989). "Let's All Beware the Latin Square." *Quality Engineering*, 1, pp. 453–465.

John, J. A., and P. Prescott (1975). "Critical Values of a Test to Detect Outliers in Factorial Experiments." *Applied Statistics*, Vol. 24, pp. 56–59.

Jones, B., and D. Montgomery (2010). "Alternatives to Resolution IV Screening Designs in 16 Runs." *International Journal of Experimental Design and Process Optimization*, 1(4), pp. 285–295.

Jones, B., and D. Montgomery (2017). "Partial replication of small two-level factorial designs." *Quality Engineering*, 29(2), pp. 190–195. DOI:10.1080/08982112.2016.1254797

Jones, B., and C. J. Nachtsheim (2011a). "A Class of Three-Level Designs for Definitive Screening in the Presence of Second-Order Effects." *Journal of Quality Technology*, 43, pp. 1–15.

Jones, B., and C. J. Nachtsheim (2011b). "Efficient Designs with Minimal Aliasing." *Technometrics*, 53, pp. 62–71.

Jones, B., and C. J. Nachtsheim (2013). "Definitive Screening Designs with Added Two-Level Categorical Factors." *Journal of Quality Technology*, 45(2), pp. 120–129.

Jones, B., and C. J. Nachtsheim (2016). "Blocking Schemes for Definitive Screening Designs." *Technometrics*, 58(1), pp. 74–83.

Jones, B., and C. J. Nachtsheim (2017). "Effective Design-Based Model Selection for Definitive Screening Designs." *Technometrics*, 59(3), pp. 319–329.

Jones, B., E. D. Schoen, and D.C. Montgomery (2016). "A Comparison of Two-Level Designs to Estimate All Main Effects and Two-Factor Interactions." *Quality Engineering*, 28(4), pp. 369–380.

Kackar, R. N. (1985). "Off-Line Quality Control, Parameter Design, and the Taguchi Method." *Journal of Quality Technology*, 17, pp. 176–188.

Kenward, Michael G. and Roger, James H. (2009) "An improved approximation to the precision of fixed effects from restricted maximum likelihood" *Computational Statistics and Data Analysis*, 53 pp. 2583–2595.

Kiefer, J. (1961). "Optimum Designs in Regression Problems II." *Annals of Mathematical Statistics*, 32, pp. 298–325.

Kiefer, J., and J. Wolfowitz (1959). "Optimum Designs in Regression Problems." *Annals of Mathematical Statistics*, 30, pp. 271–294.

Khuri, A. I., and Cornell, J. A. (1996). *Response Surfaces*, 2nd edition. Marcel Dekker, New York.

Lenth, R.V. (1989). "Quick and Easy Analysis of Unreplicated Factorials", *Technometrics*, Vol. 31 pp. 469–473.

Li, X., N. Sudarsanam, and D. D. Frey (2006). "Regularities in Data from Factorial Experiments." *Complexity*, 11(5), pp. 32–45.

Montgomery, D. C., and G. C. Runger (2011). *Applied Statistics and Probability for Engineers*. 5th edition. Wiley, New York.

Montgomery, D. C. (2020) *Design and Analysis of Experiments*, 10th edition Wiley, Hoboken N.J.

Montgomery, R. C., and G. C. Runger (2018). *Applied Statistics and Probability for Engineers*, 7th edition. Wiley, Hoboken, NJ.

Myers, R. H., D. C. Montgomery, and C. M. Anderson-Cook (2016). *Response Surface Methodology: Process and Product Optimization Using Designed Experiments*, 4th edition. Wiley, Hoboken, NJ.

Nachtsheim, C. J., and B. Jones (2018). "Design Augmentation for Response Optimization and Model Estimation." *Quality Engineering*, 30(1), pp. 38–51.

Nair, V. N. (editor) (1992). "Taguchi's Parameter Design: A Panel Discussion." Technometrics, 34, pp. 127–161.

Pignatiello, J. J., Jr., and J. S. Ramberg (1992). "Top Ten Triumphs and Tragedies of Genichi Taguchi." *Quality Engineering*, Vol. 4, pp. 211–225.

Stefansky, W. (1972). "Rejecting Outliers in Factorial Designs." *Technometrics*, Vol. 14, pp. 469–479.

Taguchi, G. (1987). *System of Experimental Design: Engineering Methods to Optimize Quality and Minimize Cost*. UNIPUB, White Plains, NY.

Taguchi, G. (1991). *Introduction to Quality Engineering*. Asian Productivity Organization, UNIPUB, White Plains, NY.

Taguchi, G., and Y. Wu (1980). *Introduction to Off-Line Quality Control. Central Japan Quality Control Association*. Nagoya, Japan.

Tukey, J. W. (1953). "The Problem of Multiple Comparisons." Unpublished Notes, Princeton University.

Woods, D. C., S. M. Lewis, J. A. Eccleston, and K. G. Russell (2006). "Designs for Generalized Linear Models with Several Variables and Model Uncertainty." *Technometrics*, 48, pp. 284–292.

Xiao, L., D. K. J. Lin, and F. Bai (2012). "Constructing Definitive Screening Designs Using Conference Matrices." *Journal of Quality Technology*, 44, pp. 1–7.

Index